T0269101

LONDON MATHEMATICAL SOCIETY LECTURE NOTE SERIES

Managing Editor: Professor N.J. Hitchin, Mathematical Institute,
University of Oxford, 24–29 St Giles, Oxford OX1 3LB, United Kingdom

The titles below are available from booksellers, or, in case of difficulty, from Cambridge University Press at www.cambridge.org.

London Mathematical Society Lecture Note Series. 301

Stable Modules and the D(2)-Problem

F. E. A. Johnson
University College London

CAMBRIDGE
UNIVERSITY PRESS

CAMBRIDGE UNIVERSITY PRESS

Cambridge, New York, Melbourne, Madrid, Cape Town,
Singapore, São Paulo, Delhi, Tokyo, Mexico City

Cambridge University Press
The Edinburgh Building, Cambridge CB2 8RU, UK

Published in the United States of America by
Cambridge University Press, New York

www.cambridge.org
Information on this title: www.cambridge.org/9780521537490

First published 2003

A catalogue record for this publication is available from the British Library

ISBN 978-0-521-53749-0 Paperback

To
C. T. C. Wall
who first considered this question.

Contents

Acknowledgements

Thanks are due to a number of people who have been generous at various times with comments, practical information and encouragement, in particular: A. J. Casson; K. W. Gruenberg both for many helpful conversations over the years and in particular for copies of some of Browning's unpublished preprints; R. J. Milgram; Ib Madsen; and Charles Thomas, whose initial suggestion it was to re-examine the problem of Poincaré 3-complexes.

Introduction

A history of the D(2)-problem

The problem with which this book is concerned arose from the attempt, during the 1960s, to classify compact manifolds by means of 'surgery' [7], [48], [71], [73]. Developing further the techniques of Thom [64], Wallace [75], [76], [77], Milnor [43], and Smale [52], a movement led notably by W. Browder, S. P. Novikov, and C. T. C. Wall made a systematic effort to understand compact manifolds in terms of homotopy theory which, by that time, was already a mature subject, with its own highly developed literature and was considered, in practice, at least under the simplifying restriction of simple connectivity, to be effectively computable [8].

Wall's particular contribution to manifold theory was to consider surgery problems in which the fundamental group is non-trivial. Perhaps one should point out that by allowing *all* finitely presented fundamental groups, one automatically turns a computable theory into a noncomputable one [6], [47]. However, even if we restrict our attention to fundamental groups which are familiar, the extent to which the resulting theory is computable is problematic. It really depends upon what is meant by 'familiar', and how well one understands the group under consideration. When describing groups by means of generators and relations, there are easily stated questions which one can ask of very familiar and otherwise tractable finite groups which at present seem completely beyond our ability to answer.

In connection with this general attack, Wall wrote two papers which merit special attention. The first of these, 'Poincaré Complexes I' [72], gives general homotopical conditions which must be satisfied by any space before it can be transformed, by surgery, into a manifold.

The second, 'Finiteness conditions for CW complexes' [68] (and despite the earlier publication date, it does seem to come later in historical development),

1

straddles the boundary between surgery and a more general attempt to describe all homotopy theory in terms of pure algebra, namely 'algebraic' or 'combinatorial' homotopy theory [80]. Wall's aim in this paper is to formulate general conditions which guarantee that a given space will be homotopically equivalent to one with certain properties. In particular, he asks what conditions it is necessary to impose before a space can be homotopy equivalent to one of dimension $\leq n$.

The obvious first condition that one looks for is that homology groups should vanish in dimensions $> n$. This is clearly a necessary condition. However, homology alone is a notoriously bad indicator of dimension as the following 'Moore space' example shows.

Let m be a positive integer; the Moore space $M(m, n)$ is formed from the n-sphere by attaching an $(n + 1)$-cell by an attaching map $S^n \to S^n$ of degree m. Then $M(m, n)$ has dimension $n + 1$. However, computing integral homology gives

$$H_k(M(m, n); \mathbf{Z}) = \begin{cases} \mathbf{Z}/m\mathbf{Z} & k = n \\ 0 & n < k \end{cases}$$

which falsely indicates the dimension as dim $= n$. In fact, the accurate indicator of dimension is integral cohomology, and in this case we get

$$H^k(M(m, n); \mathbf{Z}) = \begin{cases} \mathbf{Z}/m\mathbf{Z} & k = n + 1 \\ 0 & n + 1 < k \end{cases}$$

giving the correct answer dim$(M(m, n)) = n + 1$.

If $\tilde{X}$ denotes the universal covering of X, the assumption that $H_k(\tilde{X}; \mathbf{Z}) = 0$ for all $n < k$ is enough to guarantee that X is equivalent to a space of dimension $\leq n + 1$, but not necessarily of dimension $\leq n$. Therefore, we may pose the problem in the following form:

D(n)-problem: Let X be a complex of geometrical dimension $n + 1$. What further conditions are necessary and sufficient for X to be homotopy equivalent to a complex of dimension n?

In the simply connected case, Milnor (unpublished) had previously shown that the necessary condition $H^{n+1}(X; \mathbf{Z}) = 0$, abstracted from the Moore space example, was already sufficient. In the non-simply connected case, clearly, one still requires $H_{n+1}(\tilde{X}; \mathbf{Z}) = 0$. That being assumed, Wall showed that in dimensions $n \geq 3$, the additional condition, both necessary and sufficient, is the obvious generalization from the simply connected case, namely that $H^{n+1}(X; \mathcal{B}) = 0$ should hold for all coefficient bundles $\mathcal{B}$.

Wall also gave a 'formal' solution to the D(1)-problem, at the cost of using nonabelian sheaves $\mathcal{B}$. To an extent this was unsatisfactory, and the one-dimensional case was not cleared up completely until the Stallings–Swan proof that groups of cohomological dimension one are free [54], [61]. This left only the two-dimensional case, which we state in the following form:

D(2)-problem: Let X be a finite connected cell complex of geometrical dimension 3, and suppose that

$$H_3(\tilde{X}; \mathbf{Z}) = H^3(X; \mathcal{B}) = 0$$

for all coefficient systems $\mathcal{B}$ on X. Is it true that X is homotopy equivalent to a finite complex of dimension 2?

We shall say that a 3-complex X is *cohomologically two-dimensional* when these two conditions are satisfied. We note, and shall do so again in detail at the appropriate point, that for finite fundamental groups G, the condition $H_3(\tilde{X}; \mathbf{Z}) = 0$ is redundant, since it is implied, using the Eckmann–Shapiro Lemma, by the condition $H^3(X; \mathbf{F}) = 0$ where $\mathbf{F}$ is the standard coefficient bundle on X with fibre $\mathbf{Z}[G]$.

We have chosen to ask the question with the restriction that X be a finite complex. One can, of course, relax this condition; one can also ask a similar question phrased in terms of collapses and expansions [70]. Neither, however, will be pursued here.

The first thing to observe is that the D(2)-problem is parametrized by the fundamental group. Each finitely presented group G has its own D(2)-problem; we say that the D(2)-property *holds for* G when the above question is answered in the affirmative, and likewise *fails for* G when there is a finite 3-complex X_G with $\pi_1(X_G) \cong G$ which answers the above question in the negative.

The D(2)-problem arises in a completely natural way once one attempts, as Wall did in [72], to find a normal form for Poincaré complexes. To see how, consider a smooth closed connected n-manifold M^n, and let M_0 be the bounded manifold obtained by removing an open disc; by Morse Theory, it is easy to see that M_0 contracts on to a subcomplex of dimension $\leq n - 1$; in particular, M admits a cellular decomposition with a single-top dimensional cell. In [72] Wall attempted to mimic this construction in the context of Poincaré complexes. He showed that, as a consequence of Poincaré Duality, a finite Poincaré complex M of dimension $n + 1 \geq 4$ has, up to homotopy, a representation in the form

$$(*) \qquad M = X \cup_\alpha e^{n+1}$$

where X is finite complex satisfying $H_{n+1}(\tilde{X}; \mathbf{Z}) = H^{n+1}(X; \mathcal{B}) = 0$ for all coefficient systems $\mathcal{B}$. Wall showed, in [68], that the case $n \geq 3$ sucessfully mimics the situation for manifolds in that X is equivalent to an n-complex. When $n = 1$, it follows from the Stallings–Swan Theorem [54], [61] that X is homotopy equivalent to a one-dimensional complex. It is only in the case $n = 2$, corresponding to a Poincaré 3-complex, that we still do not know the general answer.

The two-dimensional realization problem

In the world of low-dimensional topology, there is another, older, problem which can be posed independently. If K is a finite 2-complex with $\pi_1(K) = G$ one obtains an exact sequence of $\mathbf{Z}[G]$-modules of the form

$$0 \to \pi_2(K) \to C_2(K) \xrightarrow{\partial_2} C_1(K) \xrightarrow{\partial_1} C_0(K) \xrightarrow{\epsilon} \mathbf{Z} \to 0$$

where $\pi_2(K)$ is the second homotopy group of K, and $C_n(K) = H_n(\tilde{K}^{(n)}, \tilde{K}^{(n-1)})$ is the group of cellular n-chains in the universal cover of K. Since each $C_n(K)$ is a free module over $\mathbf{Z}[G]$, this suggests that we take, as algebraic models for geometric 2-complexes, arbitrary exact sequences of the form

$$0 \to J \to F_2 \xrightarrow{\partial_2} F_1 \xrightarrow{\partial_1} F_0 \xrightarrow{\epsilon} \mathbf{Z} \to 0$$

where F_i is a finitely generated free (or, more generally, stably free) module over $\mathbf{Z}[G]$. Such objects are called algebraic 2-complexes over G, and form a category denoted by $\mathbf{Alg}_G$.

In fact, the correspondence $K \mapsto C_*(K)$ gives a faithful representation of the two-dimensional geometric homotopy relation in the algebraic homotopy category determined by $\mathbf{Alg}_G$. That is, when K, L are finite geometrical 2-complexes with $\pi_1(K) = \pi_1(L) = G$, then $K \simeq L \iff C_*(K) \simeq C_*(L)$. This has been known since the time of Whitehead [80], and perhaps even from the time of Tietze [66]. Nevertheless, it is still difficult to find explicitly in the literature in this form, and we prove it directly in Chapter 9. There is now an obvious question.

Realization Problem: Let G be a finitely presented group. Is every algebraic 2-complex

$$\left(0 \to J \to F_2 \xrightarrow{\partial_2} F_1 \xrightarrow{\partial_1} F_0 \xrightarrow{\epsilon} \mathbf{Z} \to 0\right) \in \mathbf{Alg}_G$$

geometrically realizable; that is, homotopy equivalent in the algebraic sense, to a complex of the form $C_*(K)$ where K is a finite 2-complex?

Statement of results

Firstly we show (see [28]) that for *finite fundamental groups* G, the D(2)-problem is entirely equivalent to the Realization Problem; that is:

Theorem I: When G is a finite group, the D(2)-property holds for G if and only if each algebraic 2-complex over G is geometrically realizable.

Theorem I will also be referred to as the Realization Theorem. The proof given in the text uses techniques which are specific to finite groups and does not generalize immediately to infinite fundamental groups. In particular, we make frequent use of the fact that, over a finite group, projective modules are injective relative to the class of $Z[G]$-lattices, a statement which is known to be false for even the most elementary of all infinite groups, namely the infinite cyclic group. This is not to say that the Realization Theorem, as stated, does not hold more widely. In Appendix B, we give a proof which is valid for all finitely presented groups which satisfy an additional homological finiteness condition, the so-called FL(3) condition.

Having reduced the D(2)-problem to the Realization Problem, to make progress we must now pursue the problem of realizing homotopy types of algebraic 2-complexes by geometric 2-complexes. Here we are helped by two specific technical advances which, considered together, render our task easier, at least for homotopy types over a finite fundamental group.

The first is Yoneda's Theory of module extensions [34], [82]. This was in essence known to Whitehead, as can be seen from [35]. It is rather the modern version of Yoneda Theory, expressed in terms of stable modules and derived categories, implicit in the original, but incompletely realized, that the author has found so useful. Since the systematic use of stable modules is such an obvious feature of the exposition, some words of explanation are perhaps in order.

For a module M over a ring Λ, the *stable module* $[M]$ is the equivalence class of M under the equivalence relation generated by the stabilization operation $M \mapsto M \oplus \Lambda$. We shall also need to consider a more general stability, here called *hyper-stability*,* namely $M \mapsto M \oplus P$, where P is an arbitrary finitely generated projective module.

For much of the time we work, not in the category of modules over $Z[G]$, but rather in the 'derived module category'. This is the quotient category obtained by equating 'projective $= 0$'. The objects in this category can be equated with hyper-stable classes of modules.

To make a comparison with a simpler case, Carlson's book [11] considers modular representation theory systematically from this point of view, and his

---

* Mac Lane calls this notion 'projective equivalence' ([34], p. 101, Exercise 2).

elegant account was extremely useful in the initial formulation of ideas. There all projectives are free, and objects in the derived category are indistinguishable from stable modules.

The derived category gives an *objective* form to the original Eilenberg–MacLane conception of homological algebra. Whereas they worked with 'derived functors', we work with 'derived objects'; cohomology as the derived functor of Hom is obtained simply by applying Hom to the appropriate derived objects.

Systematic use of derived objects confers some specific advantages. In the context of the cohomology of finite groups, it is 'well known and obvious' when pointed out, but has emerged from the collective subconscious only comparatively recently, [22], that cohomology functors are both representable and co-representable, in the technical sense of Yoneda's Lemma. Perhaps, given the minute analysis to which two generations have subjected the foundations of the subject, this is still less obvious than it should be. This 'geometrization' of cohomology allows a significant degree of control over the D(2)-problem, and a subsidiary aim of this book, carried out in Chapter 4, is to give an account of relative homological algebra from this point of view.

The second advance, Swan–Jacobinski cancellation theory, deals with the extent to which one can reverse the stabilization operation $M \mapsto M \oplus \mathbf{Z}[G]$. It enables us to assemble the set of all possible homotopy groups of two-dimensional complexes with given fundamental group into a tree, $\Omega_3(\mathbf{Z})$, of the following sort

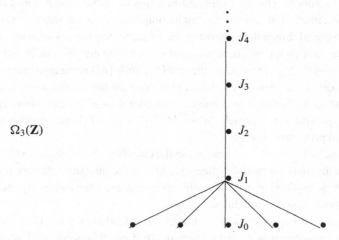

Here the vertices are modules, and one draws a line (upwards) joining two modules so $J_m \rightarrow J_m \oplus \mathbf{Z}[G] = J_{m+1}$. This idea of tree representation goes

back to Dyer and Sieradski [16]. Homotopy types with π_2 at the bottom level are called *minimal*. We say that G has the *realization property* when all algebraic 2-complexes are geometrically realizable. Our second result (again see [28]) says that it is enough to realize homotopy types at the minimal level:

Theorem II: The finite group G has the realization property if and only if all minimal algebraic 2-complexes are realizable.

This is a general condition on homotopy types. Our third result gives a criterion for realization in terms of the second homotopy group rather than a complete homotopy type. For any such module J, there is an homomorphism of groups

$$\nu^J : \mathrm{Aut}_{\mathbf{Z}[G]}(J) \to \mathrm{Aut}_{\mathcal{D}er}(J)$$

from the automorphism group in the module category to the automorphism group in the derived category. Moreover, $\mathrm{Im}(\nu^J)$ is contained in a certain subgroup of $\mathrm{Aut}_{\mathcal{D}er}(J)$ namely the kernel of the Swan map

$$S : \mathrm{Aut}_{\mathcal{D}er}(J) \to \widetilde{K}_0(\mathbf{Z}[G])$$

where $\widetilde{K}_0(\mathbf{Z}[G])$ is the projective class group. J is said to be *full* when $\mathrm{Im}(\nu^J) = \mathrm{Ker}(S)$. The module J is said to be *realizable* whenever J occurs in the form $J = \pi_2(K)$ for some finite 2-complex K with $\pi_1(K) = G$. Then we also have:

Theorem III: If each minimal module $J \in \Omega_3(\mathbf{Z})$ is both realizable and full, then G has the realization property.

Describing minimal homotopy types

For a given finite group G, obtaining an affirmative answer to the D(2)-problem involves two quite distinct steps. The first consists in giving a precise description of minimal two-dimensional algebraic homotopy types. Although possibly very difficult to implement in practice, in any particular case this step is amenable to procedures which are, in theory at least, effective. The second, for which no general procedure can be expected and which is still unsolved even for some very familiar and quite small groups, consists in determining enough minimal presentations of the group G under discussion to realize the minimal chain homotopy types.

In respect of the classification of the homotopy types of algebraic 2-complexes, the groups we find it easiest to deal with are the the finite groups of cohomological period 4. Chapter 7 is devoted to a brief exposition of what is known about them. As we point out in Chapter 11, groups of period 4 arise naturally in any discussion of Poincaré 3-complexes. For these groups, the general programme becomes much simpler. In particular, the homotopy types of minimal algebraic 2-complexes can be parametrized by more familiar objects. In Chapter 9, we prove the following classification result:

Theorem IV: Let G be a finite group of free period 4. Then there is a 1–1 correspondence

$$\widehat{\mathbf{Alg}}_G \longleftrightarrow SF(\mathbf{Z}[G])$$

where $\widehat{\mathbf{Alg}}_G$ is the set of two-dimensional algebraic homotopy types over G and $SF(\mathbf{Z}[G])$ is the set of isomorphism classes of stably free modules over $\mathbf{Z}[G]$.

Despite the intractability of the general problem of minimal group presentations mentioned above, for some groups enough is known to allow a complete solution of the D(2)-problem.

At present, Theorem IV gives the most hopeful candidate for a counterexample to the D(2)-property, for when $\mathbf{Z}[G]$ admits non-trivial stably free modules there are 'exotic' minimal 2-complexes which are not, as yet, known to be geometrically realizable. As the author has shown in [29], this occurs in the case of quaternion groups of high enough order. In fact, it follows from Swan's calculations [63] that the smallest example of this type is $Q(24)$, the quaternion group of order 24.

We point out that a special case of this classification, for the subclass of finite groups which are fundamental groups of closed 3-manifolds and which also possess the cancellation property for free modules, has also been obtained by Beyl, Latiolais and Waller [5] using a rather more geometric approach. In this case, there are no exotic minimal 2-complexes.

We note that the classification for some dihedral groups, where again there are no surprises, can be also be done directly by writing down explicit homotopy equivalences [27]. By contrast with [5], a theorem of Milnor [42] shows that dihedral groups are *not* fundamental groups of closed 3-manifolds.

It is known that if

$$\mathcal{G} = \langle X_1, \ldots, X_g; \quad W_1, \ldots, W_r \rangle$$

is a presentation of a finite group G then $g \leq r$. In the case where $g = r$, the presentation is said to be *balanced*. In terms of the D(2)-problems, finite groups which possess balanced presentations are 'smallest possible'. In connection with the problem of finding Poincaré 3-complexes of standard form, that is, having a single 3-cell, we obtain:

Theorem V: Let G be a finite group; then G has a standard Poincaré 3-form if and only if there is a finite presentation $\mathcal{G}$ of G for which $\pi_2(\mathcal{G}) \cong \mathbf{I}^*(G)$, the dual module of the augmentation ideal. Moreover, G then necessarily has free period 4, and the presentation $\mathcal{G}$ is automatically balanced.

For certain finite groups of period 4, the connection between the standard form problem and the D(2)-problem is one of equivalence.

Theorem VI: Let G be a finite group which admits a free resolution of period 4; if G has the free cancellation property then

$$G \text{ satisfies the D(2)}-\text{property} \iff G \text{ admits a balanced presentation.}$$

Earlier work on the algebraic classification problem

The first significant classification result of the type considered here is that of W. H. Cockroft and R. G. Swan [13], which classifies algebraic 2-complexes over a finite cyclic group. Taken in conjunction with Theorem I this is enough to answer the D(2)-question for finite cyclic groups in the affirmative. A general attack upon this question for finite fundamental groups was undertaken in a series of papers in the 1970s by M. N. Dyer and A. J. Sieradski. Although it does relate directly to our main concerns, one should also mention the contributions of Metzler on 2-complexes with finite abelian fundamental group [39].

However, although again we make very little direct appeal to it, without any doubt the next advance of real significance in the theory of two-dimensional homotopy was made by W. Browning in the late 1970s and early 1980s. The circumstances of Browning's career challenge the self-congratulatory assumptions on which the modern world is apt to reproach the past. In fact, Browning did not publish his results, and his work, available only in the form of his thesis [8], and some ETH pre-prints, languished in semi-obscurity for a number of years, before being accorded some of the recognition which it deserves. Happily, there is now a growing literature dealing with various aspects and generalizations of Browning's work; see for example, [20], [21], [33].

The principal result of Browning's thesis is his Stability Theorem which is a generalization from modules to chain complexes (essentially it is a complete re-proof) of the Swan–Jacobinski Theorem. Our realization results manage to avoid the technicalities of Browning's approach, though we shall review it briefly at the appropriate point in Chapter 9.

Browning's later work requires a restriction (the Eichler condition, see Chapter 3) which prevents it being applicable, except in trivial cases, to finite groups of period 4. Theorems III and IV by-pass the non-cancellation phenomena which arise when the Eichler condition fails, and are, in a sense, complementary to Browning's approach.

About this book

This book is essentially in three parts, with two appendices. In Chapters 1–3 we summarize those aspects of module theory and group representation theory that we shall need. Here the one really substantial piece of mathematics which we use systematically without any indication of proof is the Swan–Jacobinski Cancellation Theorem [25], [62]; for a proof we refer the reader to the definitive account of Curtis and Reiner [14] (vol. 2, Section 51).

Chapters 4–7 concentrate on group cohomology and module extension theory. Chapter 4 gives a systematic treatment of Yoneda Theory for our purposes; Chapters 5 and 6 specialize the general treatment to modules over group rings; Chapter 6 is particularly important since it contains the detailed classification theory by 'k-invariants' that we shall use systematically. Chapter 7 is devoted to the structure and classification of groups of periodic cohomology, which form the body of examples investigated later.

In Chapters 8–11 we consider algebraic and two-dimensional geometric homotopy theory in relation to the Realization and D(2)-problems.

Finally, in two appendices we consider briefly how the some of the arguments generalize to infinite fundamental groups. In Appendix A we show that the D(2)-property holds for finitely generated (non-abelian) free groups. In Appendix B we show that the Realization Theorem holds for finitely presented groups of type FL(3).

This work had its origin in a sequence of computations on module extensions over finite groups, with the general intention of investigating the structure theory of Poincaré 3-complexes. Begun in the Autumn of 1996, they were undertaken at first somewhat in the spirit of a diversion, for the sake of seeing what came out. However unsystematic, they nevertheless led, in short order to a perspective on the D(2)-problem which encouraged real hope of progress. Proofs of Theorems I and II followed shortly, and were announced at the British Topology Conference at Oxford in April 1997.

A number of cases of the D(2)-problem were solved on an *ad hoc* case-by-case basis in the spring and early summer of 1997. A more systematic attack required additional insight. Some of this was immediately forthcoming. At the Oxford meeting, the author was reminded, by Andrew Casson, of Milgram's finiteness obstruction computations [40], [41], and, by Ib Madsen, of the related computations of Bentzen and Madsen [4], [36]. The earlier paper of Wall [74] should also be mentioned.

The *ad hoc* calculations which began this study are no longer evident in our exposition, although some vestigial remains can be found in [27]. The key to their systematic treatment was the realization of the fundamental importance of Swan's Isomorphism Theorem (see Chapter 6). Our development of it, Theorem III, gives the required practical technique for investigating the Realization problem for groups of period 4.

It is impossible to work in this subject without appreciating the extent to which it has been shaped by the papers of R. G. Swan, in particular, [57], [58], [59], [60], [62] and [63].

Last but not least, the debt to C. T. C. Wall's papers, particularly [68], [69], [72], should be obvious. Less easy to discharge is a wider intellectual debt. As Wall's research student, I first learnt of the existence of the D(2)-problem at first hand, in June 1969 (my memory tells me it was Thursday, 19 June at about 7 p.m.). In the intervening years, working at things often far removed, I have always been aware of its continual presence, in the background, rather like the dripping of a leaky tap. And as with leaky taps, whether one turns them off or on, it is a relief at last to do something.

Chapter 1

Orders in semisimple algebras

We begin by reviewing briefly the the theory of semisimple algebras over a field, considered as a generalization of elementary linear algebra. An order is a $\mathbf{Z}$-algebra A which imbeds as a subring of a finite-dimensional $\mathbf{Q}$-algebra $A_\mathbf{Q}$. We are interested in the case where $A_\mathbf{Q}$ is semisimple. As will become apparent in subsequent chapters, when G is a finite group, the integral group ring $\mathbf{Z}[G]$ can be regarded as an order in the semisimple algebra $\mathbf{Q}[G]$.

1 Simple algebras

In the theory of linear algebra over a field k, the reader will recall that there is a basic object, the one-dimensional k-module, which we may identify with k itself. Any other (finitely generated) k-module V is isomorphic, a direct sum

$$V \cong \underbrace{k \oplus \cdots \oplus k}_{n}$$

of copies of the basic module. The number n, the *dimension* of V, is an isomorphism invariant of V. This broad outline is independent of the particular field k, which may, in fact, be taken to be any division ring, perhaps non-commutative.

An essential aspect of the basic one-dimensional module k is that it contains no proper nonzero submodule. We begin by generalizing this property.

Thus let $\mathcal{A}$ be a ring, and let M be a module over $\mathcal{A}$; M is said to be *simple* when the only $\mathcal{A}$-submodules of M are $\{0\}$ and M. The following is known as 'Schur's Lemma':

Proposition 1.1: Let M, N be simple $\mathcal{A}$-modules, and let $f : M \to N$ be a homomorphism over $\mathcal{A}$; then f is either the zero homomorphism or else f is an isomorphism.

13

Proof: Clearly $\mathrm{Ker}(f)$ is a submodule of M and $f(M)$ is a submodule of N. If f is nonzero, then simplicity of M shows that f is injective, since then $\mathrm{Ker}(f) = \{0\}$, and simplicity of N shows that f is surjective, since then $f(M) = N$. $\square$

We say that a ring $\mathcal{A}$ has *property S* when there exists a simple right $\mathcal{A}$-module Σ such that, for any finitely generated right $\mathcal{A}$-module M, there is an isomorphism $M \cong_A \Sigma^{(m)}$ for some positive integer m. In practice, it suffices to check a single condition:

Proposition 1.2: The ring $\mathcal{A}$ has *property S* if and only if there exists a simple right $\mathcal{A}$-module Σ such that for some positive integer d, there is an isomorphism of $\mathcal{A}$-right modules $\mathcal{A} \cong_A \Sigma^{(d)}$.

Proof: It clearly suffices to show ($\Longleftarrow$). Let M be a finitely generated right $\mathcal{A}$-module; then for some $n \geq 1$ there exists a surjective $\mathcal{A}$-module homomorphism $\mathcal{A}^{(n)} \to M$. Since we assume that $\mathcal{A} \cong_A \Sigma^{(d)}$, where Σ is simple, then there exists a surjective $\mathcal{A}$-module homomorphism $\varphi : \Sigma_1 \oplus \cdots \oplus \Sigma_{nd} \to M$, where each $\Sigma_i \cong \Sigma$. If $J = \{j_1, \dots, j_c\} \subset \{1, \dots, nd\}$, we denote by φ_J the restriction of φ to $\Sigma_{j_1} \oplus \cdots \oplus \Sigma_{j_c}$. Let I be a minimal subset of $\{1, \dots, nd\}$ with the property that φ_I is surjective; then, since Σ is simple, it follows easily that $\varphi_I : \Sigma_{i_1} \oplus \cdots \oplus \Sigma_{i_c} \to M$ is an isomorphism. This completes the proof. $\square$

A straightforward generalization of Schur's Lemma, left to the reader, shows that in this case the isomorphism class of the module Σ is unique; that is:

Proposition 1.3: Let $\mathcal{A}$ be a ring with property S, and let Σ_1, Σ_2 be any simple (right) $\mathcal{A}$-modules; then $\Sigma_1 \cong \Sigma_2$.

For any $\mathcal{A}$-module M, $\mathrm{End}_A(M)$ becomes a ring in which multiplication is given by composition. Schur's Lemma has the consequence that, when M is simple, every nonzero endomorphism of M over $\mathcal{A}$ is invertible; that is:

Corollary 1.4: If M is a simple $\mathcal{A}$-module then $\mathrm{End}_A(M)$ is a division ring.

As an example (as we shall see, it is really the *only* example), let D be a division ring, and consider the ring $M_n(D)$ of $n \times n$ matrices over D. There is a natural right action of $M_n(D)$ on D^n as follows

$$D^n \times M_n(D) \to D^n; \quad (\mathbf{x} \cdot \alpha)_j = \sum_{k=1}^{n} x_i \alpha_{ij}$$

where $\mathbf{x} = (x_1, \dots, x_n)$ and $\alpha = (\alpha_{ij})_{1 \leq i, j \leq n}$.

(1.5) D^n is a simple module over $M_n(D)$; moreover, $\mathrm{End}_{M_n(D)}(D^n) \cong D$.

(1.6) $M_n(D)$ has property S.

Let D_1, D_2 be division rings, and suppose that $\varphi : M_{n_1}(D_1) \to M_{n_2}(D_2)$ is a ring isomorphism. If Σ_1 is a simple right ideal of $M_{n_1}(D_1)$, then $\Sigma_2 = \varphi(\Sigma_1)$ is a simple right ideal of $M_{n_2}(D_2)$, and the restriction of φ gives an isomorphism of modules $\Sigma_1 \overset{\sim}{\to} \Sigma_2$ over φ as change of ring. It follows that $\mathrm{End}_{M_{n_1}(D_1)}(\Sigma_1) \cong \mathrm{End}_{M_{n_2}(D_2)}(\Sigma_2)$; that is

$$D_1 \cong D_2$$

Since $\dim_{D_i} M_{n_i}(D_i) = n_i^2$, we see easily that $n_1^2 = n_2^2$ and so $n_1 = n_2$; that is, we have proved:

(1.7) If D_1, D_2 are division rings, then $M_{n_1}(D_1) \cong M_{n_2}(D_2)$ if and only if $D_1 \cong D_2$ and $n_1 = n_2$.

(1.8) Let $\mathcal{A}$ be a a ring with property S; if Σ, Θ are simple (right) $\mathcal{A}$-modules, then for any positive integers e, f

$$\Sigma^{(e)} \cong \Theta^{(f)} \iff \Sigma \cong \Theta \text{ and } e = f$$

To summarize, we have:

Theorem 1.9: A ring $\mathcal{A}$ has property S if and only if $\mathcal{A} \cong M_n(D)$ for some division ring D and some integer $n \geq 1$. Moreover, in any such isomorphism, both the integer n, and the isomorphism type of D are uniquely determined by $\mathcal{A}$.

Proof: By (1.6), (1.7), (1.8) it suffices to show ($\Longrightarrow$). If $\mathcal{A}$ is a ring and $a \in \mathcal{A}$, let $\lambda_a : \mathcal{A} \to \mathcal{A}$ be the endomorphism of right $\mathcal{A}$-modules given by

$$\lambda_a(x) = ax$$

then $\lambda : \mathcal{A} \to \mathrm{End}_{\mathcal{A}}(\mathcal{A}); a \mapsto \lambda_a$, is an isomorphism of rings.

Now assume that $\mathcal{A}$ has property S. Since $\mathcal{A}$ is finitely generated over itself, by the single element $1_{\mathcal{A}}$, then by hypothesis

$$\mathcal{A} \cong \Sigma^{(n)}$$

for some simple $\mathcal{A}$-module Σ. Thus $\mathrm{End}_{\mathcal{A}}(\mathcal{A}) \cong M_n(D)$ where $D = \mathrm{End}_{\mathcal{A}}(\Sigma)$. By Schur's Lemma, D is a division ring, and the result follows since $\mathcal{A} \cong \mathrm{End}_{\mathcal{A}}(\mathcal{A})$. $\square$

However, if D is a division ring, then D^{opp} is also a division ring, and $M_n(D)^{opp} \cong M_n(D^{opp})$. It follows that this result is left–right symmetric, and property S could as well be defined in terms of left modules.

The characterization given in (1.9) is 'extrinsic', relying on properties of modules to give information about the algebra. There is a corresponding 'intrinsic' treatment, which we outline below. A ring $\mathcal{A}$ is said to be *simple* when it contains no non-trivial two-sided ideal. The basic example of a simple ring is given by:

Proposition 1.10: For any division ring D, the ring $M_n(D)$ of $n \times n$ matrices over D is simple.

Proof: Considered as a D-bimodule $M_n(D)$ is free with basis $\{E(i, j)\}_{1 \le i, j \le n}$, where $E(i, j)$ is the matrix

$$E(i, j)_{k,l} = \delta_{ik}\delta_{j,l}$$

and we write $X \in M_n(D)$ in the form

$$X = \sum r, s X_{rs} E(r, s)$$

where $X_{rs} \in D$. Let $\mathbf{I}$ be a nonzero two-sided ideal in $M_n(D)$, and $X \in \mathbf{I}$ be a nonzero element. In particular, suppose that $X_{ij} \ne 0$; then

$$E(i, j) = X_{ij}^{-1} E(i, i) X E(j, j) \in \mathbf{I}$$

Hence for all r, s

$$E(r, s) = E(r, i)E(i, j)E(j, s) \in \mathbf{I}$$

and $\mathbf{I} = M_n(D)$, since $\mathbf{I}$ contains the canonical basis. $\square$

The converse of (1.10) fails to be true without imposing some sort of finiteness condition. The most general of these, the so-called *Artinian condition*, need not concern us here, though we consider it briefly in Section 4. Instead, we impose a more straightforward finiteness condition by requiring rings to be algebras of finite dimension over a field. We then have:

Proposition 1.11: Let $\mathcal{A}$ be an algebra of finite dimension over a field k; then the following conditions are equivalent:

  (i) $\mathcal{A}$ has property S;
 (ii) $\mathcal{A}$ is simple;
(iii) $\mathcal{A} \cong M_n(D)$ for some division algebra D of finite dimension over k.

Proof: We have shown already that (i) $\iff$ (iii) and that (iii) $\Rightarrow$ (ii), so it suffices to show that (ii) $\Rightarrow$ (i).

Since $\mathcal{A}$ is finite dimensional over k, so is every right ideal of $\mathcal{A}$. Let M be a nonzero right ideal in $\mathcal{A}$ of lowest possible dimension, and put

$$J = \sum_{a \in \mathcal{A}} aM$$

Clearly J is a nonzero two-sided ideal; since $\mathcal{A}$ is simple it follows that $J = \mathcal{A}$. Moreover, since $\mathcal{A}$ is finite dimensional over k there exists a finite subset $\{a_1, \ldots, a_n\}$ such that $\mathcal{A} = a_1 M + \cdots + a_n M$.

Suppose that $\{a_1, \ldots, a_n\}$ is a minimal subset with this property. Since M is simple it is straightforward to see that we have a direct sum

$$\mathcal{A} = a_1 M \oplus \cdots \oplus a_n M$$

Thus $\mathcal{A}$ has property S by (1.2). This completes the proof. $\qquad\square$

2 Semisimple modules

The classification of Section 1 extends in a straightforward fashion to modules over a direct product $M_{d_1}(D_1) \times \cdots \times M_{d_m}(D_m)$ where $D_1, \ldots, D_m$ are division rings. In fact, the classification of modules over any (finite) direct product of rings reduces to that of the classification over the individual factors. Thus suppose that $\mathcal{A}$ is a direct product of rings $\mathcal{A} = \mathcal{A}_1 \times \cdots \times \mathcal{A}_m$; we denote by $e_1, \ldots, e_m$ the canonical central idempotents

$$e_1 = (1, 0, \ldots, 0); \quad e_2 = (0, 1, 0, \ldots, 1); \ldots; \quad e_m = (0, 0, \ldots, 1)$$

Clearly

(2.1) $$e_r . e_s = \delta_{rs} . e_r$$

If M is an $\mathcal{A}$-module, we obtain a collection $M_1, \ldots, M_m$ of $\mathcal{A}$-modules thus

$$M_r = M . e_r = \{x . e_r : x \in M\}$$

It is straightforward to check that M is the internal direct sum

(2.2) $$M = M_1 \dotplus \cdots \dotplus M_m$$

If $\varphi : B \to A$ is a ring homomorphism, and M is a module over A, we obtain a module $\varphi^*(M)$ over B in which the underlying abelian group of $\varphi^*(M)$ is that of M, but where the B-action is given by

$$m \cdot b = m \cdot \varphi(b)$$

Although the canonical injection $i_r : \mathcal{A}_r \to \mathcal{A}$ is not a ring homomorphism in the strict sense (the identity goes to a central idempotent rather than the identity), the formalism nevertheless applies to give a module $\widetilde{M}_r = i_r^*(M)$ over the direct factor $\mathcal{A}_r$. The projection $\pi_r : \mathcal{A} \to \mathcal{A}_r$ is a *bona fide* ring homomorphism, and we see easily that

(2.3) $M_r = \pi_r^*(\widetilde{M}_r)$

From the identity

$$\text{Id} = i_1 \circ \pi_1 + \cdots + i_m \circ \pi_m$$

we obtain:

Theorem 2.4: Let $\mathcal{A} = \mathcal{A}_1 \times \cdots \times \mathcal{A}_m$ be a direct product of rings; then any $\mathcal{A}$-module M can be expressed as a direct sum

$$M = \pi_1^*(\widetilde{M}_1) \oplus \cdots \oplus \pi_m^*(\widetilde{M}_m)$$

It follows that

(2.5) Let M, N be modules over $\mathcal{A} = \mathcal{A}_1 \times \cdots \times \mathcal{A}_m$; then

$$M \cong_\mathcal{A} N \iff M_s \cong_{\mathcal{A}_s} N_s \text{ for each } s$$

Provided the context is clear, we shall ignore the distinctions between M_r, $\widetilde{M}_r$ and $\pi_r^*(\widetilde{M}_r)$, and regard M_r as a module over both $\mathcal{A}$ and $\mathcal{A}_r$.

The problem of classifying modules over a product $\mathcal{A}_1 \times \cdots \times \mathcal{A}_m$ is now reduced to the corresponding problem for each of the factors. We shall say that a nonzero module M over $\mathcal{A}_1 \times \cdots \times \mathcal{A}_m$ is *supported on the rth factor* when $M_s = 0$ for $s \neq r$. From (2.4) and (2.5), it follows that:

(2.6) If $\mathcal{A} = \mathcal{A}_1 \times \cdots \times \mathcal{A}_m$ is a direct product of rings, then any simple (nonzero) $\mathcal{A}$-module is supported on some *unique* factor.

A right $\mathcal{A}$-module M is said to be *semisimple* when it can be written in the form $M = \bigoplus_{i \in I} N_i$ for some collection $(N_i)_{i \in I}$ of simple $\mathcal{A}$-modules N_i, and *finitely semisimple* when $M = \bigoplus_{i=1}^m N_i$ for some *finite* collection $(N_i)_{1 \leq i \leq n}$ of simple $\mathcal{A}$-modules. Here we shall only be concerned with finitely semisimple modules.

If N is any module over $\mathcal{A}$, we denote by $N^{(d)}$ the d-fold direct sum

$$N^{(d)} = \underbrace{N \oplus \cdots \oplus N}_{d}$$

A module of this form is said to be *isotypic*, or, more precisely, *isotypic of type* (N,d). It is straightforward to see that $\text{End}_A(N^{(d)})$ is naturally represented as a matrix ring

$$\text{End}_A(N^{(d)}) \cong M_d(\text{End}_A(M))$$

If M is an $\mathcal{A}$-module, by an isotypic decomposition for M is a direct sum $M = \oplus_{i=1}^{m} M_i$, where each M_i is isotypic of type (N_i, d_i), and where the *isomorphism types* N_i are pairwise distinct; that is

$$M \cong N_1^{(d_1)} \oplus \cdots \oplus N_m^{(d_m)}$$

where $N_1, \ldots, N_m$ are simple and $N_i \not\cong N_j$ for $i \neq j$. By collecting together isomorphic summands, we see that:

Proposition 2.7: An $\mathcal{A}$-module M is finitely semisimple if and only if it admits an isotypic decomposition.

It is an easy consequence of Schur's Lemma that in such an isotypic decomposition, $\mathrm{Hom}_{\mathcal{A}}(N_i^{(d_i)}, N_j^{(d_j)}) = 0$ whenever $i \neq j$. It follows that:

Proposition 2.8: For a module M admitting the isotypic decomposition $M \cong N_1^{(d_1)} \oplus \cdots \oplus N_m^{(d_m)}$ there is an isomorphism of rings

$$\mathrm{End}_{\mathcal{A}}(M) \cong M_{d_1}(D_1) \times \cdots \times M_{d_m}(D_m)$$

where D_i is the division ring $D_i = \mathrm{End}_{\mathcal{A}}(N_i)$.

We say that the ring $\mathcal{A}$ is *right semisimple* (resp. *left semisimple*) when it is semisimple as a *right* (resp. *left*) module over itself. Evidently $\mathcal{A}$ is right semisimple if and only if $\mathcal{A}^{opp}$ is left semisimple.

When $\mathcal{A}$ is right semisimple, then $\mathcal{A}$ is automatically finitely semisimple as a module over itself, for, if $f : \mathcal{A} \to \oplus_{i \in I} M_i$ is an isomorphism of right $\mathcal{A}$-modules where each M_i is simple, then the index set I must be finite. In fact, if $\pi_r : \oplus_{i \in I} M_i \to M_r$ and $i_r : M_r \to \oplus_{i \in I} M_i$ denote, respectively, the projection on to the rth summand and the inclusion of the rth summand, and, if $J = \{r \in I : \pi_r(f(1_{\mathcal{A}})) \neq 0\}$, then J is finite by definition of $\oplus$; it clearly suffices to show that $J = I$.

Write $A = f^{-1}(\oplus_{i \in J} M_i)$, $B = f^{-1}(\oplus_{i \in I \setminus J} M_i)$. Then A, B are right ideals of $\mathcal{A}$ with $A \cap B = \{0\}$; moreover, $1_{\mathcal{A}} \in A$. If $I \setminus J \neq \emptyset$, we may choose $b \in B$ such that $b \neq 0$. Since A is a right ideal, $b = 1_{\mathcal{A}}.b \in A$. Hence $A \cap B$ contains a nonzero element b; contradiction. Thus $I = J$, and for any ring $\mathcal{A}$ we have

(2.9) $\mathcal{A}$ is right semisimple $\iff$ $\mathcal{A}$ is right finitely semisimple.

A similar argument holds for left modules.

Let $\mathcal{A}$ be a right semisimple ring; then, by (2.8), $\mathrm{End}_{\mathcal{A}}(\mathcal{A})$ is isomorphic as a ring, to a direct product $M_{d_1}(D_1) \times \cdots \times M_{d_m}(D_m)$, where each D_i is a division ring. However, for any ring $\mathcal{A}$, there is a ring isomorphism $\mathrm{End}_{\mathcal{A}}(\mathcal{A}) \cong \mathcal{A}$ given by the evaluation map $\mathrm{ev} : \mathrm{End}_{\mathcal{A}}(\mathcal{A}) \to \mathcal{A}$, $\mathrm{ev}(\alpha) = \alpha(1_{\mathcal{A}})$. It follows that, when $\mathcal{A}$ is right semisimple, $\mathcal{A}$ is isomorphic to a direct product

$$\mathcal{A} \cong M_{d_1}(D_1) \times \cdots \times M_{d_m}(D_m)$$

where $D_1, \ldots, D_m$ are division rings. It follows from the results of Section 1 both that the converse is true and that the result is left–right symmetric; that is:

Theorem 2.10: The following conditions on $\mathcal{A}$ are equivalent:

(i) $\mathcal{A}$ is right semisimple;
(ii) $\mathcal{A}$ is left semisimple;
(iii) every finitely generated left $\mathcal{A}$-module is semisimple;
(iv) every finitely generated right $\mathcal{A}$-module is semisimple;
(v) $\mathcal{A}$ is is isomorphic to a direct product

$$\mathcal{A} \cong M_{d_1}(D_1) \times \cdots \times M_{d_m}(D_m)$$

where $D_1, \ldots, D_m$ are division rings.

Let $\mathcal{A}$ be an algebra of finite dimension over a field k; we say that $\mathcal{A}$ is *semisimple* when it satisfies any of the equivalent conditions of (2.10). (We note that the general notion of 'semisimple ring', namely that the Jacobson radical should be zero, allows, in the absence of any finiteness condition, for more complicated examples than the above. However, in our development, we shall neither meet nor need anything more general than finite products of matrix rings as above.) From (2.4), (2.5), and the results of Section 1, we obtain the following classification theorem:

Theorem 2.11: Let M be a finitely generated module over $M_{d_1}(D_1) \times \cdots \times M_{d_1}(D_m)$ where each D_i is a division ring; then

$$M \cong \pi_1^*(\Sigma_1)^{(\mu_1)} \oplus \cdots \oplus \pi_m^*(\Sigma_m)^{(\mu_m)}$$

where Σ_i is a simple module over $M_{d_i}(D_i)$, and $\pi_i : \mathcal{A} \to M_{d_i}(D_i)$ is the projection on to the ith factor; with the given ordering of the factors on $\mathcal{A}$, the sequence $(\mu_1, \ldots, \mu_m)$ is uniquely determined by M. Furthermore, the simple modules over $\mathcal{A}$ are precisely those of the form $\pi_i^*(\Sigma_i)$.

The elucidation of the structure of finite-dimensional semisimple algebras is due to Wedderburn [79]. By a *Wedderburn decomposition* of a finite-dimensional k-algebra $\mathcal{A}$ we mean an isomorphism of k-algebras $\mathcal{A} \cong M_{d_1}(D_1) \times \cdots \times M_{d_m}(D_m)$ where each D_i is a division algebra. It is not difficult to see that such a product structure is *essentially unique*; that is:

Theorem 2.12: Let $\mathcal{A}$ be a finite-dimensional k-algebra possessing two product structures

$$\mathcal{A} \cong \prod_{i=1}^{m} M_{d_i}(D_i) \cong \prod_{\iota=1}^{\mu} M_{\delta_\iota}(\Delta_\iota)$$

then $\mu = m$, and for some permutation σ of $\{1, \ldots, m\}$, $\delta_i = d_{\sigma(i)}$ and $\Delta_i \cong D_{\sigma(i)}$. Thus up to a permutation, we may speak of *the Wedderburn decomposition of* $\mathcal{A}$. In the special case where the base field k is *algebraically closed*, every finite-dimensional division algebra over k is actually isomorphic to k, so that the Wedderburn decomposition then simplifies to:

Theorem 2.13: Let $k = \bar{k}$ be an algebraically closed field and $\mathcal{A}$ a finite-dimensional k-algebra; then there is an isomorphism of k-algebras $\mathcal{A} \cong \prod_{i=1}^{m} M_{d_i}(\bar{k})$ for some sequence of positive integers $(d_i)_{1 \leq i \leq m}$, unique up to permutation.

3 Projective modules and semisimplicity

As we have seen in Section 1, the theory of modules over a field, or more generally over a simple finite-dimensional algebra, is straightforward. Over a general ring $\mathcal{A}$ however, the situation is usually far more complicated.

One still has the obvious analogue of vector spaces; these are the so-called 'free modules' of the form

$$\underbrace{\mathcal{A} \oplus \cdots \oplus \mathcal{A}}_{n}$$

More general, but almost as convenient, is the notion of *projective module*. An $\mathcal{A}$-module M is said to be projective when it is a direct summand of a free module; that is, when there exists a module N such that $M \oplus N$ is a free over $\mathcal{A}$. Since a direct sum of free modules is free, we see that:

(3.1) A direct sum of projective modules is projective.

Suppose that each short exact sequence $0 \to Q \to M \to P \to 0$ splits. Let $(e_\lambda)_{\lambda \in \Lambda}$ be a set which generates P over $\mathcal{A}$, and let $e : F_\Lambda(\mathcal{A}) \to P$

$$e(f) = \sum_{\lambda \in \Lambda} f(\lambda).e_\lambda$$

be the canonical epimorphism from the free module on Λ. Then

$$0 \to \mathrm{Ker}(e) \to F_\Lambda(\mathcal{A}) \to P \to 0$$

is exact, so that $P \oplus \mathrm{Ker}(e) \cong F_\Lambda(\mathcal{A})$, and P is projective.

Conversely, suppose that P is projective. Let N be an $\mathcal{A}$-module such that, for some set X, $P \oplus N \cong F_X(\mathcal{A})$. Making the identification $P \oplus N = F_X(\mathcal{A})$, for each $\hat{x} \in X$ choose $p(x) \in P$, and $t(x) \in N$ such that

$$p(x) + t(x) = \hat{x}$$

Given an exact sequence of $\mathcal{A}$-modules $\mathcal{E} = (0 \to Q \to M \xrightarrow{\pi} P \to 0)$, for each $x \in X$ choose an element $s(x) \in M$ such that $\pi(s(x)) = p(x)$. Then the correspondence $\hat{x} \mapsto s(x)$ defines a homomorphism of $\mathcal{A}$-modules $s : F_X(\mathcal{A}) \to M$, and, by restriction to P, a homomorphism $s : P \to M$; s splits $\mathcal{E}$ on the right, so that:

(3.2) An $\mathcal{A}$- module P is projective if and only if each short exact sequence of the form $0 \to Q \to M \to P \to 0$ splits.

This has the following consequence:

Proposition 3.3: Let $\mathcal{A}$ be a finite-dimensional algebra over a field k; if every finitely generated module over $\mathcal{A}$ is projective, then $\mathcal{A}$ is semisimple.

Proof: Let M be a finitely generated $\mathcal{A}$-module; we must show that M is a direct sum of simples. Since M is automatically a finite-dimensional vector space over k we may proceed by induction on $\dim_k(M)$.

If $\dim_k(M) = 1$, then M is simple, and there is nothing to prove. Suppose proved for all modules L such that $\dim_k(L) < m$, and let $\dim_k(M) = m$. If M is simple, there is nothing to prove. If M is not simple, then it has a proper nonzero submodule M_1. However, by the projectivity hypothesis and (3.2), the exact sequence

$$0 \to M_1 \to M \to M/M_1 \to 0$$

splits, so that $M \cong M_1 \oplus M/M_1$. However, $\dim_k(M_1) < m$ and $\dim_k(M/M_1) < m$, so that, by induction, both M_1 and M/M_1 are direct sums of simple modules. Thus M is also a direct sum of simples, and this completes the proof. $\square$

The converse to this is also true, as we now proceed to show. Consider first the case where $\mathcal{A} = M_n(D)$ where D is a division ring. Let $\mathbf{R}(i) \subset M_n(D)$ be the right ideal consisting of matrices concentrated in the ith-row thus

$$\mathbf{R}(i) = \left\{ \sum_r X_{ir} E(i, r); \quad X_{ir} \in D \right\}$$

We note that there are isomorphisms $\mathbf{R}(1) \cong \mathbf{R}(2) \cong \cdots \cong \mathbf{R}(n)$ over $M_n(D)$, and, as we have already seen, any finitely generated $M_n(D)$-module M is isomorphic to a direct sum of copies of $\mathbf{R}(1)$

$$M \cong \mathbf{R}(1)^{(m)}$$

Since

$$M_n(D) \cong \mathbf{R}(1) \oplus \mathbf{R}(2) \oplus \cdots \oplus \mathbf{R}(n)$$

we see that $\mathbf{R}(1)$ is projective, and so we obtain:

Proposition 3.4: If D is a division ring, then any finitely generated module over $M_n(D)$ is projective.

More generally, by (2.6), if $D_1, \ldots, D_m$ are division rings, then any simple right (resp. left) module over a finite product $M_{d_1}(D_1) \times \cdots \times M_{d_m}(D_m)$ is isomorphic to a simple right (resp. left) ideal supported on a single factor, and from the argument already given, is necessarily projective. Since, by (2.11), any finitely generated module over $M_{d_1}(D_1) \times \cdots \times M_{d_m}(D_m)$ is a direct sum of simple modules, then we have:

Proposition 3.5: Let $\mathcal{A}$ be a finite-dimensional semisimple algebra over a field k; then every finitely generated module over $\mathcal{A}$ is projective.

From (3.3) and (3.5), together with the results of Section 2, we obtain one final charactization of semisimplicity:

Theorem 3.6: Let $\mathcal{A}$ be a finite-dimensional algebra over a field k. The following conditions on $\mathcal{A}$ are equivalent:

(i) every finitely generated right $\mathcal{A}$-module is projective;
(ii) every finitely generated left $\mathcal{A}$-module is projective;
(iii) $\mathcal{A}$ is isomorphic to a product $\mathcal{A} \cong \prod_{i=1}^{m} M_{d_i}(D_i)$ where each D_i is a finite-dimensional division algebra over k, and each d_i is a positive integer.

4 The Jacobson radical

Let $\mathcal{A}$ be a finite-dimensional algebra over k; a two-sided ideal $\mathbf{a}$ in $\mathcal{A}$ is said to be *nilpotent* when $\mathbf{a}^n = 0$ for some $n \geq 1$. If $\mathbf{a}, \mathbf{b}$ are nilpotent two-sided ideals in $\mathcal{A}$, then so is $\mathbf{a} + \mathbf{b}$. In fact

$$(\mathbf{a} + \mathbf{b})^N \subset \sum_{r=0}^{N} \mathbf{a}^r \mathbf{b}^{N-r}$$

In fact, if $\mathbf{a}^N = 0$ and $\mathbf{b}^M = 0$, then $(\mathbf{a} + \mathbf{b})^{N+M} = 0$. In particular:

Proposition 4.1: Let $\mathcal{A}$ be a finite-dimensional algebra over a field k; then $\mathcal{A}$ has a maximal two-sided nilpotent ideal, the radical rad($\mathcal{A}$).

We shall need the following also:

Lemma 4.2: Let $\mathcal{A}$ be a finite dimensional algebra over a field k such that $\mathrm{rad}(\mathcal{A}) = 0$; then any simple right ideal $Q \subset \mathcal{A}$ contains a nonzero element e such that $e^2 = e$.

Proof: If $a \in Q$ we have a homomorphism of $\mathcal{A}$-modules $\hat{a} : Q \to Q$ given by $\hat{a}(x) = ax$. Since $\mathrm{Im}(\hat{a})$ is a right ideal of $\mathcal{A}$ and Q is simple then *either*

(I) $\mathrm{Im}(\hat{a}) = \{0\}$ *or*
(II) $\hat{a} : Q \to Q$ is an isomorphism for at least one $a \in Q$.

Put

$$T = \sum_{b \in \mathcal{A}} bQ$$

then T is a two-sided ideal which contains Q and hence is nonzero. If (I) above holds, then $T^2 = 0$, and thus $\mathrm{rad}(\mathcal{A}) \neq 0$, contradiction. Hence there exists $a \in Q$ such that $\hat{a} : Q \to Q$ is an isomorphism. In particular, there exists $e \in Q$ such that $ae = a$. Evidently $e \neq 0$. Also $ae^2 = a$, and so $e^2 - e \in \mathrm{Ker}(\hat{a}) \subset Q$. However $\mathrm{Ker}(\hat{a})$ is a right ideal in $\mathcal{A}$, and by choice of a is contained properly in Q. Since Q is simple, $\mathrm{Ker}(\hat{a}) = \{0\}$, and $e^2 = e$ as claimed. $\square$

As a corollary, we get:

Theorem 4.3: If $\mathcal{A}$ is a finite-dimensional algebra over a field k, then

$$\mathcal{A} \text{ is semisimple} \iff \mathrm{rad}(\mathcal{A}) = 0$$

Proof: Suppose $\mathcal{A}$ is a finite-dimensional semisimple algebra over k. Let $\mathcal{A} = \mathcal{A}_1 \oplus \cdots \oplus \mathcal{A}_m$ be the Wedderburn decomposition of $\mathcal{A}$ into simple two-sided ideals $\mathcal{A}_1, \ldots, \mathcal{A}_m$. Since each $\mathcal{A}_i$ is simple, it is straightforward to see that any nonzero two-sided ideal $\mathbf{a}$ has the form $\mathbf{a} = \mathcal{A}_{i_1} \oplus \cdots \oplus \mathcal{A}_{i_n}$ for some unique sequence $i_1 < \cdots < i_n$. However, the simple summands $\mathcal{A}_1, \ldots, \mathcal{A}_m$ have the property that

$$\mathcal{A}_i \mathcal{A}_j = \begin{cases} \mathcal{A}_i & \text{if } i = j \\ 0 & \text{if } i \neq j \end{cases}$$

from which it follows that

$$\mathbf{a}^2 = \mathbf{a}$$

in particular, no nonzero two-sided ideal of $\mathcal{A}$ is nilpotent, and $\mathrm{rad}(\mathcal{A}) = 0$.

To prove the converse, we first prove, by induction on $\dim_k(P)$, that any right ideal P is a direct sum of simple right ideals

$$P = P_1 \oplus \cdots \oplus P_n$$

If $\dim_k(P) = 1$ there is nothing to prove. Thus suppose that $\dim_k(P) = N$ and that the result is established for all right ideals of dimension $< N$.

If P is simple, there is nothing to prove. If P is not simple, choose a nonzero right ideal $P_1 \subset P$ of lowest possible dimension. Then P_1 is simple, so that, by (4.2), there exists $e \in P_1$ such that $e^2 = e$. Then P decomposes as a direct sum of right ideals

$$P = eP \oplus (1 - e)P$$

Now $e \in P_1 \subset P$ and $e^2 = e$; from the simplicity of P_1 it now follows that $eP = P_1$. However, $(1 - e)P$ is a right ideal with $\dim_k((1 - e)P) < N$, so that, inductively, $(1 - e)P$ decomposes as a direct sum of simple right ideals

$$(1 - e)P = P_2 \oplus \cdots \oplus P_n$$

giving the required decomposition into simples

$$P = P_1 \oplus \cdots \oplus P_n$$

The theorem as stated now follows on, taking $P = \mathcal{A}$. $\qquad\square$

The finiteness condition we have imposed, namely that of being a finite-dimensional algebra over a field, is not the most general that can be imagined. For example, a finite product of semisimple algebras over different fields is still semisimple. The most general finiteness condition in our context is the so-called *Artinian* or *descending chain condition*. A ring $\mathcal{A}$ satisfies the *right Artinian condition* when each descending sequence $\cdots \subset P_{n+1} \subset P_n \subset \cdots \subset P_1 \subset P_0$ of right ideals has the property that for some N, $P_r = P_N$ for $N \leq r$. We shall not need, and do not prove, the appropriate generalization of (4.3), which is:

Theorem 4.4: Let $\mathcal{A}$ be a ring; then $\mathcal{A}$ is semisimple (that is, semisimple as right module over itself) if and only if $\mathcal{A}$ is right Artinian and $\mathrm{rad}(\mathcal{A}) = 0$.

There is a corresponding, and equivalent, statement for *left Artinian* rings. It is perhaps worth pointing out that *the Artinian condition by itself is not left–right symmetric*; nevertheless, in conformity with (2.10), the combined condition 'Artinian + rad = 0' *is* left–right symmetric.

5 Nondegenerate algebras

Let R be a commutative integral domain of characteristic zero with field of fractions k; by an R-order, we mean an R-algebra whose underlying R-module is finitely generated and free. In the special case where $R = k$, a k-order is simply an k-algebra of finite dimension.

When Λ is an R-order, and $x \in \Lambda$, we denote by $\hat{x}$ the homomorphism of right modules $\hat{x} : \Lambda \to \Lambda$ given by

$$\hat{x}(y) = xy$$

There is a symmetric bilinear form $\beta_\Lambda : \Lambda \times \Lambda \to \Lambda$ given by

$$\beta_\Lambda(x, y) = \text{Tr}(\hat{x}\hat{y})$$

We say that Λ is *nondegenerate* as an R-algebra when β_Λ is nondegenerate as a symmetric bilinear form; that is, when the correlation

$$\beta_\Lambda^* : \Lambda \to \Lambda^*; \quad \beta_\Lambda^*(x)(y) = \beta_\Lambda(x, y)$$

is injective, where Λ^* denotes the R-dual of Λ.

Proposition 5.1: Let k be a field of characteristic zero, and let $\mathcal{A}$ be a finite-dimensional algebra over k; if $\mathcal{A}$ is simple, then $\mathcal{A}$ is nondegenerate.

Proof: Put

$$\mathcal{A}^\perp = \{x \in \mathcal{A} : \beta_\mathcal{A}(x, y) = 0 \text{ for all } y \in \mathcal{A}\}$$

It is straightforward to see that $\mathcal{A}^\perp$ is a two-sided ideal in $\mathcal{A}$. Since $\mathcal{A}$ is simple, then either $\mathcal{A}^\perp = \mathcal{A}$ or $\mathcal{A}^\perp = \{0\}$. However, $\beta_\mathcal{A}(1, 1) = \dim_k(\mathcal{A}) \neq 0$ so that $1_\mathcal{A} \notin \mathcal{A}^\perp$, and $\mathcal{A}^\perp \neq \mathcal{A}$. Thus $\mathcal{A}^\perp = \{0\}$, and $\mathcal{A}$ is nondegenerate. $\quad\square$

Nondegeneracy is preserved under a direct product:

Proposition 5.2: Let R be a commutative integral domain of characteristic zero, and let Λ_1, Λ_2 be R-algebras whose underlying R-modules are finitely generated and free; then

$$\Lambda_1 \times \Lambda_2 \text{ is nondegenerate} \iff \Lambda_1, \Lambda_2 \text{ are each nondegenerate.}$$

Proof: It is straightforward to see that there is an isomorphism of symmetric bilinear forms

$$\left(\Lambda_1 \times \Lambda_2, \beta_{\Lambda_1 \times \Lambda_2}\right) \cong \left(\Lambda_1, \beta_{\Lambda_1}\right) \perp \left(\Lambda_2, \beta_{\Lambda_2}\right) \quad\square$$

In consequence, we obtain:

Theorem 5.3: Let k be a field of characteristic zero, and let $\mathcal{A}$ be a finite-dimensional algebra over k; then

$$\mathcal{A} \text{ is nondegenerate} \iff \mathcal{A} \text{ is semisimple}$$

Proof: If $\mathcal{A}$ is semisimple, then $\mathcal{A}$ is isomorphic to a finite direct product

$$\mathcal{A} \cong \mathcal{A}_1 \times \cdots \times \mathcal{A}_m$$

where each $\mathcal{A}_i$ is simple, and hence, by (5.1), nondegenerate. Thus $\mathcal{A}$ is nondegenerate by (5.2).

To establish the converse, it suffices to show that $\mathrm{rad}(\mathcal{A}) = 0$ whenever $\mathcal{A}^\perp = 0$. Thus suppose that $\mathbf{a}$ is nonzero two-sided ideal of $\mathcal{A}$ such that $\mathbf{a}^{n+1} = 0$ for some $n \geq 1$. Put $m = \max\{r : \mathbf{a}^r \neq 0\}$, and let $\{e_1^m, \ldots, e_{d_m}^m\}$ be a basis for $\mathbf{a}^m$. For each k in the range $0 \leq k \leq m - 1$, let $\{e_1^k, \ldots, e_{d_k}^k\}$ be a linearly independent set in $\mathbf{a}^k$ which projects to a basis in $\mathbf{a}^k/\mathbf{a}^{k+1}$, where by convention we take $\mathbf{a}^0 = \mathcal{A}$. Then

$$\{e_1^m, \ldots, e_{d_m}^m, e_1^{m-1}, \ldots, e_{d_2}^2, e_1^1, \ldots, e_{d_1}^1, e_1^0, \ldots, e_{d_0}^0\}$$

is a basis for $\mathcal{A}$. Moreover, if $x \in \mathbf{a}$, it is straightforward to check that, with respect to this basis, $\hat{x}$ is represented by a matrix in triangular block form whose diagonal is identically zero. Consequently $\mathrm{Tr}(\hat{x}) = 0$ for all $x \in \mathbf{a}$. However, if $x \in \mathbf{a}$, then $xy \in \mathbf{a}$ for all $y \in \mathcal{A}$, so that

$$\beta_{\mathcal{A}}(x, y) = \mathrm{Tr}(\hat{x}\hat{y})) = \mathrm{Tr}(\widehat{(xy)}) = 0 \quad \text{for all } x \in \mathbf{a}$$

and so $\mathbf{a} \subset \mathcal{A}^\perp$. In particular, $\mathrm{rad}(\mathcal{A}) \subset \mathcal{A}^\perp$.

However, if $\mathcal{A}$ is nondegenerate, then $\mathrm{rad}(\mathcal{A}) = \mathcal{A}^\perp = 0$, and hence $\mathcal{A}$ is semisimple. This completes the proof. $\square$

We consider the effect of 'extension of scalars'. Let R be a commutative integral domain of characteristic zero.

Proposition 5.4: Let $R \subset S$ be a ring extension; if Λ is an R-order, then $\Lambda_S = \Lambda \otimes_R S$ is an S-order, and

$$\Lambda \text{ is nondegenerate} \iff \Lambda_S \text{ is nondegenerate}$$

Corollary 5.5: Let R be a commutative integral domain with field of fractions k, and let Λ be an R-order; then

$$\Lambda \text{ is nondegenerate} \iff \Lambda_k \text{ is semisimple}$$

6 The discriminant

If $\{\epsilon_1, \ldots, \epsilon_n\}$ is an R basis for Λ, the discriminant $\text{Disc}(\epsilon_1, \ldots, \epsilon_n)$ is defined by

$$\text{Disc}(\epsilon_1, \ldots, \epsilon_n) = \det((\beta_{ij})_{1 \leq i, j \leq n})$$

where $\beta_{ij} = \beta_\Lambda(\epsilon_i, \epsilon_j) = \text{Tr}(\hat{\epsilon}_i \hat{\epsilon}_j)$. If $\{E_1, \ldots, E_n\}$ is also an R basis for Λ, and, if $A = (\alpha_{ij})_{1 \leq i, j \leq n}$ is the matrix of the change of basis

$$\epsilon_i = \sum_{j=1}^{n} \alpha_{ij} E_j$$

then $\text{Disc}(\epsilon_1, \ldots, \epsilon_n)$, $\text{Disc}(E_1, \ldots, E_n)$ are related by

(6.1) $$\text{Disc}(\epsilon_1, \ldots, \epsilon_n) = \det(A)^2 \text{Disc}(E_1, \ldots, E_n)$$

It follows easily that:

Proposition 6.2: The following conditions on an R-order Λ are equivalent:

(i) Λ is nondegenerate;
(ii) $\text{Disc}(\epsilon_1, \ldots, \epsilon_n) \neq 0$ for *some* R-basis $\{\epsilon_1, \ldots, \epsilon_n\}$ of Λ;
(iii) $\text{Disc}(\epsilon_1, \ldots, \epsilon_n) \neq 0$ for *every* R-basis $\{\epsilon_1, \ldots, \epsilon_n\}$ of Λ.

This gives an absolute invariant of R-orders; if Λ is an R-order and $\{\epsilon_1, \ldots, \epsilon_n\}$ is an R-basis for Λ, we define the *discriminant*, $\text{Disc}(\Lambda)$, to be the image of $\text{Disc}(\epsilon_1, \ldots, \epsilon_n)$ in the multiplicative monoid $R/(R^*)^2$ obtained as the quotient of R by the subgroup of squares of units.

As an example, we compute β_Λ in the case where $\Lambda = M_n(R)$; let $(E(i, j))_{1 \leq i, j \leq n}$ be the canonical basis of $M_n(R)$

$$E(i, j)_{kl} = \delta_{ik} \delta_{jl}$$

so that $X \in M_n(R)$ is represented in the form

$$X = \sum_{ij} X_{ij} E(i, j)$$

We wish to compute the trace $\text{Tr}(\hat{X})$ for $X \in M_n(R)$; observe that the matrix X has a trace in the usual sense

$$\text{tr}(X) = \sum_{k} X_{kk}$$

In fact, we have:

Proposition 6.3: If $X \in M_n(R)$ then

$$\operatorname{Tr}(\hat{X}) = \sum_{kl} X_{kk} = n \operatorname{tr}(X)$$

Proof:

$$\hat{X}(E(k, l)) = \sum_{ij} \delta_{jk} X_{ij} E(il)$$

$$= \sum_{r} X_{rk} E(r, l)$$

With respect to the basis $(E(i, j))_{1 \le i, j \le n}$, the matrix $\mu(X)_{kl,rs}$ of $\hat{X}$ is given by

$$X \cdot E(k, l) = \sum_{rs} \mu(X)_{kl,rs} E(r, s)$$

and a straightforward comparison shows that

$$\mu(X)_{kl,rs} = \begin{cases} X_{rk} & \text{if } s = l \\ 0 & \text{if } s \ne l \end{cases}$$

In particular, $\mu(X)_{kl,kl} = X_{kk}$, so that

$$\operatorname{Tr}(\hat{X}) = \sum_{kl} X_{kk} = n \operatorname{tr}(X)$$

as desired.

One can now compute the discriminant $\operatorname{Disc}(E(i, j)_{1 \le i, j \le n})$ as follows:

Proposition 6.4:

$$\operatorname{Disc}(E(i, j))_{1 \le i, j \le n} = (-1)^{\frac{n(n-1)}{2}} n^{n^2}$$

Proof: Write $\beta = \beta_{M_n(R)}$. Then it follows easily from (6.3) that

$$\beta(E(i, j), E(k, l)) = n \, \delta_{jk} \operatorname{tr}(E(il))$$

so that

$$\beta(E(i, j), E(k, l)) = \begin{cases} n \text{ if } i = l \text{ and } j = k \\ 0 \qquad \text{otherwise} \end{cases}$$

If we put $B(X, Y) = \beta(X, Y^T)$, we see that

$$B(E(i, j), E(k, l)) = \begin{cases} n \text{ if } (i, j) = (k, l) \\ 0 \quad \text{ otherwise} \end{cases}$$

so that B is represented by a diagonal $n^2 \times n^2$ matrix with the constant entry 'n' on the diagonal. Thus

$$\det[(B(E(i, j), E(k, l))_{ij,kl}] = n^{n^2}$$

It follows that

$$\det[(\beta(E(i, j), E(k, l))_{ij,kl}] = \sigma n^{n^2}$$

where σ is the sign of the 'transpose' permutation $(i, j) \mapsto (j, i)$ on $\{1, \ldots, n\} \times \{1, \ldots, n\}$; this permutation is obviously the composition of $\frac{n(n-1)}{2}$ transpositions, so that $\sigma = (-1)^{\frac{n(n-1)}{2}}$, which is the required result. $\qquad\square$

It follows already from (5.1) that $M_n(R)$ is nondegenerate when R is a commutative integral domain of characteristic zero. The above computation gives an alternative proof of this result.

Let $\mathcal{A}$ be a simple algebra whose centre is the field $\mathbf{E}$; we may make the identification $\mathcal{A} \otimes_{\mathbf{E}} \bar{\mathbf{E}} \cong M_n(\bar{\mathbf{E}})$. If $X \in \mathcal{A}$, then $\mathrm{Tr}(X) = n\mathrm{tr}(X \otimes 1)$. It follows that, in this case, we can define a *reduced discriminant* $\mathrm{disc}(\mathcal{A}) \in \mathbf{E}/(\mathbf{E}^*)^2$ as follows: let $\{\epsilon_1, \ldots, \epsilon_n\}$ be a basis for $\mathcal{A}$ over $\mathbf{E}$, and put

$$\mathrm{disc}(\epsilon_1, \ldots, \epsilon_n) = \det((b_{ij})_{1 \le i, j \le n})$$

We take $\mathrm{disc}(\mathcal{A})$ to be the class in $\mathbf{E}/(\mathbf{E}^*)^2$ of $\mathrm{disc}(\epsilon_1, \ldots, \epsilon_n)$ for any $\mathbf{E}$-basis $\{\epsilon_1, \ldots, \epsilon_n\}$ for $\mathcal{A}$.

This device eliminates the high powers of n which occur in the computation of (6.4). We obtain instead the following result for the basis $\{E(i, j)\}_{1 \le i, j \le n}$ of $M_n(\mathbf{E})$.

Proposition 6.5:

$$\mathrm{disc}\,(E(i, j)_{1 \le i, j \le n}) = (-1)^{\frac{n(n-1)}{2}}$$

7 Z-orders

In the case where $R = \mathbf{Z}$, $(\mathbf{Z}^*)^2$ is the trivial group, and we identify $\mathbf{Z}/(\mathbf{Z}^*)^2$ with $\mathbf{Z}$; that is:

Proposition 7.1: If Λ is a $\mathbf{Z}$-order, then the discriminant $\mathrm{Disc}(\Lambda)$ is a well-defined integer.

We also have the following:

Proposition 7.2: Let Λ be a **Z**-order; then the following conditions on Λ are equivalent:

  (i) $\text{Disc}(\Lambda) \neq 0$;
 (ii) Λ is nondegenerate;
(iii) $\Lambda_{\mathbf{Q}}$ is nondegenerate;
(iv) $\Lambda_{\mathbf{E}}$ is nondegenerate for any extension field $\mathbf{E}$ of $\mathbf{Q}$.

Let Λ, Ω be **Z**-orders in the same **Q**-algebra $\mathcal{A}$, and suppose that $\Lambda \subset \Omega$; then, since Λ and Ω both span $\mathcal{A}$ over **Q**, $\text{rk}_{\mathbf{Z}}(\Lambda) = \text{rk}_{\mathbf{Z}}(\Omega)$ and the index $j = |\Omega/\Lambda|$ is finite. It follows from (6.1) that:

$$(7.3) \qquad\qquad \text{Disc}(\Lambda) = j^2 \text{Disc}(\Omega)$$

from which we easily deduce

$$(7.4) \qquad\qquad \text{disc}(\Lambda) = j^2 \text{disc}(\Omega)$$

We say that an integer $j \geq 1$ is a *co-index* for the **Z**-order Λ when there exists a pair (Λ', φ) such that Λ' is a **Z**-order and $\varphi : \Lambda \to \Lambda'$ is an injective ring homomorphism such that $|\Lambda'/\varphi(\Lambda)| = j$; j is a *proper* co-index when $j \geq 2$. Λ is said to be a *maximal order* when its only co-index is $j = 1$. The reduced discriminant gives the following practical criterion for detecting when an order is maximal.

Proposition 7.5: Let $\mathcal{A}$ be a central simple **Q**-algebra, and let $\Lambda \subset \mathcal{A}$ be a **Z**-order; if $\text{disc}(\Lambda) = \pm 1$, then Λ is maximal.

Clearly a **Z**-order Λ is maximal when given any injective ring homomorphism $\varphi : \Lambda \to \Lambda'$ into a **Z**-algebra Λ', *either* φ is an isomorphism or $\text{rk}_{\mathbf{Z}}(\Lambda') > \text{rk}_{\mathbf{Z}}(\Lambda)$.

In practice, maximal **Z**-orders often appear in a slightly disguised form: let **E** be a finite algebraic extension field of **Q**, and let $R \subset \mathbf{E}$ be the subring of algebraic integers in **E**. Let $\mathcal{A}$ be a simple **E**-algebra, and, let $\Lambda \subset \mathcal{A}$ be an order over R. We can regard Λ as a **Z**-order by restricting scalars from R to **Z**. If $\mathcal{A}$ is a central simple **E**-algebra, then $\mathcal{A} \otimes_{\mathbf{E}} \bar{\mathbf{E}} \cong M_n(\bar{\mathbf{E}})$ and, for any R-basis $\{\epsilon_1, \ldots, \epsilon_{n^2}\}$ for Λ, the reduced discriminant $\text{disc}(\epsilon_1, \ldots, \epsilon_{n^2})$ is defined and takes values in R, so that the reduced discriminant $\text{disc}(\Lambda)$ is also well defined as an element of $R/(R^*)^2$. Then we have:

Proposition 7.6: Let **E** be a finite algebraic extension field of **Q**, and let $R \subset \mathbf{E}$ be the subring of algebraic integers in **E**. Let $\mathcal{A}$ be a central simple **E**-algebra, and let $\Lambda \subset \mathcal{A}$ be an order over R. If $\mathrm{disc}(\Lambda)$ is represented by a unit in R^*, then Λ is maximal considered as a $\mathbf{Z}$-order.

We denote by $c(\Lambda)$ the set of co-indices for Λ. We say that Λ has the *bounded co-index property* when $c(\Lambda)$ is a finite subset of $\mathbf{Z}_+$. It follows easily from (7.3) that:

Proposition 7.7: A nondegenerate $\mathbf{Z}$-order has the bounded co-index property.

The converse is also true as we now see:

Proposition 7.8: Let Λ be a $\mathbf{Z}$-order; if Λ is degenerate, then $c(\Lambda)$ is infinite.

Proof: We first show that there exists an injective ring homomorphism $\varphi : \Lambda \to \Lambda_1$ where Λ_1 is a $\mathbf{Z}$-order with $\mathrm{rk}_{\mathbf{Z}}(\Lambda_1) = \mathrm{rk}_{\mathbf{Z}}(\Lambda)$ such that $|\Lambda_1/\varphi(\Lambda)| > 1$. The following construction is due to Fadeev [18]; see also Roggenkamp and Huber-Dyson [50].

Assume without loss that Λ is imbedded in $A = \Lambda_{\mathbf{Q}}$ via $\lambda \mapsto \lambda \otimes 1$, and suppose that Λ is degenerate, so that $\mathrm{rad}(A) \neq 0$. Put $\mathbf{a} = \mathrm{rad}(A)$ and let $n = \max\{r \geq 1 : \mathbf{a}^r \neq 0\}$. For each k, $1 \leq k \leq n$, put $L_k = \Lambda \cap \mathbf{a}^k$. It is straightforward to see that L_n is a full sublattice in $\mathbf{a}^k$, and that Λ/L_n is torsion free. Put

$$\Lambda_1 = \Lambda + \frac{1}{2}L_1 + \frac{1}{2^2}L_2 + \cdots + \frac{1}{2^n}L_n$$

Then clearly Λ_1 is a subring of A. Moreover, Λ_1 is a $\mathbf{Z}$-lattice in A, and $\Lambda \subset \Lambda_1$. Since $\Lambda \otimes \mathbf{Q} \subset \Lambda_1 \otimes \mathbf{Q} \subset A = \Lambda \otimes \mathbf{Q}$, then $\Lambda \otimes \mathbf{Q} = \Lambda_1 \otimes \mathbf{Q}$, and so $\mathrm{rk}_{\mathbf{Z}}(\Lambda_1) = \mathrm{rk}_{\mathbf{Z}}(\Lambda)$, and so Λ_1/Λ is finite.

Suppose that $\Lambda_1 = \Lambda$. Then multiplication by 2^n gives $L_n \subset 2\Lambda$, and so $L_n \subset (2\Lambda) \cap \mathbf{a}^n$. However, since Λ/L_n is torsion free, we see that $(2\Lambda) \cap \mathbf{a}^n = 2L_n$. Hence $L_n \subset 2L_n$, and so $L_n = 2L_n$. This is a contradiction, since $\mathbf{L}_n$ is a $\mathbf{Z}$-lattice. Thus the inclusion $\Lambda \subset \Lambda_1$ is strict.

This is now enough to establish the conclusion, since $\Lambda_1 \otimes \mathbf{Q} \cong \Lambda \otimes \mathbf{Q}$ so that Λ_1 is also degenerate, and we may apply the construction to Λ_1 to get a proper finite index imbedding $\Lambda_1 \subset \Lambda_2$. Repeating, we get a sequence of proper finite index imbeddings $\varphi_k : \Lambda_{k-1} \to \Lambda_k$. Put $j_k = |\Lambda_k/\varphi_k(\Lambda_{k-1})|$, where $\Lambda_0 = \Lambda$, and $\varphi_0 = \varphi$. Put $\pi_n = \prod_{k=0}^n j_k$. On composing the imbeddings φ_k we see that $(\pi_n)_{0 \leq n}$ is an increasing sequence contained within $c(\Lambda)$. This completes the proof. $\square$

Theorem 7.9: The following conditions on a $\mathbf{Z}$-order Λ are equivalent:

(i) Λ is nondegenerate;
(ii) Λ has the bounded co-index property;

(iii) there exists a finite index imbedding $\varphi : \Lambda \to \Lambda_{max}$ into a maximal order Λ_{max};

(iv) Λ_Q is semisimple.

Proof: (i) and (ii) are equivalent, by (7.3) and (7.8). (i) and (iv) are also equivalent, by (5.3).

We show that (iii) $\implies$ (i). Suppose therefore that $\varphi : \Lambda \to \Lambda_{max}$ is an imbedding into a maximal order Λ_{max}, with finite co-index j. It follows from Fadeev's Theorem that Λ_{max} is nondegenerate, so that $\text{Disc}(\Lambda_{max}) \neq 0$. However, $\text{Disc}(\Lambda) = j^2 \text{Disc}(\Lambda_{max})$, so that Λ is nondegenerate. This proves (iii) $\implies$ (i).

Finally, it is clear that (ii) $\implies$ (iii). For if $c(\Lambda)$ is bounded, let $\varphi : \Lambda \to \Lambda_1$ be a finite index imbedding into an order Λ_1 for which the co-index j achieves the maximal value $\max\{c(\Lambda)\}$. Then Λ_1 is necessarily maximal, otherwise, if $\psi : \Lambda_1 \to \Lambda_2$ is a finite index imbedding with co-index $k > 1$, then $\psi \circ \varphi : \Lambda \to \Lambda_2$ is a finite index imbedding with co-index $jk > j = \max\{c(\Lambda)\}$, contradicting the definition of $\max\{c(\Lambda)\}$. This completes the proof. $\square$

The study of maximal orders in a general semisimple algebra reduces to that of maximal orders in the simple factors; let $\mathcal{A}$ be a finite-dimensional semisimple algebra over $\mathbf{Q}$ with Wedderburn decomposition

$$\mathcal{A} = \mathcal{A}_1 \oplus \cdots \oplus \mathcal{A}_m$$

where $\mathcal{A}_1, \ldots, \mathcal{A}_m$ are the simple two-sided ideals of $\mathcal{A}$. Let $\Lambda \subset \mathcal{A}$ be a maximal order; then Λ is a direct sum

$$\Lambda = \Lambda_1 \oplus \cdots \oplus \Lambda_m$$

where $\Lambda_i \subset \mathcal{A}_i$ is a maximal order.

8 Examples

If R is a commutative ring and $a, b \in R^*$, we denote by $(\frac{a,b}{R})$ the *quaternion algebra* over R, with R-basis $\{1, i, j, k\}$, and multiplication given by

$$i^2 = a \cdot 1; \quad j^2 = b \cdot 1; \quad ij = -ji = k$$

When $R = \mathbf{K}$ is a field, $(\frac{a,b}{K})$ is a simple $\mathbf{K}$-algebra, and is a division algebra if and only if the equation

$$ax_1^2 + bx_2^2 - abx_3^2 = 0$$

has no nonzero solution $\mathbf{x} = (x_1, x_2, x_3) \in \mathbf{K}^3$. We denote by $\mathbf{H}$ the standard quaternion algebra

$$\mathbf{H} = \left(\frac{-1, -1}{\mathbf{R}} \right)$$

over $\mathbf{R}$.

Following Swan [59], we exhibit two distinct maximal orders in a quaternionic division algebra $\mathbf{Q}$-algebra $(\frac{-1,-1}{K})$, where K is a a totally real number field of degree 4 over $\mathbf{Q}$. Our examples will be quaternionic conjugates of Swan's.

Let $\mathcal{A}$ be an algebra over a commutative ring R; we denote the *group of units* of $\mathcal{A}$ by U($\mathcal{A}$). In the case where $\mathcal{A} = \mathbf{H}$, U($\mathbf{H}$) is the product Spin(3) $\times$ $\mathbf{R}_+$, where Spin(3), the unit sphere in $\mathbf{H}$, is the simply connected covering of the group SO(3) of rotations in $\mathbf{R}^3$.

If $R = \mathbf{Q}$ and Λ is a $\mathbf{Z}$-order in $\mathcal{A}$ we denote by $U_0(\Lambda)$ the intersection of U(Λ) with the the *commutator subgroup* of U($\mathcal{A}$)

$$U_0(\Lambda) = U(\Lambda) \cap [U(\mathcal{A}), U(\mathcal{A})]$$

The ring homomorphism $\mathcal{A} \to \mathcal{A} \otimes_{\mathbf{Q}} \mathbf{R}; x \mapsto x \otimes 1$ induces an imbedding U($\mathcal{A}$) $\to$ U($\mathcal{A} \otimes_{\mathbf{Q}} \mathbf{R}$) and so we get a natural imbedding of $U_0(\Lambda)$ into the commutator subgroup $[U(\mathcal{A} \otimes_{\mathbf{Q}} \mathbf{R}), U(\mathcal{A} \otimes_{\mathbf{Q}} \mathbf{R})]$. However, U($\mathcal{A} \otimes_{\mathbf{Q}} \mathbf{R}$) has a natural structure as a Lie group, in which the commutator subgroup $[U(\mathcal{A} \otimes_{\mathbf{Q}} \mathbf{R}), U(\mathcal{A} \otimes_{\mathbf{Q}} \mathbf{R})]$ is a closed subgroup and so is itself a real Lie group. Moreover, this imbedding is discrete; that is:

Proposition 8.1: Let Λ be a $\mathbf{Z}$-order in a finite-dimensional algebra $\mathcal{A}$ over $\mathbf{Q}$; then $U_0(\Lambda)$ imbeds as a discrete subgroup of $[U(\mathcal{A} \otimes_{\mathbf{Q}} \mathbf{R}), U(\mathcal{A} \otimes_{\mathbf{Q}} \mathbf{R})]$.

We take $\mathcal{A}$ to be a quaternion algebra $\mathcal{A} = (\frac{-1,-1}{\mathbf{K}})$, where $\mathbf{K}$ is a totally real algebraic extension of finite degree d over $\mathbf{Q}$. In this case there is an isomorphism

$$\mathcal{A} \otimes_{\mathbf{Q}} \mathbf{R} \cong \underbrace{\mathbf{H} \times \cdots \times \mathbf{H}}_{d}$$

then

$$[U(\mathcal{A} \otimes_{\mathbf{Q}} \mathbf{R}), U(\mathcal{A} \otimes_{\mathbf{Q}} \mathbf{R})] \cong \underbrace{\text{Spin}(3) \times \cdots \times \text{Spin}(3)}_{d}$$

which is compact. Since $U_0(\Lambda)$ imbeds as a discrete subgroup in a compact group, it is finite; to summarize:

Proposition 8.2: Let $\mathbf{K}$ be a totally real algebraic extension of finite degree d over $\mathbf{Q}$, let $\mathcal{A}$ be a quaternion algebra of the form $\mathcal{A} = (\frac{-1,-1}{\mathbf{K}})$, and let $\Lambda \subset \mathcal{A}$ be a $\mathbf{Z}$-order; then $U_0(\Lambda)$ is finite.

Note that, if Λ is a $\mathbf{Z}$-order in the quaternion algebra $\mathcal{A} = (\frac{-1,-1}{\mathbf{K}})$, then, since K is totally real, $U(\Lambda)$ actually imbeds as a group of units in $U(\mathbf{H})$. In particular, the finite group $U_0(\Lambda)$ imbeds as a subgroup of $U(\mathbf{H})$.

Now the isomorphism types of finite subgroups of $U(\mathbf{H})$ are known ([82], p. 88); with three exceptions, a finite subgroup of $\mathbf{H}^*$ is either cyclic or generalized quaternion. The exceptions are:

  (i) T^*, the binary tetrahedral group of order 24;
 (ii) O^*, the binary octahedral group of order 48; and
(iii) I^*, the binary icosahedral group of order 120.

We proceed to our examples: put $\zeta = \exp(\frac{2\pi i}{16})$ and put $K = \mathbf{Q}(\zeta + \bar{\zeta})$; then K is a totally real number field of degree 4 over $\mathbf{Q}$ whose ring of integers R takes the form

$$R = \mathbf{Z}[x]/(x^4 - 4x^2 + 2)$$

Observe that the minimal polynomial $x^4 - 4x^2 + 2$ over $\mathbf{Q}$ factorizes as

$$x^4 - 4x^2 + 2 = (x - \tau)(x + \tau)(x - \tilde{\tau})(x + \tilde{\tau})$$

where $\tau = \sqrt{2 + \sqrt{2}}$ and $\tilde{\tau} = \sqrt{2 - \sqrt{2}}$. Moreover, the Galois group $\mathrm{Gal}(K/\mathbf{Q})$ is cyclic of order 4 and a generator acts on the roots of $x^4 - 4x^2 + 2 = (x - \tau)(x + \tau)(x - \tilde{\tau})(x + \tilde{\tau})$ by means of

$$\tau \mapsto -\tilde{\tau}; \quad \tilde{\tau} \mapsto \tau$$

Since K is totally real, the quaternion algebra $A = (\frac{-1,-1}{K})$ is a division algebra of dimension 16 over $\mathbf{Q}$. Next define $\alpha, \beta \in A$ by

$$\alpha = \frac{\tilde{\tau} + i\tau}{2}; \quad \beta = \frac{(\tau - \tilde{\tau})(1 - j)}{2}$$

Then it is easy to compute

$$\alpha\beta = \frac{(\tau^2 - 3) + i + (3 - \tau^2)j - ij}{2} \in A$$

We put $\Lambda = \mathrm{span}_R\{1, \alpha, \beta, \alpha\beta\}$.

Proposition 8.3: Λ is a subring of A.

Proof: Since Λ is an R-submodule of A, it suffices to show that Λ is closed with respect to multiplication. The product $\alpha\beta$ is in Λ by definition, and the products $\alpha^2, \beta^2, \beta\alpha$ all belong to Λ as we see from the following identities

$$\alpha^2 = -1 + (\tau^3 - 3\tau)\alpha$$
$$\beta^2 = (4 - \tau^2) + (4\tau - \tau^3)\beta$$
$$\beta\alpha = (3 - \tau^2) + (4\tau - \tau^3)\alpha + (\tau^3 - 3\tau)\beta - \alpha\beta$$

Straightforward substitions now reveal that $\alpha^2\beta$, $\alpha\beta^2$ and $\alpha\beta\alpha$ all belong to Λ.

For $\xi \in \Lambda$ we denote by $\rho_\xi : A \to A$ the mapping

$$\rho_\xi(x) = x\xi$$

To show that Λ is a subring of A, it suffices to show that $\rho_\xi(\Lambda) \subset \Lambda$ for any $\xi \in \Lambda$. Since the products $\alpha^2, \beta\alpha, \alpha\beta\alpha$ all belong to Λ, we see easily that $\rho_\alpha(\Lambda) \subset \Lambda$. Likewise, $\alpha\beta, \beta^2, \alpha\beta^2$ all belong to Λ, so $\rho_\beta(\Lambda) \subset \Lambda$. Clearly

$$\rho_{\alpha\beta}(\Lambda) \subset \Lambda$$

since $\rho_{\alpha\beta} = \rho_\beta \circ \rho_\alpha$. Upon writing an arbitrary element $\xi \in \Lambda$ in the form $\xi = c_0 \cdot 1 + c_1 \cdot \alpha + c_2 \cdot \beta + c_3 \cdot \alpha\beta$, with $c_i \in R$, we see that

$$\rho_\xi = c_0 \cdot \mathrm{Id} + c_1 \cdot \rho_\alpha + c_2 \cdot \rho_\beta + c_3 \cdot \rho_{\alpha\beta}$$

Hence $\rho_\xi(\Lambda) \subset \Lambda$, and Λ is closed under multiplication.   □

α, β are units in Λ, since $\alpha^{16} = \beta^{16} = 1$; taking Q to be the subgroup of $U(\Lambda)$ generated by α, β, we obtain:

Proposition 8.4: Q is contained in $U_0(\Lambda)$ and is isomorphic to the generalized quaternion group $Q(32)$.

Proof: Note that the commutator quotient group of $U(A)$ is a torsion free abelian group. It follows that every element of finite order in $U(\Lambda)$ is contained in $U_0(\Lambda)$. The verification that $Q \cong Q(32)$ is straightforward.   □

Now put $\Theta = \mathrm{span}_R\{1, \xi, \eta, \xi\eta\}$, where $\xi = \frac{(1-i)}{\sqrt{2}}$ and $\eta = \frac{(1-j)}{\sqrt{2}}$; then $\xi\eta = \frac{(1-i-j+ij)}{2}$. Observe that $i \in \Theta$, since

$$i = 1 - \sqrt{2}\xi$$

likewise $j, ij \in \Theta$, and a somewhat easier computation than (8.3) shows:

Proposition 8.5: Θ is a subring of A.

In fact, straightforward computation shows that the reduced discriminant equals -1 for Λ and Θ, so that:

Proposition 8.6: Λ and Θ are both maximal orders in A.

However:

Proposition 8.7: $\Lambda \not\cong \Theta$.

Proof: Observe that the elements ξ, η are units in Θ, since $\xi^8 = \eta^8 = 1$. Consequently $\xi\eta \in U_0(\Theta)$, since we have

$$\xi\eta = \eta\xi^{-1}\eta^{-1}\xi$$

A straightforward calculation shows $(\xi\eta)^3 = -1$; hence $(\xi\eta)^2$ has order 3, and the finite group $U_0(\Theta)$ has order divisible by 3. However, since $U_0(\Lambda)$ contains a subgroup isomorphic to $Q(32)$ it follows that $U_0(\Lambda)$ has order divisible by 32.

Suppose that $\Theta \cong \Lambda$, then $U_0(\Lambda) \cong U_0(\Theta)$ has order divisible by 96. This eliminates the possibility that $U_0(\Lambda)$ is one of the exceptional groups I^*, O^*, T^*.

Since $U_0(\Lambda)$ is non-abelian, it must therefore be a generalized quaternion group, and so must have an element of order 48. In particular, Λ must contain a copy of the field $\mathbf{Q}[x]/(c_{48}(x))$, where $c_d(x)$ denotes the dth cyclotomic polynomial, that is, the product $c_d(x) = \prod(x - \zeta)$, where ζ ranges over the *primitive dth-roots of unity*. In this case, $c_{48}(x) = x^{16} - x^8 + 1$, and so

$$\dim_{\mathbf{Q}}(\mathbf{Q}[x]/(c_{48}(x))) = 16 = \dim_{\mathbf{Q}}(\Lambda \otimes_{\mathbf{Z}} \mathbf{Q}) = A$$

It follows that $A \cong \mathbf{Q}[x]/(c_{48}(x))$. But this is a contradiction since A is non-abelian. Thus $\Theta \not\cong \Lambda$ as claimed. $\qquad\square$

Chapter 2

Representation theory of finite groups

We review briefly the representation theory of a finite group G over a field k of characteristic zero; expressed differently, the module theory of the group algebra $k[G]$. The case $k = \mathbf{Q}$ gives a first approximation to the module theory of $\mathbf{Z}[G]$.

9 Group representations

Let G be a group and $\mathbf{A}$ be a commutative ring; by a $(G, \mathbf{A})$-representation, or $\mathbf{A}$-representation if the group G is understood, we mean a pair (V, ρ), where V is an $\mathbf{A}$-module and $\rho : G \to GL_{\mathbf{A}}(V)$ is a group homomorphism. If (V, ρ), (W, σ) are $(G, \mathbf{A})$-representations, then by a $(G, \mathbf{A})$-morphism $\Psi : (V, \rho) \to (W, \sigma)$ we mean an $\mathbf{A}$-linear map $\Psi : V \to W$ such that $\Psi(\rho(g)(\underline{v})) = \sigma(g)(\Psi(\underline{v}))$ for all $g \in G$, $\underline{v} \in V$.

If $(V_i, \rho_i)_{i \in I}$ are $(G, \mathbf{A})$-representations, we define the direct sum representation $\bigoplus_{i \in I}(V_i, \rho_i)$ thus

$$\bigoplus_{i \in I}(V_i, \rho_i) = \left(\bigoplus_{i \in I} V_i, \bigoplus_{i \in I} \rho_i \right)$$

where $\bigoplus_{i \in I} \rho_i : G \to GL_{\mathbf{A}}(\bigoplus_{i \in I} V_i)$ is the homomorphism $(\bigoplus_{i \in I} \rho_i)(g)(v)_j = \rho_j(g)(v_j)$.

When V is free module of rank n over $\mathbf{A}$, $GL_{\mathbf{A}}(V)$ is isomorphic to $GL_n(\mathbf{A})$, the group of invertible $n \times n$ matrices over $\mathbf{A}$, and a $(G, \mathbf{A})$ representation can be interpreted in explicit coordinate form as a homomorphism $\rho : G \to GL_n(\mathbf{A})$.

Recall the group algebra construction; if G is a group and $\mathbf{A}$ is a commutative ring, the group algebra $\mathbf{A}[G]$ consists of all $\mathbf{A}$-valued functions with finite support defined on G; it is naturally an $\mathbf{A}$-module under pointwise addition and scalar multiplication.

If $g \in G$, we denote by $\hat{g}$ the element of $\mathbf{A}[G]$ defined thus

$$\hat{g}(x) = \begin{cases} 1 & \text{if } x = g \\ 0 & \text{if } x \neq g \end{cases}$$

The set $\{\hat{g}\}_{g \in G}$ is a basis for $\mathbf{A}[G]$ over $\mathbf{A}$, leading to the canonical representation of elements in $\mathbf{A}[G]$ as finite sums of the form

$$\alpha = \sum_{g \in G} \alpha_g \hat{g}$$

in particular, $\mathbf{A}[G]$ is a free module over $\mathbf{A}$. Moreover, $\mathbf{A}[G]$ acquires the structure of an $\mathbf{A}$-algebra in which the multiplication is given by

$$(\alpha * \beta)(g) = \sum_{h \in G} \alpha(gh^{-1})\beta(h)$$

Here the multiplicative identity is $\hat{1}$, so that the inclusion $\mathbf{A} \to \mathbf{A}[G]$ is $\lambda \mapsto \lambda.\hat{1}$. If (V, ρ) is an $\mathbf{A}$-representation of G, we associate with (V, ρ) a right $\mathbf{A}[G]$-module $V(\rho)$ whose underlying $\mathbf{A}$-module is V, and on which $\mathbf{A}[G]$ acts by means of

$$\mathbf{v} \bullet \left(\sum_{g \in G} a_g \hat{g} \right) = \sum_{g \in G} a_g \rho(g^{-1})(\mathbf{v})$$

Conversely, if V is a finite-dimensional right $\mathbf{A}[G]$-module, we associate with V a finite-dimensional $\mathbf{A}$-representation $\rho_V : G \to GL_{\mathbf{A}}(V)$ by means of

$$\rho_V(g)(\mathbf{v}) = \mathbf{v} \cdot \hat{g}^{-1}$$

This correspondence between $\mathbf{A}$-representations of G and right modules over $\mathbf{A}[G]$ is clearly 1–1.

Any group ring $\mathbf{A}[G]$ admits a *canonical involution* $\tau : \mathbf{A}[G] \to \mathbf{A}[G]$ given by

$$\tau \left(\sum_{g \in G} a_g \hat{g} \right) = \sum_{g \in G} a_g \hat{g}^{-1}$$

When (V, ρ) is an $\mathbf{A}$-representation, an $\mathbf{A}$-submodule $W \subset V$ is said to be (G, ρ)-invariant (or just G-invariant, if ρ is understood) when, for all $g \in G$, $\rho(g)(W) \subset W$. Since $1_V = \rho(g)\rho(g^{-1})$, this is equivalent to requiring that $\rho(g)(W) = W$. A G-invariant subspace $\tilde{V}$ of V defines the subrepresentation $(\tilde{V}, \tilde{\rho})$ of (V, ρ) by requiring that $\tilde{\rho} : G \to GL_k(\tilde{V})$ be the homomorphism $\tilde{\rho}(g)(v) = \rho(g)(v)$.

Being a module over itself, $\mathbf{A}[G]$ gives rise, via this correspondence, to a representation of particular importance, the regular representation $\rho_{\text{reg}} : G \to GL_{\mathbf{A}}(\mathbf{A}[G])$; that is

$$\rho_{\text{reg}}(g)(\mathbf{v}) = g.\mathbf{v}$$

Any representation $\rho : G \to GL_{\mathbf{A}}(V)$ extends to a homomorphism of $\mathbf{A}$-algebras, denoted by the same symbol, $\rho : \mathbf{A}[G] \to \text{End}_{\mathbf{A}}(V)$; thus

$$\rho\left(\sum_{g \in G} a_g g\right) = \sum_{g \in G} a_g \rho(g).$$

We make no notational distinction between appearances of ρ as a group representation and as an algebra representation.

Suppose that $\mathbf{A} = k$ is a field; a k-representation (V, ρ) of G; (V, ρ) is said to be *irreducible over* k when the only G-invariant k-linear subspaces of V are $\{0\}$ and V itself, and *completely reducible over* k when there is an isomorphism in $\langle G, k \rangle$

$$\Psi : (V, \rho) \to \bigoplus_{i \in I}(V_i, \rho_i)$$

where each (V_i, ρ_i) is irreducible over k. We see easily that:

(9.1) The correspondence between k-representations of G and right modules over $k[G]$ preserves direct sums, and irreducible (G, k) representatations correspond to to simple $k[G]$-modules.

10 Maschke's Theorem

A classical result of Maschke reduces the study of (G, k)-representations to the Wedderburn theory of semisimple algebras developed in Chapter I (although historically, the development went the other way).

Theorem 10.1: (Maschke): Let G be a finite group and let k be a field whose characteristic does not divide the order of G; then every $k[G]$-module is projective.

Proof: Let $0 \to A \xrightarrow{i} B \xrightarrow{p} C \to 0$ be an exact sequence of $k[G]$-modules. Forgetting the G-actions, this is an exact sequence of k-modules, and so splits. In particular, there exists a k-linear map $s : C \to B$ such that $p \circ s = 1_C$. The map $t : C \to B$ defined by $t(x) = \sum_{g \in G} s(xg)g^{-1}$ is clearly k-linear, and the computation

$$t(xh) = \sum_{g \in G} s((xh)g)g^{-1} = \left\{\sum_{g \in G} s((xh)g)g^{-1}h^{-1}\right\} h$$

$$= \left\{\sum_{g \in G} s(x(hg))(hg)^{-1}\right\} h = t(x)h$$

for $x \in C$, $h \in G$, shows that t is also $k[G]$-linear. Moreover, for all $x \in C$

$$p \circ t(x) = p \left(\sum_{g \in G} s(xg)g^{-1} \right) = \sum_{g \in G} (p \circ s)(xg)g^{-1}$$

so that $p \circ t(x) = \sum_{g \in G} xgg^{-1} = |G| \cdot x$. However, $|G|.1$ is invertible in k, and the map $\sigma : C \to B$ given by

$$\sigma(x) = \frac{t(x)}{|G|}$$

is $k[G]$-linear, and $p \circ \sigma = 1_C$. Thus C is projective since each exact sequence of $k[G]$-modules ending in C splits, and this completes the proof. $\quad\square$

Applying Theorem (3.5) to (10.1) above, we see that:

Corollary 10.2: Let G be a finite group and let k be a field whose characteristic does not divide $|G|$; then $k[G]$ is semisimple.

Corollary (10.2) is false if the characteristic of k is allowed to divide the order of G. As an example, we may take $\mathbf{F}_2[C_2]$ where $\mathbf{F}_2$ is the field with two elements. In this case

$$\mathbf{F}_2[C_2] = \{0, 1, T, 1 + T\}$$

and

$$\mathcal{I} = \{0, 1 + T\}$$

is a (two-sided) ideal which does not possess a complementary ideal. Since $k[G]$ is finite dimensional over k, (10.2) has the following consequences.

Corollary 10.3: Let G be a finite group and let k be a field whose characteristic does not divide the order of G; then every finitely generated $k[G]$-module is semisimple.

Corollary 10.4: When G is a finite group, and k is a field of characteristic coprime to the order of G, there is an isomorphism of k-algebras

$$k[G] \cong M_{n_1}(D_1) \times \cdots \times M_{n_m}(D_m)$$

for some positive integer $m(= m(G, k))$ and some sequence $(D_i, n_i)_{1 \leq i \leq m}$ of positive integers n_i, and division algebras D_i over k, determined uniquely up to order and isomorphism by G and k.

We have noted, in (9.1), that the correspondence between k-representations and $k[G]$-modules preserves direct sums; moreover, irreducible representations correspond to simple modules. From Maschke's Theorem we see that:

Proposition 10.5: Let G be a finite group, and let k be a field of characteristic coprime to $|G|$; then all (G, k)-representations are completely reducible.

Indeed, this was the original formulation of Maschke's result [37].

Suppose that the characteristic of k is coprime to $|G|$. From the uniqueness of the Wedderburn decomposition

$$k[G] \cong M_{n_1}(D_1) \times \cdots \times M_{n_m}(D_m)$$

each simple right module over $k[G]$ can be identified with a simple right ideal $N_i \cong D_i^{n_i}$ in $M_{n_1}(D_1)$. Hence each simple *right* ideal N of $k[G]$ is isomorphic to N_i, for some unique i, $1 \leq i \leq m$. The trivial one-dimensional representation of G occurs with multiplicity $= 1$ in $k[G]$, allowing us to write the Wedderburn decomposition in the form

$$k[G] \cong k \times M_{n_2}(D_2) \times \cdots \times M_{n_m}(D_m)$$

Here we are taking $D_1 = k$ and $n_1 = 1$ for the factor of the trivial representation.

So far our results have only assumed that k is a field of characteristic coprime to the order of G. From now on, without further mention, the field k is assumed to be of characteristic zero. When $k = \bar{k}$ is algebraically closed, the only division algebra of finite dimension over $\bar{k}$ is $\bar{k}$ itself, so that the Wedderburn decomposition takes the form

$$\bar{k}[G] \cong M_{d_1}(\bar{k}) \times M_{d_2}(\bar{k}) \cdots \times M_{d_m}(\bar{k})$$

where $d_1 = 1$. Here the number m of simple factors is equal to the number of conjugacy classes of G, and the integers d_i are called the *degrees of the absolutely simple representations* of G. Clearly

$$\sum_{i=1}^{m} d_i^2 = |G|$$

11 Division algebras over Q

If G is a finite group and $\mathbf{K}$ is a field of characteristic zero, the theorems of Maschke and Wedderburn imply the existence of a product decomposition of $\mathbf{K}[G]$ in the form

$$\mathbf{K}[G] \cong M_{n_1}(D_1) \times \cdots \times M_{n_m}(D_m)$$

where $D_1, \ldots, D_m$ are finite-dimensional division algebras over $\mathbf{K}$; as above, $n_1 = 1$ and $D_1 = \mathbf{K}$.

Eventually, we wish to drop the condition that $\mathbf{K}$ be a field and study modules over $\mathbf{Z}[G]$; there are three cases of increasing interest, and difficulty, which may be seen as successive approximations to $\mathbf{Z}[G]$. When $k = \mathbf{C}$, the only finite-dimensional division algebra over $\mathbf{C}$ is $\mathbf{C}$ itself, and the complex Wedderburn decomposition of a finite group G then assumes the form

$$\mathbf{C}[G] \cong \mathbf{C} \times M_{d_2}(\mathbf{C}) \cdots \times M_{d_m}(\mathbf{C})$$

where the number m of simple factors equals the number of conjugacy classes of G.

Slightly more complicated is the case where $\mathbf{K} = \mathbf{R}$; then the real Wedderburn decomposition takes the form

$$\mathbf{R}[G] \cong \mathbf{R} \times M_{n_2}(D_2) \times \cdots \times M_{n_m}(D_m)$$

where for $i \geq 2$, D_i is either $\mathbf{R}$, $\mathbf{C}$ or $\mathbf{H}$. As we shall see, in Chapter 3, when discussing the *Eichler condition*, a particular difficulty arises in the case where $\mathbf{H}$ is a direct factor of $\mathbf{R}[G]$, that is when $D_i = \mathbf{H}$ and $n_i = 1$ for some i,

A rather closer approximation to $\mathbf{Z}[G]$ is $\mathbf{Q}[G]$; in this case, there are infinitely many isomorphically distinct finite-dimensional division algebras over $\mathbf{Q}$. The degree of complication, though infinite, is not completely arbitrary however, and there is at least one *a priori* restriction which is imposed on the type of the division algebra D_i occurring in a Wedderburn decomposition. In fact, for any totally real field $\mathbf{K}$, the canonical involution

$$\tau(a) = \sum_{g \in \Phi} a_g g^{-1} \quad \text{where} \quad a = \sum_{g \in \Phi} a_g g$$

on $\mathbf{K}[G]$ is positive in the sense that $a\tau(a)$ is a positive element of $\mathbf{K}$ whenever $a \neq 0$. A theorem of Albert ([1]) (see also [51]) now implies an isomorphism of involuted $\mathbf{K}$-algebras

$$(\mathbf{K}[G], \tau) \cong \left(M_{n_1}(D_1), \hat{\tau}_1 \right) \times \cdots \times \left(M_{n_m}(D_m), \hat{\tau}_m \right)$$

where each D_i is a finite-dimensional division algebra over $\mathbf{K}$, and τ_i is a positive involution on D_i. In particular, the division algebras $\mathcal{D}_i$ which occur in a rational Wedderburn decomposition

$$\mathbf{Q}[G] \cong \mathbf{Q} \times M_{n_2}(D_2) \times \cdots \times M_{n_m}(D_m)$$

admit positive involutions. We proceed to sketch Albert's classification of these positive division algebras. First recall the construction of cyclic algebras; fix the following notation:

$\mathcal{R}$: a commutative ring,

n: an integer ≥ 2,

s: a ring automorphism of $\mathcal{R}$ satisfying $s^n = \text{Id}$,

a: a nonzero element of $\mathcal{R}$ such that $s(a) = a$.

The cyclic algebra $C_n(\mathcal{R}, s, a)$ is the two-sided free $\mathcal{R}$-module of rank n, with basis $([X^r])_{0 \le r \le n-1}$, subject to the relations

$$[X^r]\lambda = s^r(\lambda)[X^r] \quad (\lambda \in \mathcal{R})$$

$C_n(\mathcal{R}, s, a)$ is an algebra over the fixed point ring $\mathbf{E} = \{x \in \mathcal{R} : s(x) = x\}$ with multiplication determined by

$$[X][X^r] = \begin{cases} [X^{r+1}] & 0 \le r < n - 1 \\ a[X^0] & r = n - 1 \end{cases}$$

We note that when $\mathcal{R} = \mathbf{K}$ is a field and s has order n (rather than merely satisfying $s^n = \mathrm{Id}$) then the fixed point field ${}^s\mathbf{K} = \mathbf{E}$ is actually the centre of $C_n(\mathbf{K}, s, a)$; indeed, in some definitions it is required that s has order n precisely to guarantee this outcome. However, we find it more useful to work with the weaker condition $s^n = \mathrm{Id}$. This construction is natural with respect to direct products; that is, we have:

Proposition 11.1:

$$C_n(\mathcal{R}_1 \times \mathcal{R}_2, s_1 \times s_2, (a_1, a_2)) \cong C_n(\mathcal{R}_1, s_1, a_1) \times C_n(\mathcal{R}_2, s_2, a_2)$$

Albert's classification of simple $\mathbf{Q}$-algebras which admit a positive involution is as follows: let $\mathbf{K}$ be a subfield of $\mathbf{R}$, and let $\mathbf{A}$ be a finite-dimensional semisimple $\mathbf{K}$-algebra; by an *algebra involution* τ on $\mathbf{A}$, we mean an isomorphism of $\mathbf{A}$ with its opposite algebra such that $\tau^2 = 1_A$; τ is said to be *positive* when $\mathrm{tr}_{\mathbf{K}}(x\,\tau(x)) > 0$ for all nonzero $x \in \mathbf{A}$, where '$\mathrm{tr}_{\mathbf{K}}$' denotes '*reduced trace*'. If $\mathbf{A}$ is expressed as a sum of simple ideals $\mathbf{A} = \mathbf{A}_1 \oplus \mathbf{A}_2 \oplus \cdots \oplus \mathbf{A}_n$, then by uniqueness of the Wedderburn decomposition, an algebra involution τ induces an involution τ_* on the index set $\{1, 2, \ldots, n\}$ by the condition that

$$\tau(\mathbf{A}_i) = \mathbf{A}_{\tau_*(i)}$$

If $\tau_*(i) \ne i$, it is immediate that, for any $x \in \mathbf{A}_i$, $x\tau(x) = 0$. Hence the positivity condition forces $\tau_* = \mathrm{Id}$; that is:

Proposition 11.2: Let τ be a positive involution on finite-dimensional semisimple K-algebra $\mathcal{A}$; then there is an isomorphism of involuted K-algebras

$$(\mathbf{A}, \tau) = (\mathbf{A}_1, \tau_1) \oplus (\mathbf{A}_2, \tau_2) \oplus \cdots \oplus (\mathbf{A}_n, \tau_n)$$

in which each $(\mathbf{A}_i, \tau_i)$ is a simple positively involuted algebra.

If D is a finite-dimensional division algebra over $\mathbf{K}$, an involution σ on D extends to an involution $\hat{\sigma}$ on $M_n(D)$ thus

$$\hat{\sigma}((x_{ij})) = (\sigma(x_{ji}))$$

with transposed indices as indicated. By the Skolem–Noether Theorem, each involution on $M_n(D)$ has this form. Moreover, $\hat{\sigma}$ is positive if and only if σ is positive.

An involution τ of a simple algebra $\mathbf{A}$ is said to be *of the first kind* when it restricts to the identity on the centre $\mathcal{Z}$ of $\mathbf{A}$; otherwise, τ is said to be *of the second kind*. A quaternion algebra $(\frac{a,b}{\mathbf{E}})$ admits two essentially distinct involutions of the first kind, namely *conjugation*, c, and *reversion*, r, defined thus

$$c(x_0 + x_1 i + x_2 j + x_3 k) = x_0 - x_1 i - x_2 j - x_3 k$$
$$r(x_0 + x_1 i + x_2 j + x_3 k) = x_0 + x_1 i - x_2 j + x_3 k$$

Albert showed that a positively involuted division algebra (D, τ), of finite dimension over $\mathbf{Q}$, falls into one of four classes; here $\mathbf{E}$ and $\mathbf{K}$ are algebraic number fields:

I $D = \mathbf{E}$ is totally real and $\tau = 1_{\mathbf{E}}$;

II $D = (\frac{a,b}{\mathbf{E}})$, where $\mathbf{E}$ is totally real, a is totally positive, b is totally negative, and τ is reversion;

III $D = (\frac{a,b}{\mathbf{E}})$, where $\mathbf{E}$ is a totally real, a and b are both totally negative, and τ is conjugation;

IV $D = \mathcal{C}_m(\mathbf{K}, s, a)$, where s is an automorphism of $\mathbf{K}$, of order m, whose fixed point field $\mathbf{E}$ is an imaginary quadratic extension, $\mathbf{E} = \mathbf{E}_0(\sqrt{b})$, of a totally real field $\mathbf{E}_0$, and $a \in \mathbf{E}$; moreover, if $\mathbf{L}$ is a maximal totally real subfield of $\mathbf{K}$, there exists a totally positive element $d \in \mathbf{L}$ such that $N_{\mathbf{E}/\mathbf{E}_0}(a) = N_{\mathbf{L}/\mathbf{E}_0}(d)$.

12 Examples of Wedderburn decompositions

In what follows, we give the rational Wedderburn decompositions for some familiar finite groups, in particular, cyclic, dihedral and quaternionic groups.

Cyclic groups

We write the cyclic group C_n in the form

$$C_n = \langle x \mid x^n = 1 \rangle$$

denoting by $\mathbf{Q}[x]$ the polynomial ring in 1-variable x, we may identify $\mathbf{Q}[C_n]$ with the quotient $\mathbf{Q}[x]/(x^n - 1)$. For each positive integer d we put

$$\Omega(d) = \{\zeta \in \mathbf{C} : \zeta^d = 1 \text{ and } \mathrm{ord}(\zeta) = d\}$$

and put

$$c_d(x) = \prod_{\zeta \in \Omega(d)} (x - \zeta)$$

It is well known that each $c_d(x)$ is an irreducible polynomial over $\mathbf{Q}$, so that

$$\mathbf{Q}(d) = \mathbf{Q}[x]/(c_d(x))$$

is a field. Moreover, the factorization of $x^n - 1$ into $\mathbf{Q}$-irreducible factors is given by

$$x^n - 1 = \prod_{d|n} c_d(x)$$

so that the decomposition of $\mathbf{Q}[C_n] \cong \mathbf{Q}[x]/(x^n - 1)$ into simple factors, that is the *rational* Wedderburn decomposition of C_n, is given by

(12.1) $$\mathbf{Q}[C_n] \cong \bigoplus_{d|n} \mathbf{Q}(d).$$

If ζ_d is a primitive dth-root of unity, the fixed field of $\mathbf{Q}(d)$ under complex conjugation is $\mathbf{Q}(\mu_d)$, where $\mu_d = \zeta_d + \bar{\zeta}_d$.

The dihedral group of order 6

The smallest non-abelian group is D_6, the dihedral group of order 6. We write $D_6 = \{1, x, x^2, y, xy, x^2 y\}$ with the relations $x^3 = 1$, $y^2 = 1$ and $yx = x^2 y$. There are three conjugacy classes

$$\{1\}; \quad \{x, x^2\}; \quad \{y, xy, x^2 y\}$$

and consequently three isomorphism classes of $\mathbf{C}$-representations, (V_i, ρ_i), $(1 \leq i \leq 3)$, with $\dim(V_i) = d_i$. Then $d_1^2 + d_2^2 + d_3^2 = 6$, so that, up to order, we may suppose that $d_1 = d_2 = 1$ and $d_3 = 2$. Let ρ_1 be the trivial representation, then we may take

$$\rho_2(x) = 1; \quad \rho_2(y) = -1$$

and

$$\rho_3(x) = \begin{pmatrix} 0 & 1 \\ 1 & -1 \end{pmatrix}; \quad \rho_3(y) = \begin{pmatrix} 0 & 1 \\ 1 & 0 \end{pmatrix}$$

In particular, every simple representation of D_6 is defined over $\mathbf{Z}$, and hence over $\mathbf{Q}$, and the rational Wedderburn decomposition is

$$(12.2) \qquad \mathbf{Q}[D_6] \cong \mathbf{Q} \times \mathbf{Q} \times M_2(\mathbf{Q})$$

It follows that for any field k of characteristic 0

$$k[D_6] \cong k \times k \times M_2(k)$$

The general dihedral group

For each $n \geq 3$, the dihedral group D_{2n} is defined by the presentation

$$D_{2n} = \langle \xi, \eta \mid \xi^n = \eta^2 = 1, \quad \eta \xi \eta^{-1} = \xi^{-1} \rangle$$

If $\mathcal{R}$ is a commutative ring then for any $n \geq 2$, the group ring $\mathcal{R}[D_{2n}]$ can be described as a cyclic algebra in terms of the group ring of C_n thus

$$\mathcal{R}[D_{2n}] \cong C_2(\mathcal{R}[C_n], \widehat{}, 1)$$

where $\widehat{}: \mathcal{R}[C_n] \to \mathcal{R}[C_n]$ is the involution given on group elements by

$$\widehat{g} = g^{-1}$$

We specialize to the case where $\mathcal{R} = \mathbf{Q}$; as we have seen

$$\mathbf{Q}[C_n] \cong \prod_{d \mid n} \mathbf{Q}(d)$$

Under the isomorphism $\mathbf{Q}[C_n] \cong \mathbf{Q}[x]/(x^n - 1) \cong \prod_{d \mid n} \mathbf{Q}(d)$ the canonical involution $\widehat{}: \mathbf{Q}[C_n] \to \mathbf{Q}[C_n]$ induces an involution $\gamma_d : \mathbf{Q}(d) \to \mathbf{Q}(d)$, which is the identity for $d = 1, 2$, and complex conjugation otherwise; for any $n \geq 2$, we obtain

$$(12.3) \qquad \mathbf{Q}[D_{2n}] \cong \prod_{d \mid n} C_2(\mathbf{Q}(d), \gamma_d, 1)$$

Moreover, it is straightforward to see that

$$C_2(\mathbf{Q}(d), \gamma_d, 1) \cong \begin{cases} \mathbf{Q} \times \mathbf{Q} & d = 1, 2 \\ M_2(\mathbf{Q}(\mu_d)) & d \geq 3 \end{cases}$$

Thus, for any $n \geq 2$, we obtain the following complete decomposition formula

$$\mathbf{Q}[D_{2n}] \cong \begin{cases} \mathbf{Q} \times \mathbf{Q} \times \prod_{d \mid n, d \geq 3} M_2(\mathbf{Q}(\mu_d)) & n \text{ odd} \\ \mathbf{Q} \times \mathbf{Q} \times \mathbf{Q} \times \mathbf{Q} \times \prod_{d \mid n, d \geq 3} M_2(\mathbf{Q}(\mu_d)) & n \text{ even} \end{cases}$$

The quaternion group of order 8

$Q(8)$, the quaternion group of order 8, is given by the following presentation

$$Q(8) = \langle X, Y : X^4 = 1; \quad X^2 = Y^2 = (XY)^2; \quad XY = YX^3 \rangle$$

The elements are represented in the normal forms $X^r, X^s Y$ where $r, s = 0, 1, 2, 3$, and there are five conjugacy classes

$$\{1\}; \quad \{X^2\}; \quad \{X, X^3\}; \quad \{Y, X^2 Y\}; \quad \{XY, X^3 Y\}$$

Clearly $d_1^2 + d_2^2 + d_3^2 + d_4^2 + d_5^2 = 8$, so that, up to order, we may suppose $d_1 = d_2 = d_3 = d_4 = 1$ and $d_5 = 2$. The four inequivalent one-dimensional representations are $\tau_1, \tau_2, \tau_3, \tau_4$, where τ_1 is the trivial representation and

$$\tau_2(X) = -1; \quad \tau_2(Y) = 1;$$
$$\tau_3(X) = 1; \quad \tau_3(Y) = -1;$$
$$\tau_4(X) = -1; \quad \tau_4(Y) = -1.$$

In particular, each τ_i is defined over $\mathbf{Q}$.

Over any field $\mathbf{F}$ of characteristic zero, $Q(8)$ admits an irreducible representation σ in the unit group $(\frac{-1,-1}{\mathbf{F}})^*$ by means of

$$\sigma(X) = i; \quad \sigma(Y) = j$$

and we get the decomposition

(12.4) $$\mathbf{F}[G] \cong \mathbf{F} \times \mathbf{F} \times \mathbf{F} \times \mathbf{F} \times \left(\frac{-1, -1}{\mathbf{F}} \right)$$

In the cases $\mathbf{F} = \mathbf{Q}$ or $\mathbf{R}$, $(\frac{-1,-1}{\mathbf{F}})$ is a division algebra. When $\mathbf{F} = \mathbf{C}$, $(\frac{-1,-1}{\mathbf{F}}) \cong M_2(\mathbf{C})$, giving the complex Wedderburn decomposition

$$\mathbf{C}[Q(8)] \cong \mathbf{C} \times \mathbf{C} \times \mathbf{C} \times \mathbf{C} \times M_2(\mathbf{C})$$

The general quaternion group

For any integer $n \geq 3$, the generalized quaternion group $Q(4n)$ is defined by the presentation

$$Q(4n) = \langle a, b \mid a^{2n} = b^2, \quad aba = b \rangle$$

Denoting the canonical generators of D_{2n} by ξ, η, there is a non-trivial central extension

$$1 \to C_2 \to Q(4n) \xrightarrow{\psi_n} D_{2n} \to 1$$

defined by the correspondence $\psi_n(a) = \xi$; $\psi_n(b) = \eta$. This will enable us give an expression for $\mathbf{Q}[Q(4n)]$ directly in terms of $\mathbf{Q}[D_{2n}]$.

We denote by $\mathbf{Q}\langle n \rangle$ the quotient

$$\mathbf{Q}\langle n \rangle = \mathbf{Q}[x]/(x^n + 1)$$

and by σ_n the involution on $\mathbf{Q}\langle n \rangle$, namely

$$\sigma_n(x^r) = -x^{n-r}$$

Theorem 12.5: For each $n \geq 2$, there is an algebra isomorphism

$$\mathbf{Q}[Q(4n)] \cong \mathbf{Q}[D_{2n}] \oplus \mathcal{C}_2(\mathbf{Q}\langle n \rangle, \sigma_n, -1)$$

Proof: Taking the presentation

$$Q(4n) = \langle a, b \mid a^{2n} = b^2, \quad aba = b \rangle$$

we let z denote the central element $z = a^m = b^2$, and denote by θ the translation map

$$\theta : Q(4n) \to Q(4n); \quad \theta(\mathbf{v}) = z\mathbf{v}$$

Then $\theta^2 = \text{Id}$, so that we have a decomposition

$$Q(4n) = J_+ \oplus J_-$$

as a direct sum of two-sided ideals, where J_+, J_- denote respectively the $+1$ and -1 eigenspaces of θ. We may take bases for J_+, J_- as follows

$$J_+ = \text{span}_{\mathbf{Q}} \left\{ a^r b^s \left(\frac{1+z}{2} \right) : 0 \leq r \leq n-1, 0 \leq s \leq 1 \right\}$$

$$J_- = \text{span}_{\mathbf{Q}} \left\{ a^r b^s \left(\frac{1-z}{2} \right) : 0 \leq r \leq n-1, 0 \leq s \leq 1 \right\}$$

It is easy to check that ψ_n induces an algebra isomorphism $\psi_n : J_+ \to \mathbf{Q}[D_{2n}]$. In fact

$$\psi_n \left(a^r b^s \left(\frac{1+z}{2} \right) \right) = \xi^r \eta^s$$

ψ_n is automatically multiplicative, and we have an algebra decomposition

$$\mathbf{Q}[Q(4n)] \cong \mathbf{Q}[D_{2n}] \oplus J_-$$

To complete the proof, it suffices to show that

$$J_- \cong \mathcal{C}_2(\mathbf{Q}\langle n \rangle, \sigma_n, -1)$$

Put

$$S = \text{span}_{\mathbf{Q}} \left\{ a^r \left(\frac{1-z}{2} \right) : 0 \le r \le n-1 \right\}$$

We view S as a ring in which $\frac{1-z}{2}$, the central idempotent of $\mathbf{Q}[Q(4n)]$ generating J_- as an ideal, acts as the identity. It is straightforward to see that, as rings, $S \cong \mathbf{Q}\langle n \rangle$. J_- is a free module of rank 2 over S, generated by $\{X^0, X^1\}$, where $X^0 = (\frac{1-z}{2})$ and $X^1 = b(\frac{1-z}{2})$. However, conjugation by X^1 on S corresponds to the action of σ_n on $\mathbf{Q}\langle n \rangle$, and $(X^1)^2 = -X^0$. Thus

$$J_- \cong C_2(\mathbf{Q}\langle n \rangle, \sigma_n, -1)$$

as required.   □

The Wedderburn decomposition of $\mathbf{Q}[D_{2n}]$ has already been given. We can analyse the structure of the summand $C_2(\mathbf{Q}\langle n \rangle, \sigma_n, -1)$ further as follows; from the identities

$$x^{2n} - 1 = \prod_{d|2n} c_d(x)$$

and

$$x^n - 1 = \prod_{d|n} c_d(x)$$

it follows by uniqueness of factorization that

$$x^n + 1 = \prod_{d|2n, d \nmid n} c_d(x)$$

Hence

(12.6) $$\mathbf{Q}\langle n \rangle \cong \prod_{d|2n, d \nmid n} \mathbf{Q}(d)$$

Under the isomorphism $\mathbf{Q}[C_n] \cong \mathbf{Q}[x]/(x^n - 1) \cong \prod_{d|n} \mathbf{Q}(d)$ the canonical involution $\widehat{} : \mathbf{Q}[C_n] \to \mathbf{Q}[C_n]$ induces an involution $\gamma_d : \mathbf{Q}(d) \to \mathbf{Q}(d)$ which is the identity for $d = 1, 2$, and complex conjugation otherwise; we obtain:

Corollary 12.7: For each $n \ge 2$, there is an algebra isomorphism

$$\mathbf{Q}[Q(4n)] \cong \mathbf{Q}[D_{2n}] \times \prod_{d|2n, d \nmid n} C_2(\mathbf{Q}(d), \gamma_d, -1)$$

Writing

$$\zeta_d = \cos\left(\frac{2\pi r}{d}\right) + i\ \sin\left(\frac{2\pi r}{d}\right)$$

for r coprime to d, we have:

Proposition 12.8: $(\zeta_d - \bar{\zeta}_d)^2 = -4\ \sin(\frac{2\pi r}{d})^2$.

$C_2(\mathbf{Q}(d), \gamma, -1)$ is a free module of rank 2 over $\mathbf{Q}(\zeta_d)$. In the case $d = 2$, it is straightforward to check that $C_2(\mathbf{Q}(2), \gamma, -1)$ is isomorphic to the field $\mathbf{Q}\sqrt{-1}$. By contrast, when $d \geq 3$, $C_2(\mathbf{Q}(d), \gamma_d, -1)$ is a free module of rank 4 over $\mathbf{Q}(\mu_d)$. As a basis over $\mathbf{Q}(\mu_d)$ we may take $1 = X^0$; $i = (\zeta_d - \bar{\zeta}_d)$; $j = X^1$; $k = (\zeta_d - \bar{\zeta}_d)X^1$. It is easily checked that

$$ij = -ji = k$$

and that

$$i^2 = -4s(d)^2; \quad j^2 = -1$$

where $s(d) = \sin(\frac{2\pi r}{d})$. We have proved

(12.9) $$C_2(\mathbf{Q}(d), \gamma_d, -1) = \left(\frac{-4s(d)^2, -1}{\mathbf{Q}(\mu_d)}\right)$$

Finally, we get the complete decomposition formula for $\mathbf{Q}[Q(4n)]$.

$\mathbf{Q}[Q(4n)]$

$$\cong \begin{cases} \mathbf{Q} \times \mathbf{Q} \times \prod_{d|n, d \geq 3} M_2(\mathbf{Q}(\mu_d)) \times \mathbf{Q}(i) \times \prod_{d|2n, d\nmid n, d \geq 3} \left(\frac{-4s(d)^2, -1}{\mathbf{Q}(\mu_d)}\right) & n \text{ odd} \\ \mathbf{Q} \times \mathbf{Q} \times \mathbf{Q} \times \mathbf{Q} \times \prod_{d|n, d \geq 3} M_2(\mathbf{Q}(\mu_d)) \times \prod_{d|2n, d\nmid n} \left(\frac{-4s(d)^2, -1}{\mathbf{Q}(\mu_d)}\right) & n \text{ even} \end{cases}$$

Here $\left(\frac{-4s(d)^2, -1}{\mathbf{Q}(\mu_d)}\right)$ is a rational division algebra with the property that

$$\mathcal{D}_d \otimes_{\mathbf{Q}} \mathbf{R} \cong \underbrace{\mathbf{H} \times \cdots \times \mathbf{H}}_{v(d)}$$

where $v(d)$ is the degree of $\mathbf{Q}(\mu_d)$ over $\mathbf{Q}$.

Chapter 3
Stable modules and cancellation theorems

Modules M, N over a ring Λ, are said to be stably equivalent when $M \oplus \Lambda^m \cong N \oplus \Lambda^n$. In the case when Λ is a nondegenerate $\mathbf{Z}$-order, the fundamental cancellation theorem of Swan and Jacobinski gives conditions under which one may argue in the opposite direction, from stable equivalence to isomorphism. As was pointed out by Dyer and Sieradski [16], stable modules have the natural structure of a *directed tree*, by drawing an arrow $N_1 \twoheadrightarrow N_2$ between modules N_1, N_2 whenever $N_2 \cong N_1 \oplus \Lambda$.

We also give an example of Swan to show that, even in the simplest case of projective modules, cancellation may fail if the hypotheses of the Swan–Jacobinski Theorem are not satisfied.

13 Schanuel's Lemma

We begin with a basic result from module theory, which is usually known as 'Schanuel's Lemma' [58].

Proposition 13.1: Let $0 \to D \xrightarrow{i} P \xrightarrow{f} A \to 0$ and $0 \to D' \xrightarrow{i} P' \xrightarrow{f} A \to 0$ be short exact sequences of Λ-modules in which P and P' are projective; then

$$D \oplus P' \cong D' \oplus P$$

Proof: Form the *fibre product*

$$Q = P \underset{f,g}{\times} P' = \{(x, y) \in P \times P' : f(x) = g(y)\}$$

There is a short exact sequence $0 \to D' \to Q \xrightarrow{\pi} P \to 0$ where $\pi(x, y) = x$. Since P is projective, the sequence splits, so that $Q \cong D' \oplus P$. Likewise, the short exact sequence $0 \to D \to Q \xrightarrow{\pi'} P' \to 0$ with $\pi'(x, y) = y$ also splits, since P' is projective, and $Q \cong D \oplus P'$. Now $Q \cong D' \oplus P \cong D \oplus P'$ as claimed. □

14 The structure of stable modules

Until further notice, Λ will denote a **Z**-*order*. By a Λ-*lattice*, we mean a Λ-module whose underlying abelian group is finitely generated and torsion free. We denote by $\mathcal{F}(\Lambda)$ the class of all Λ-lattices. It can be regarded as a full subcategory of the category of Λ-modules; that is, if M, N are Λ-lattices

$$\text{Hom}_{\mathcal{F}(\Lambda)}(M, N) = \text{Hom}_{\Lambda}(M, N)$$

We begin by considering the various cancellation properties possessed by Λ-lattices. Here we are not so much concerned to investigate the detailed structure of individual modules as to draw general combinatorial conclusions from the ability to cancel in certain ways. We put

$$\text{Stab}(\mathcal{F}(\Lambda)) = \mathcal{F}(\Lambda)/\sim$$

where the stability relation '$\sim$' is defined by

$$M \sim N \Longleftrightarrow M \oplus \Lambda^n \cong N \oplus \Lambda^m \text{ for some } m, n$$

Say that a Λ-lattice M has the *cancellation property* when, for any $N \in \mathcal{F}(\Lambda)$ such that $\text{rk}_{\mathbf{Z}}(M) \leq \text{rk}_{\mathbf{Z}}(N)$

$$N \oplus \Lambda^m \cong M \oplus \Lambda^n \Longrightarrow N \cong M \oplus \Lambda^{n-m}$$

It is a straightforward observation that:

Proposition 14.1: If $M \in \mathcal{F}(\Lambda)$ has the cancellation property, then $M \oplus \Lambda^b$ has the cancellation property for each $b \geq 0$.

We shall say that the module $M \in \mathcal{F}(\Lambda)$ has the *weak cancellation property* when $M \oplus \Lambda$ has the cancellation property. It follows trivially from (14.1) that:

Proposition 14.2: If $M \in \mathcal{F}(\Lambda)$ has the cancellation property, then M has the weak cancellation property.

The converse to (14.2) is definitely false. Specific counterexamples will be discussed in Section 17 and Section 18. We shall say that a Λ-lattice M is *minimal*, when $\text{rk}_{\mathbf{Z}}(M) \leq \text{rk}_{\mathbf{Z}}(N)$ for all $N \in [M]$.

Proposition 14.3: Let $M \in \mathcal{F}(\Lambda)$; then the following conditions are equivalent:

(i) every non-minimal module $N \in [M]$ has the cancellation property;
(ii) every module $N \in [M]$ has the weak cancellation property;
(iii) every minimal module $M_0 \in [M]$ has the weak cancellation property;
(iv) at least one minimal module $M_0 \in [M]$ has the weak cancellation property.

Proof: We first show that (i) $\implies$ (ii). Thus suppose (i) holds; if $N \in [M]$ is non-minimal, then, by hypothesis, it has the cancellation property and hence the weak cancellation property by (14.2); if N is minimal, then $N \oplus \Lambda$, being non-minimal, has the cancellation property. Either way, N has the weak cancellation property, which is the desired conclusion. Moreover, it is clear that (ii) $\implies$ (iii) $\implies$ (iv). Thus we must show that (iv) $\implies$ (iii) $\implies$ (ii) $\implies$ (i).

(iv) $\implies$ (iii) Suppose that $M_0, N \in [M]$ are both minimal, and that $M_0 \oplus \Lambda$ has the cancellation property. We must show that $N \oplus \Lambda$ has the cancellation property. Now $\mathrm{rk}_{\mathbf{Z}}(N) = \mathrm{rk}_{\mathbf{Z}}(M_0)$, and since N and M_0 are stably equivalent it follows that

$$M_0 \oplus \Lambda^a \cong N \oplus \Lambda^a$$

for some $a \geq 0$. If $a = 0, 1$, there is nothing to prove; if $a \geq 2$, then

$$M_0 \oplus \Lambda \oplus \Lambda^{a-1} \cong N \oplus \Lambda \oplus \Lambda^{a-1}$$

so that, since $M_0 \oplus \Lambda$ has the cancellation property

$$M_0 \oplus \Lambda \cong N \oplus \Lambda$$

and so $N \oplus \Lambda$ has the cancellation property as required.

(iii) $\implies$ (ii) Suppose that $M_0 \in [M]$ is minimal and that $N \in [M]$ is non-minimal, then $\mathrm{rk}_{\mathbf{Z}}(M_0) < \mathrm{rk}_{\mathbf{Z}}(N)$, and, since N and M_0 are stably equivalent

$$N \oplus \Lambda^a \cong M_0 \oplus \Lambda^b$$

for some a, b with $a < b$. If $a = 0$, there is nothing to prove. So suppose that $1 \leq a$, then $1 \leq a \leq b - 1$, and

$$N \oplus \Lambda^a \cong M_0 \oplus \Lambda \oplus \Lambda^{b-1}$$

Since $M_0 \oplus \Lambda$ has the cancellation property

$$N \cong M_0 \oplus \Lambda \oplus \Lambda^{b-a-1} \cong M_0 \oplus \Lambda^{b-a}$$

and so, by (14.1), (14.2), N has the weak cancellation property as required.

(ii) $\implies$ (i) Let $N \in [M]$ be non-minimal, and choose a minimal representative $M_0 \in [M]$; then

$$N \oplus \Lambda^c \cong M_0 \oplus \Lambda^b \oplus \Lambda^c$$

for some b, c with $0 < b$. By hypothesis, $M_0 \oplus \Lambda$ has the cancellation property, so that $M_0 \oplus \Lambda^b$ has the cancellation property by (14.1). Hence $N \cong M_0 \oplus \Lambda^b$, and so N also has the cancellation property. This completes the proof. $\square$

If $M \in \mathcal{F}(\Lambda)$, we say that the stable module $[M]$ has the *weak cancellation property* when any, and hence all, of the four conditions of (14.3) are satisfied. From the proof of (14.3) we may draw the following conclusions:

Proposition 14.4: Suppose that the stable module $[M]$ has the weak cancellation property for $M \in \mathcal{F}(\Lambda)$, and let $M_0 \in [M]$ be minimal:

(i) if $N \in [M]$ is minimal, then $N \oplus \Lambda \cong M_0 \oplus \Lambda$;
(ii) if $N \in [M]$ is not minimal, then $N \cong M_0 \oplus \Lambda^n$ for some $n \geq 1$.

When $M \in \mathcal{F}(\Lambda)$, the stable module $[M]$ has a natural combinatorial structure which we represent as a *directed graph* in which the vertices are the modules $N \in [M]$, and where we draw an arrow $N \to N \oplus \Lambda$. It is clear that there are no loops in the corresponding undirected graph, so that $[M]$ is a tree. In the cases which interest us, the number of prongs is finite and the structure of the graph is more specific. Represent the natural numbers **N** as a tree with one end in which there is no branching away from the main stem. By a *fork*, we shall mean a tree **T** for which there exists a surjective map of graphs $\lambda_\mathbf{T} : \mathbf{T} \to \mathbf{N}$ with the property that for all $r \geq 1$, the segments $[r, r + 1]$ in **N** are covered precisely once. The name 'fork' conveys the essence of the definition, namely, that there should be no branching above level 1. It is easy to check that, for any fork **T** there is a unique surjective 'level function' $\mathbf{T} \to \mathbf{N}$, so that $\lambda_\mathbf{T}$ is intrinsic

(14.5)

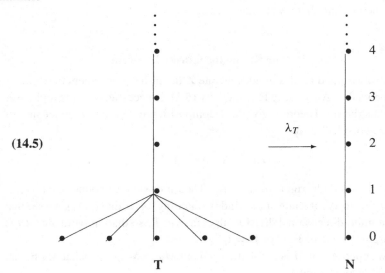

λ_T

T   N

Proposition 14.6: Let $M \in \mathcal{F}(\Lambda)$; then the stable module $[M]$ has the weak cancellation property if and only if $[M]$ is a fork.

Proof: The implication ($\Longrightarrow$) follows from (14.4), and the converse is trivial.   □

In the special case when $[M] \cong \mathbf{N}$ we say that $[M]$ is *straight*

(14.7)   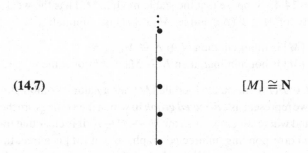 $[M] \cong \mathbf{N}$

The stable module $[M]$ is said to have the *strong cancellation property* when *every* $N \in [M]$ has the cancellation property. A straightforward chase of definitions now shows:

Proposition 14.8: Let $M \in \mathcal{F}(\Lambda)$; then $[M]$ has the strong cancellation property if and only if $[M]$ is straight.

In Chapter 9, we shall encounter a number of important cases where stable modules do indeed have the strong cancellation property, and whose underlying tree structures are thereby straight.

15 The Swan–Jacobinski Theorem

Λ is now assumed to be a nondegenerate $\mathbf{Z}$-order. It is convenient to regard Λ as imbedded in $\Lambda_{\mathbf{R}} = \Lambda \otimes \mathbf{R}$, which, by (5.3), is necessarily semisimple over $\mathbf{R}$. Wedderburn's Theorem gives a decomposition of $\Lambda_{\mathbf{R}}$ into a direct sum of simple two-sided ideals

$$\Lambda_{\mathbf{R}} \cong \Lambda_1 \oplus \cdots \oplus \Lambda_n$$

Let V_i be a simple right ideal in Λ_i. Then any simple module over $\Lambda_{\mathbf{R}}$ is isomorphic to V_i for some unique index i. Putting $D_i = \mathrm{End}_{\Lambda_{\mathbf{R}}}(V_i)$, we see that D_i is a finite-dimensional division algebra over $\mathbf{R}$, so that, in particular, D_i is isomorphic to one of $\mathbf{R}$, $\mathbf{C}$ or $\mathbf{H} = (\frac{-1,-1}{\mathbf{R}})$.

Now suppose that M is a Λ-lattice and put $M_{\mathbf{R}} = M \otimes \mathbf{R}$, so that $M_{\mathbf{R}}$ has an isotypic decomposition

$$M_{\mathbf{R}} \cong V_1^{(f_1)} \oplus \cdots \oplus V_n^{(f_n)}$$

where, to allow for the case that some multiplicities might be zero, we adhere to the convention that $V^{(0)} = 0$. Note that the ring $\text{End}_{\Lambda_{\mathbf{R}}}(M_{\mathbf{R}})$ is also a semisimple algebra, and decomposes as a direct sum of two-sided ideals, thus

$$\text{End}_{\Lambda_{\mathbf{R}}}(M_{\mathbf{R}}) \cong \mathcal{B}_1 \oplus \cdots \oplus \mathcal{B}_n$$

where $\mathcal{B}_i = M_{f_i}(D_i)$ is the ring of $f_i \times f_i$ matrices over D_i; here, of course, the convention takes the form that $M_0(D) = 0$. We say that M is an *Eichler lattice* when no simple factor $\mathcal{B}_i$ is isomorphic to $\mathbf{H}$. The following special case of the Swan–Jacobinski Theorem will suffice for our purposes: (compare [14], vol. 2 (51.28), p. 324).

Theorem 15.1: Suppose that M is an Eichler lattice over Λ with the property that $M \cong M_0 \oplus \Lambda$ for some module M_0. Then M has the cancellation property.

From the point of view of cancellation properties, it is convenient to recast the definition. If M is a Λ-lattice, we shall say that M is *pre-Eichler* when $M \oplus \Lambda$ is Eichler. The Swan–Jacobinski Theorem then implies the following, which is the conclusion from this discussion that we use most frequently:

Theorem 15.2: Let M be a Λ-lattice; if M is a pre-Eichler lattice, then M has the weak cancellation property.

The question of whether a Λ-lattice M satisfies the Eichler condition depends only upon the isomorphism type of $M \otimes \mathbf{R}$. As a right module over itself, $\Lambda_{\mathbf{R}}$ decomposes as a direct sum

$$\Lambda_{\mathbf{R}} \cong V_1^{(e_1)} \oplus \cdots \oplus V_n^{(e_n)}$$

where each V_i is simple over $\Lambda_{\mathbf{R}}$ and $V_i \not\cong V_j$ for $i \neq j$. Moreover, the isomorphism types of $V_1, \ldots, V_n$, and their multiplicities $e_1, \ldots, e_n$, are uniquely determined up to order. We say that the *order Λ satisfies the Eichler condition* when, considered as a Λ-module, Λ is an Eichler lattice. Clearly one now has:

Proposition 15.3: Let Λ be a semisimple order and let

$$\Lambda_{\mathbf{R}} \cong \Lambda_1 \oplus \cdots \oplus \Lambda_n$$

be the Wedderburn decomposition of $\Lambda_{\mathbf{R}}$ into a direct sum of simple-two-sided ideals. Then Λ satisfies the Eichler condition if and only if no simple two-sided ideal Λ_i is isomorphic to $\mathbf{H}$.

From the above discussion, we see immediately that:

Proposition 15.4: If Λ satisfies the Eichler condition, then every $M \in \mathcal{F}(\Lambda)$ satisfies the Eichler condition.

One may express the Eichler condition for Λ more explicitly; if the Wedderburn decomposition of $\Lambda_{\mathbf{R}}$ takes the form

$$\Lambda_{\mathbf{R}} \cong M_{d_1}(\mathbf{R}) \times \cdots \times M_{d_a}(\mathbf{R}) \times M_{e_1}(\mathbf{C}) \times \cdots \times M_{e_b}(\mathbf{C}) \times M_{f_1}(\mathbf{H}) \times \cdots \times M_{f_c}(\mathbf{H})$$

then Λ satisfies the Eichler condition when either $c = 0$ or each $f_i \geq 2$.

Under the hypothesis that Λ satisfies the Eichler condition, it follows that each $M \in \mathcal{F}(\Lambda)$ has the weak cancellation property. Though satisfyingly general, this statement is, however, of limited utility, since in some cases, which we shall encounter, although Λ itself may fail the Eichler condition, nevertheless the lattices of most interest to us may still possess the weak cancellation property. To see how this may arise, we need to analyse the situation further.

Let

$$\Lambda_{\mathbf{R}} \cong V_1^{(a_1)} \oplus \cdots \oplus V_n^{(a_n)}$$

be the decomposition of $\Lambda_{\mathbf{R}}$ into isotypic modules; we say that the simple module V_i is *quaternionic* when $D_i = \mathrm{End}_{\Lambda_{\mathbf{R}}}(V_i) \cong \mathbf{H}$; V_i is then said to be *good* when $a_i \geq 2$; otherwise, when $a_i = 1$, V_i is said to be *bad*.

When M is a Λ-lattice we write $M_{\mathbf{R}} = M \otimes \mathbf{R}$; moreover then we have

$$M_{\mathbf{R}} \cong V_1^{(b_1)} \oplus \cdots \oplus V_n^{(b_n)}$$

where we adhere to the convention that $V^{(0)} = 0$. In this case, $\mathrm{End}_{\Lambda_{\mathbf{R}}}((M \oplus \Lambda)_{\mathbf{R}})$ decomposes as a direct sum of two-sided ideals, thus

$$\mathrm{End}_{\Lambda_{\mathbf{R}}}((M \oplus \Lambda)_{\mathbf{R}}) \cong \mathcal{A}_1 \oplus \cdots \oplus \mathcal{A}_n$$

where $\mathcal{A}_i$ is the ring $\mathcal{A}_i = M_{a_i + b_i}(D_i)$ of $(a_i + b_i) \times (a_i + b_i)$ matrices over D_i. Then $M \oplus \Lambda$ satisfies the Eichler condition precisely when $a_i + b_i \neq 1$ for each simple quaternionic module V_i. Since $a_i \geq 1$ and $b_i \geq 0$, the possibility that $a_i + b_i = 0$ does not occur, and $M \oplus \Lambda$ satisfies the Eichler condition precisely when $a_i + b_i \geq 2$ for each simple quaternionic module V_i. This is automatically satisfied when V_i is good; when V_i is bad, we have $a_i = 1$ and for such modules we require $b_i \geq 1$, that is:

Theorem 15.5: Let Λ be a semisimple order and let $M \in \mathcal{F}(\Lambda)$; then M has the weak cancellation property provided *either*

(i) Λ satisfies the Eichler condition *or*
(ii) each bad simple quaternionic $\Lambda_{\mathbf{R}}$-module has multiplicity ≥ 1 in $M_{\mathbf{R}}$.

We say that Λ has the *cancellation property property for free modules* when

$$M \oplus \Lambda \cong \Lambda^{(k)} \oplus \Lambda \Longrightarrow M \cong \Lambda^{(k)}$$

Since any finitely generated free module belongs to the stable module $[\Lambda]$ containing Λ itself, it follows easily from (15.3) and (15.5) that:

Proposition 15.6: The following conditions are equivalent for any semisimple order Λ:

  (i) Λ has the cancellation property for free modules;
 (ii) the module Λ has the cancellation property;
(iii) the stable module $[\Lambda]$ has the strong cancellation property;
(iv) each finitely generated stably free module over Λ is actually free.

From this we see that:

Proposition 15.7: If Λ has the Eichler property, then Λ has the cancellation property for free modules.

16 Finiteness of $\tilde{K}_0(\mathbf{Z}[G])$

In connection with the stability relation, there is a classical invariant of rings obtained from equivalence classes of projective Λ-modules; thus let $\mathbf{P}(\Lambda)$ denote the collection of finitely generated projective modules over Λ. Then $\mathbf{P}(\Lambda)$ is closed with respect to direct sum. Consequently, the set

$$\mathbf{P}(\Lambda)/\sim$$

of stable classes in $\mathbf{P}(\Lambda)$ forms a commutative monoid with addition derived from $\oplus$

$$[P_1] + [P_2] = [P_1 \oplus P_2]$$

in which the class of any free module represents zero. Since for any projective P there is a projective Q such that $P \oplus Q$ is free, this monoid is actually a group, the *reduced projective class group* of Λ, and denoted classically by $\mathcal{C}(\Lambda)$, but nowadays, in view of its status within algebraic K-theory, more usually by $\tilde{K}_0(\Lambda)$.

For any finite group G, $\tilde{K}_0(\mathbf{Z}[G])$ is finite. This was first proved by Swan in [57]. Here we content ourselves with pointing out the main landmarks in the proof.

The main result needed is Swan's fundamental observation that projective modules over $\mathbf{Z}[G]$ are locally free. We shall subsequently need to use a special case of this, which we note now:

Theorem 16.1: Let G be a finite group, and let P be a finitely generated projective module over $\mathbf{Z}[G]$; then for some $\delta \geq 1$

$$P \otimes \mathbf{Q} \cong \mathbf{Q}[G]^\delta$$

For a proof see [57], and also [14], Vol. I, Section 32.

Swan also gives (see [59]) the following representation theorem for projective modules over a semisimple order:

Theorem 16.2: Let Λ be an order in a semisimple **Q**-algebra; then each finitely generated projective module P over Λ can be represented in the form

$$P \cong J \oplus \Lambda^n$$

where J is a projective ideal in Λ.

Nowadays, (16.2) is proved using Swan's general local freeness theorem in conjunction with the Swan–Jacobinski Theorem. In particular, it applies in the case $\Lambda = \mathbf{Z}[G]$ where G is finite. An immediate consequence is that any class in $\tilde{K}_0(\mathbf{Z}[G])$ can be represented by a module J satisfying $\mathrm{rk}_\mathbf{Z}(J) = |G|$. However, by the Jordan-Zassenhaus Theorem ([14] Vol. I, Section 24) there are only finitely many isomorphism classes of $\mathbf{Z}[G]$-lattices of a given finite rank. As an immediate consequence we have:

Corollary 16.3: If G is a finite group, then the reduced projective class group $\tilde{K}_0(\mathbf{Z}[G])$ is finite.

This has the following obvious but useful consequence:

Corollary 16.4: If P is a finitely generated projective module over $\mathbf{Z}[G]$, then for some $n \geq 1$, $P^{(n)} = \underbrace{P \oplus \cdots \oplus P}_{n}$ is stably free over $\mathbf{Z}[G]$.

Although we make no use of the fact, we note that in the case where Λ is a *maximal order* every Λ-lattice is Λ-projective ([14] vol I (26.12)).

17 Non-cancellation and an example of Swan

Let Λ be a ring. We say that two right (resp. left) ideals I, J of Λ are *co-isomorphic* when Λ/I and Λ/J are isomorphic as right (resp. left) Λ-modules.

Elementary considerations show that isomorphism does not imply co-isomorphism; for example, when $\Lambda = \mathbf{Z}$, any two nonzero ideals are isomorphic, whilst co-isomorphic ideals are necessarily identical. In general, co-isomorphism does not imply isomorphism either. Nevertheless a weaker statement is true. Say that two ideals I, J are *stably isomorphic* when $I \oplus \Lambda \cong J \oplus \Lambda$ as Λ-modules. Then Schanuel's Lemma gives immediately

(17.1)   co-isomorphic ideals are stably isomorphic.

We describe in detail an example, due to Eichler–Swan, of co-isomorphic left ideals which are not isomorphic. First let Λ be the maximal order constructed in Chapter 1, Section 8. To recall the details briefly, put $K = \mathbf{Q}(\zeta + \bar{\zeta})$, where $\zeta = \exp(\frac{2\pi i}{16})$. Then K is a totally real number field of degree 4 over $\mathbf{Q}$ whose ring of integers R takes the form

$$R = \mathbf{Z}[x]/(x^4 - 4x^2 + 2)$$

Let A be the quaternion division algebra

$$A = \left(\frac{-1, -1}{K}\right)$$

Then $\Lambda = \operatorname{span}_R\{1, \alpha, \beta, \alpha\beta\}$, where

$$\alpha = \frac{-\bar{\tau} + i\tau}{2}; \quad \beta = \frac{(\tau - \bar{\tau})(1 + j)}{2}$$

We have seen already that Λ is a maximal order in A. Observe that 17 factorizes in R, thus

$$17 = p_1 p_2 p_3 p_4$$

where

$$p_1 = 1 + 2\tau; \quad p_2 = 1 - 2\bar{\tau}; \quad p_3 = 1 - 2\tau; \quad p_4 = 1 + 2\bar{\tau}$$

Likewise, $-1 + 4j$ factorizes in Λ as

$$-1 + 4j = a_1 a_2 a_3 a_4$$

where

$$a_1 = 1 - \frac{(j-1)}{\bar{\tau}}; \quad a_2 = 1 - \frac{(j-1)}{\tau}; \quad a_3 = 1 + \frac{(j-1)}{\bar{\tau}}; \quad a_4 = 1 + \frac{(j-1)}{\tau}$$

In fact each $a_i \in \Lambda$ as we have

$$a_1 = (1 - 2\tau + \tau^3) + (1 - \tau^2)\beta$$
$$a_2 = (1 + 4\tau - \tau^3) - \beta$$
$$a_3 = (1 + 2\tau - \tau^3) + (\tau^2 - 1)\beta$$
$$a_4 = (1 - 4\tau + \tau^3) + \beta$$

Proposition 17.2: $p_i \in \Lambda(a_i)$ for each i.

Proof: If $\eta = \eta_0 + \eta_1 j$, where $\eta_i \in K$, we write $\bar{\eta} = \eta_0 - \eta_1 j$. We may express this in terms of β rather than j thus: write $\xi = \xi_0 + \xi_1 \beta$, where $\xi_i \in K$; then

$$\bar{\xi} = \xi_0 + (4\tau - \tau^3)\xi_1 - \xi_1 \beta$$

62 *Stable Modules and the D(2)-Problem*

In particular, if $\xi_0, \xi_1 \in R$, then both ξ and $\bar{\xi} \in \Lambda$. We define $u_i \in R$ for $i = 1, 2, 3, 4$ by

$$u_1 = 3 + 4\tau - \tau^2 - \tau^3;$$
$$u_2 = -1 + 2\tau + \tau^2 - \tau^3;$$
$$u_3 = 3 - 4\tau - \tau^2 + \tau^3;$$
$$u_4 = -1 - 2\tau + \tau^2 + \tau^3.$$

A straightforward computation reveals that, for each i

$$p_i = (u_i \bar{a}_i) a_i$$

and this completes the proof. □

For any rational prime q, we denote by $\mathbf{F}_q$ the field with q elements.

Proposition 17.3: $R/(p_i) \cong \mathbf{F}_{17}$ for $i = 1, 2, 3, 4$.

Proposition 17.4: $\Lambda/(p_i) \cong M_2(\mathbf{F}_{17})$ for $i = 1, 2, 3, 4$.

It follows that any two simple modules over $\Lambda/(p_i)$ are isomorphic. One may be more precise; if V_i is a simple module over $\Lambda/(p_i)$, then $\dim_{R/(p_i)}(V_i) = 2$, and:

Proposition 17.5: Let U_i be a nonzero module over $\Lambda/(p_i)$; then

$$\dim_{R/(p_i)}(U_i) < 4 \Longleftrightarrow U_i \cong_{\Lambda/(p_i)} V_i$$

Since $p_i \in \Lambda(a_i)$, then $\Lambda/\Lambda(a_i)$ is a module over $\Lambda/(p_i)$, and hence over $R/(p_i)$. The sequence of Λ-submodules

$$\Lambda(p_i) \subset \Lambda(a_i) \subset \Lambda$$

gives an exact sequence of Λ-modules

$$0 \to \Lambda(a_i)/\Lambda(p_i) \to \Lambda/\Lambda(p_i) \to \Lambda/\Lambda(a_i) \to 0$$

However, the inclusion $\Lambda(a_i) \subset \Lambda$ is proper, so that $\Lambda/\Lambda(a_i) \neq 0$. Likewise, the inclusion $\Lambda(a_i)/\Lambda(p_i)$ is also proper, so that $\Lambda(a_i)/\Lambda(p_i)$ is also nonzero, and, consequently

$$\dim_{R/(p_i)}(\Lambda/\Lambda(a_i)) < \dim_{R/(p_i)}(\Lambda/\Lambda(p_i)) = 4$$

Thus from (17.5), we see that:

Proposition 17.6: $\Lambda/\Lambda(a_i) \cong_{\Lambda/(p_i)} V_i$.

We define

$$\Omega_1 = \Lambda(\tilde{\tau}\alpha - \bar{\beta}, \beta\alpha - \tilde{\tau})$$

From the identity $p_1 = (\tau^2 - 1)\{\bar{\alpha}(\tilde{\tau}\alpha - \bar{\beta}) - (\beta\alpha - \tilde{\tau})\}$, we see that $p_1 \in \Omega_1$.

Proposition 17.7: Ω_1 and $\Lambda(a_1)$ are co-isomorphic left ideals in Λ; in fact, we have

$$\Lambda/\Omega_1 \cong \Lambda/\Lambda(a_1) \cong V_1$$

Proof: The sequence of Λ-submodules $(p_1) \subset \Omega_1 \subset \Lambda$ gives an exact sequence of Λ-modules

$$0 \to \Omega_1/(p_1) \to \Lambda/(p_1) \to \Lambda/\Omega_1 \to 0$$

However, the inclusion $\Omega_1 \subset \Lambda$ is proper; for example, $\alpha \notin \Omega_1$, so that $\Lambda/\Omega_1 \neq 0$. Likewise, since $\beta \notin (p_1)$, the inclusion $(p_1) \subset \Omega_1$ is also proper, so that $\Omega_1/(p_1)$ is also nonzero, and, consequently

$$\dim_{R/(p_1)}(\Lambda/\Omega_1) < \dim_{R/(p_1)}(\Lambda/(p_1)) = 4$$

Hence from (17.5) we see that $\Lambda/\Omega_1 \cong V_1$. The result now follows since, by (17.6), $\Lambda/\Lambda(a_1) \cong V_1$. □

From Schanuel's Lemma, we see that $\Omega_1 \oplus \Lambda \cong \Lambda(a_1) \oplus \Lambda$. Moreover, since $\Lambda(a_1) \cong \Lambda$, we also obtain

(17.8) $$\Omega_1 \oplus \Lambda \cong \Lambda \oplus \Lambda$$

However:

Proposition 17.9: $\Omega_1 \not\cong \Lambda$ as a left Λ-module.

Proof: Put $\Omega_2 = \Lambda(\tilde{\tau}, \beta)$, and observe that the mapping $\varphi : \Omega_2 \to \Omega_1$ given by

$$\varphi(x) = x(\alpha - \tilde{\tau}^{-1}\bar{\beta})$$

is an isomorphism of Λ-modules. It suffices, therefore, to show that Ω_2 is not isomorphic to Λ. Putting

$$\Sigma = \{\sigma \in A : \Omega_2\sigma \subset \Omega_2\}$$

it is clear that Σ is an order in A. If $\Omega_2 \cong \Lambda$, then there must exist exist $x \in \Omega$ such that $\Omega_2 = \Lambda x$. If $\sigma \in \Sigma$, then, in particular

$$x\sigma \in \Lambda x$$

Now x is necessarily nonzero, so x^{-1} exists in A, since A is a division algebra. Thus $\sigma \in x^{-1}\Lambda x$, and $\Sigma \subset x^{-1}\Lambda x$. However, it is evident that $x^{-1}\Lambda x \subset \Sigma$. Thus $\Sigma(\Omega) = x^{-1}\Lambda x \cong \Lambda$. If $\Theta \subset A$ is the maximal order of Section 8, it is easy to check that $\Theta \subset \Sigma$, and, since Θ is a maximal order, it follows that $\Sigma = \Theta$. However, $\Theta \not\cong \Lambda$ by (8.7), so the result follows from the argument above. □

Observe that one may produce examples of co-isomorphic *right ideals* by applying a suitable anti-involution. A convenient choice is the anti-involution $\theta : A \to A$ given by

$$\theta(x_0 + x_1 i + x_2 j + x_3 k) = x_0 - x_1 i + x_2 j - x_3 k$$

Since $\theta(j) = j$, we see that $\theta(a_1) = a_1$ and $\theta(\beta) = \beta$. Putting $\Delta = \theta(\Lambda)$, we see easily that Δ is a maximal order in A, and that:

Proposition 17.10: The right ideals $(a_1)\Delta$ and $(\tau, \beta)\Delta$ in Δ satisfy:

(i) $(a_1)\Delta \oplus \Delta \cong (\tau, \beta)\Delta \oplus \Delta$ and
(ii) $(a_1)\Delta \not\cong (\tau, \beta)\Delta$.

18 Non-cancellation over group rings

Swan's example to show non-cancellation over a semisimple order can be elaborated to show non-cancellation holds over some integral group rings. The first case discovered was $\mathbf{Z}[Q(32)]$.

First recall the structure of $\mathbf{Q}[Q(4n)]$; for each $n \geq 2$, there is, by (8.5), an algebra isomorphism

$$\mathbf{Q}[Q(4n)] \cong \mathbf{Q}[D_{2n}] \oplus \mathcal{C}_2(\mathbf{Q}\langle n \rangle, \sigma_n, -1)$$

When n is even

$$\mathcal{C}_2(\mathbf{Q}\langle n \rangle, \sigma_n, -1) \cong \prod_{d|2n, d\nmid h} \left(\frac{-4s(d)^2, -1}{\mathbf{Q}(\mu_d)} \right)$$

where $\mu_d = \zeta_d + \bar{\zeta}_d$, and $\zeta_d = e^{\frac{2\pi i}{d}}$, and where $s(d) = sin(\frac{2\pi}{d})$. In the case $n = 8$, there is a single factor, and we have

$$\mathcal{C}_2(\mathbf{Q}\langle 8 \rangle, \sigma_8, -1) \cong \left(\frac{-4s(16)^2, -1}{K} \right)$$

where, continuing with the notation of Section 17, $K = \mathbf{Q}(\mu_{16})$. However,

$4s(16)^2 = 2 + \sqrt{2} = \tau^2$ is a square in K. Thus

$$\mathbf{Q}[Q(32)] \cong \mathbf{Q}[D_{16}] \oplus \left(\frac{-1, -1}{K}\right)$$

Under this product structure, $\mathbf{Z}[Q(32)]$ imbeds in $\mathbf{Z}[D_{16}] \times \Lambda$, where $\Lambda \subset (\frac{-1,-1}{K})$ is the maximal order of Section 17. Moreover, if $\Gamma \subset \Lambda$ is the image of $\mathbf{Z}[Q(32)]$ under projection, then

$$\Gamma \cong \mathbf{Z}[Q(32)]/(y^2 + 1)$$

Since the reduced discriminants of the $\mathbf{Z}$-algebras $\mathbf{Z}[Q(32)]$ and $\mathbf{Z}[D_{16}]$ are both powers of 2, and since the reduced discriminant of Λ is -1, it follows that the index of Γ in Λ is also a power of 2.

Let $J \subset \mathbf{Z}[Q(32)]$ be the left ideal $J = \mathbf{Z}[Q(32)](y + 4)$, and put $T = \mathbf{Z}[Q(32)]/J$. In $\mathbf{Z}[Q(32)]$ we have

$$-255 = y^4 - 256 = (y^2 + 16)(y + 4)(y - 4)$$

so that J has finite index in $\mathbf{Z}[Q(32)]$. The prime factorization of 255 is

$$255 = 3 \cdot 5 \cdot 17$$

so that T decomposes as $T = M \oplus N$, where M consists of 17-torsion, and N is a direct sum of 3 and 5-torsion.

Since y^2 is in the centre of $Q(32)$, $y^2 \equiv 16 = (-4)^2$ on $\mathbf{Z}[Q(32)]/J$. Since $16 \equiv -1 \pmod{17}$ and $16 \equiv 1 \pmod{q}$, when $q = 3, 5$, we see that M and N are respectively the -1 and $+1$-eigenspaces of y^2 in T. It follows that M is a module over $\Gamma = \mathbf{Z}[Q(32)](y^2+1)$ and N is a module over $\mathbf{Z}[D_{16}] = \mathbf{Z}[Q(32)](y^2-1)$. Since T is a quotient of $\mathbf{Z}[Q(32)](y^2 - 1)$, it follows also that M is a quotient of Γ. Moreover, since the index of Γ in Λ is a power of 2, and hence coprime to 17, we may identify M with $\Lambda/\Lambda(y + 4)$, and so regard M as a module over Λ. Likewise N is a quotient of $\mathbf{Z}[D_{16}]$; we denote the respective projections by $\pi_M : \Lambda \to M$ and $\pi_N : \mathbf{Z}[D_{16}] \to N$. We saw in Section 17 that there is a factorization in Λ

$$-1 + 4j = a_1 a_2 a_3 a_4$$

where

$$a_1 = 1 - \frac{(j-1)}{\tilde{\tau}}; \ a_2 = 1 - \frac{(j-1)}{\tau}; \ a_3 = 1 + \frac{(j-1)}{\tilde{\tau}}; \ a_4 = 1 + \frac{(j-1)}{\tau}$$

Since $-1 + 4j = j(j + 4)$, we see easily that

$$M = \Lambda/\Lambda(y + 4) \cong \Lambda/\Lambda(a_1) \oplus \Lambda/\Lambda(a_2) \oplus \Lambda/\Lambda(a_3) \oplus \Lambda/\Lambda(a_4)$$

Let $\pi_i : \Lambda \to \Lambda / \Lambda(a_i)$ be the projection, so that $\pi_M = \pi_1 \times \pi_2 \times \pi_3 \times \pi_4$. However, we have seen that $\Lambda / \Lambda(a_1) \cong \Lambda / \Omega_1$. Let $\pi_1' : \Lambda \to \Lambda / \Omega_1$ and put $\pi_M' = \pi_1' \times \pi_2 \times \pi_3 \times \pi_4$; then π_M' is also a projection $\pi_M' : \Lambda \to M$

Finally, we denote by $\pi : \mathbf{Z}[Q(32)] \to T = M \oplus N$ the canonical projection and by $\pi' : \mathbf{Z}[Q(32)] \to M \oplus N$ the restriction of $\pi_M' \times \pi_N : \Lambda \times \mathbf{Z}[D_{16}] \to M \oplus N$ to $\mathbf{Z}[Q(32)]$ (identified with its image in $\Lambda \times \mathbf{Z}[D_{16}]$ under the canonical imbedding). Again since the index of Γ in Λ is coprime to the exponent of M, $\pi' : \mathbf{Z}[Q(32)] \to M \oplus N$ is surjective. Put $J = \mathrm{Ker}(\pi')$. Then by Schanuel's Lemma

$$\mathrm{Ker}(\pi') \oplus \mathbf{Z}[Q(32)] \cong \mathrm{Ker}(\pi) \oplus \mathbf{Z}[Q(32)]$$

However the mapping $\mathbf{Z}[Q(32)] \to \mathrm{Ker}(\pi)$; $x \mapsto x(y+4)$ gives an isomorphism $\mathrm{Ker}(\pi) \cong \mathbf{Z}[Q(32)]$. Thus

$$\Psi \oplus \mathbf{Z}[Q(32)] \cong \mathbf{Z}[Q(32)] \oplus \mathbf{Z}[Q(32)]$$

where $\Psi = \mathrm{Ker}(\pi')$. However, Ψ is not isomorphic to $\mathbf{Z}[Q(32)]$ since by extending scalars to Λ we get $\Psi \otimes_{\mathbf{Z}[Q(32)]} \Lambda \cong \Omega_1$ and Ω_1 is not isomorphic to Λ.

The above example of non-cancellation is the original one given by Swan [59]. As before, one obtains a corresponding statement for right ideals by applying an anti-involution, for example the canonical anti-involution, on $\mathbf{Z}[Q(32)]$. It follows from the Swan–Jacobinski Theorem that the presence of a quaternionic order is an essential feature of non-cancellation.

Perhaps the most basic cancellation property is the cancellation property for free modules. Subsequently, Swan ([63]) gave a systematic treatment of non-cancellation in integral group rings, in which he shows there are precisely seven exceptional binary polyhedral groups which *do* possess the cancellation property for free modules, namely the binary tetrahedral, octahedral, and icosahedral groups T^*, O^*, I^*, and four quaternion groups Q_8, Q_{12}, Q_{16}, Q_{20}. We say that the remaining binary polyhedral groups are *typical*. Swan shows that a finite group G which has a typical binary polyhedral group as a quotient fails to possess the cancellation property for free modules.

The converse is not true; for example, $Q_8 \times C_2$ fails to possess the cancellation property for free modules, and this is the smallest such example.

We shall consider non-cancellation phenomena again in Chapter 9.

Chapter 4

Relative homological algebra

In [83], Yoneda showed how to formulate the classification of n-fold module extensions in cohomological terms. Here we present a version of Yoneda's Extension Theory which is appropriate for our problem.

We begin by introducing the notion of a *tame class*. This, approximately, is a class of modules closed with respect to short exact sequences, which contains all finitely generated projective modules, and relative to which projective modules are injective. From the outset, we formulate matters within the 'derived module category' of a tame class C; that is, the quotient category obtained by setting projective modules equal to zero.

The derived category approach permits the explicit construction of a sequence of 'derived functors'

$$\mathbf{D}_n : \mathcal{D}\mathrm{er}(C) \to \mathcal{D}\mathrm{er}(C)$$

for $n \geq 0$: M and $\mathbf{D}_n(M)$ are connected by an exact sequence

$$0 \to \mathbf{D}_n(M) \to P_{n-1} \to \cdots \to P_0 \to M \to 0$$

where each P_r is projective. Cohomology is then introduced as the 'nth derived functor' of Hom by means of

$$\mathcal{H}^n(M, N) = \mathrm{Hom}_{\mathcal{D}\mathrm{er}}(\mathbf{D}_n(M), N)$$

We show that this formulation is equivalent, for a tame class, to the the traditional Eilenberg–Maclane definition via the homology of the chain complex obtained by applying Hom to a projective resolution. The above then becomes, in effect, a 'corepresentation formula'.

The relative injectivity of projectives shows that each $\mathbf{D}_n$ is a self-equivalence of categories, and has an inverse functor $\mathbf{D}_{-n}$ which gives a corresponding 'representation formula'

$$\mathcal{H}^n(M, N) = \mathrm{Hom}_{\mathcal{D}er}(M, \mathbf{D}_{-n}(N))$$

Yoneda's Theorem can also be expressed, relative to such a projective n-stem, as follows

$$\mathbf{P} = (0 \to \mathbf{D}_n(M) \to P_{n-1} \to \cdots \to P_0 \to M \to 0)$$

an arbitrary extension

$$\mathbf{A} = (0 \to J \to A_{n-1} \to \cdots \to A_0 \to M \to 0)$$

with $A_r \in \mathcal{C}$ is classified by $\mathrm{End}_{\mathcal{D}er}(J) \cong \mathrm{End}_{\mathcal{D}er}(M)$.

The objects of the derived category can be identified with 'hyper-stable modules'; that is, with equivalence classes of modules $M \in \mathcal{C}$ under the relation

$$M \sim\!\!\sim N \iff M \oplus P_1 \cong N \oplus P_2$$

where P_1, P_2 are finitely generated projectives. For later applications, this relation is too coarse for us, the appropriate stability notion being that previously considered in Chapter 3, namely

$$M \sim N \iff M \oplus F_1 \cong N \oplus F_2$$

where F_1, F_2 are finitely generated free, or, what is equivalent, stably free. It is thus necessary to refine the construction $M \mapsto \mathbf{D}_n(M)$. To any module $M \in \mathcal{C}$, we associate a well-defined stable module $\Omega_n(M)$ $(n \geq 0)$ by requiring that M is connected to some $J \in \Omega_n(M)$ by an exact sequence

$$\mathbf{F} = (0 \to J \to F_{n-1} \to \cdots \to F_0 \to M \to 0)$$

where each F_r is finitely generated free. When G is a finite group, the stable modules $\Omega_n(\mathbf{Z})$ over $\mathbf{Z}[G]$ will subsequently assume a primary significance.

19 The derived category of a tame class

Λ will denote an associative ring with unity. We say that a Λ-homomorphism $f : M_1 \to M_2$ *factors through a projective module*, written '$f \approx 0$', when f can be written as a composite $f = \beta \circ \alpha$, for some projective module P, and some Λ-homomorphisms $\alpha : M_1 \to P$ and $\beta : P \to M_2$. The relation '$\approx$' is additive; that is:

Proposition 19.1: Let $f, g : M \to N$ be Λ-homomorphisms; if $f \approx 0$ and $g \approx 0$, then $f + g \approx 0$.

Proof: Let $f = \alpha \circ \beta$ be a factorization through the projective P and $g = \gamma \circ \delta$ be a factorization through the projective Q; then

$$f + g = (\alpha, \gamma) \begin{pmatrix} \beta \\ \delta \end{pmatrix}$$

is a factorization of $f + g$ through the projective $P \oplus Q$. □

Proposition 19.2: Let $f : M \to N$ be a Λ-homomorphism; if $f \approx 0$, then $-f \approx 0$.

We extend $\approx$ to a binary relation on $\text{Hom}_\Lambda(M, N)$ by means of

$$f \approx g \Longleftrightarrow f - g \approx 0$$

With the definition so extended, $\approx$ is an equivalence relation. Moreover, it is compatible with composition; that is:

Proposition 19.3: Let $f, f' : M_0 \to M_1$, $g, g' : M_1 \to M_2$ be Λ-homomorphisms; if $f \approx f'$ and $g \approx g'$, then $g \circ f \approx g' \circ f'$.

For any class C of Λ modules, there is a well-defined category, the *derived module category* $\mathcal{D}\text{er}(C)$ of C, whose objects are (right) Λ-modules, and in which, for any two objects M, N, the set of morphisms $\text{Hom}_{\mathcal{D}\text{er}(C)}(M, N)$ is given by

$$\text{Hom}_{\mathcal{D}\text{er}}(M, N) = \text{Hom}_\Lambda(M, N)/\approx$$

It should cause no confusion to call $\mathcal{D}\text{er}(C)$ simply the 'derived category' of C (see note at end of Section 22). The additivity property of $\approx$ shows that:

Proposition 19.4: For any class C, and any $M, N \in C$, $\text{Hom}_{\mathcal{D}\text{er}(C)}(M, N)$ has the natural structure of an abelian group.

Suppose that C is a class of Λ-modules. A module $J \in C$ is said to be *injective relative to* C when any short exact sequence of the form

$$0 \to J \to M_1 \to M_2 \to 0$$

with $M_1, M_2 \in C$ splits. The class C is said to be *tame* provided it is closed under isomorphism, that is, if $M \cong M'$ and $M \in C$, then $M' \in C$, and it also satisfies the following properties $T(0) - T(4)$:

$T(0)$: Each $M \in C$ is finitely generated over Λ.
$T(1)$: C contains all finitely generated projective Λ-modules.

$\mathcal{T}(2)$: If $0 \to K \to M \to Q \to 0$ is an exact sequence of Λ-modules and $Q \in \mathcal{C}$, then $M \in \mathcal{C} \iff K \in \mathcal{C}$

$\mathcal{T}(3)$: If $M \in \mathcal{C}$, then M is injective relative to $\mathcal{C} \iff M$ is projective.

$\mathcal{T}(4)$: If $M \in \mathcal{C}$, then there exists an exact sequence of Λ-modules

$$0 \to M \to F \to L \to 0$$

where $L \in \mathcal{C}$ and F is finitely generated free over Λ.

A module $J \in \mathcal{C}$ is said to be *strongly injective relative to* $\mathcal{C}$ when, given any short exact sequence of the form $0 \to K \overset{j}{\to} L \overset{p}{\to} M \to 0$ with $K, L, M \in \mathcal{C}$, and any Λ-homomorphism $\varphi : K \to J$, there exists a Λ-homomorphism $\bar{\varphi} : L \to J$ such that $\bar{\varphi} \circ j = \varphi$.

Lemma 19.5: If $\mathcal{C}$ is a class of Λ-modules satisfying property $\mathcal{T}(2)$, then $J \in \mathcal{C}$ is injective relative to $\mathcal{C}$ if and only if it is strongly injective relative to $\mathcal{C}$.

Proof: ($\implies$) Suppose that J is injective relative to $\mathcal{C}$. Let

$$\mathcal{E} = \left(0 \to D \overset{i}{\to} L \overset{p}{\to} M \to 0\right)$$

be an exact sequence in $\mathcal{C}$, and let $\varphi : D \to J$ be a homomorphism. We must produce a homomorphism $\Phi : L \to J$ such that $\varphi = \Phi \circ i$. Let Δ be the submodule

$$\Delta = \{(i(x), -\varphi(x)) : x \in D\}$$

of $L \oplus J$; then the sequence

$$\varphi_*(\mathcal{E}) = \left(0 \to J \overset{j}{\to} L' \overset{\pi}{\to} M \to 0\right)$$

is exact, where $L' = (L \oplus J)/\Delta$, $j : J \to L'$ is the homomorphism $j(x) = [i(x), 0]$, and $\pi : L' \to M$ is the homomorphism

$$\pi[y, q] = p(y)$$

$\varphi_*(\mathcal{E})$ is an exact sequence in $\mathcal{C}$, by property $\mathcal{T}(2)$, since $J, M \in \mathcal{C}$. Since J is injective relative to $\mathcal{C}$, there exists a homomorphism $\rho : L' \to J$ which splits $\varphi_*(\mathcal{E})$ on the left. Let $\psi : L \to L'$ be the homomorphism $\psi(y) = [y, 0]$, and put $\Phi = \rho \circ \psi$. From the commutativity of

$$\begin{pmatrix} 0 \to D \xrightarrow{i} L \xrightarrow{p} M \to 0 \\ \downarrow \varphi \quad \downarrow \psi \quad \downarrow \mathrm{Id}_M \\ 0 \to J \xrightarrow{j} L' \xrightarrow{\pi} M \to 0 \end{pmatrix}$$

and the fact that $\rho \circ j = \mathrm{Id}_J$, it follows immediately that $\varphi = \Phi \circ i$ as required, proving ($\Longrightarrow$).

($\Longleftarrow$) If J is strongly injective relative to $\mathcal{C}$ and

$$\mathcal{E} = \left(0 \to D \xrightarrow{i} L \xrightarrow{p} M \to 0\right)$$

is an exact sequence in $\mathcal{C}$ then the mapping $\rho : L \to J$ extending $\mathrm{Id}_J : J \to J$ splits $\mathcal{E}$ on the left, showing that J is injective relative to $\mathcal{C}$. This completes the proof. $\qquad\square$

In the derived category of a tame class $\mathcal{C}$, it is not necessary to postulate that the projective modules through which morphisms are factorized need to be finitely generated. Moreover, since projective modules are direct summands of free modules which are also projective, the condition '$f \approx 0$' is entirely equivalent to the requirement that $f : M_1 \to M_2$ factors through a free module. In summary, we have:

Proposition 19.6: Let M be a finitely generated Λ-module, and let $f : M \to N$ be a Λ-homomorphism; then the following conditions are equivalent:

 (i) f factors through a projective module;
 (ii) f factors through a finitely generated projective module;
 (iii) f factors through a free module;
 (iv) f factors through a finitely generated free module.

We extend these considerations to exact sequences; denote by $\mathbf{Ext}_{\mathcal{C}}^1$ the collection of exact sequences of Λ-modules and homomorphisms

$$\mathbf{E} = (0 \to E_+ \to E_0 \to E_- \to 0)$$

in which the modules E_+, E_0 and E_- are all in $\mathcal{C}$; $\mathbf{Ext}_{\mathcal{C}}^1$ can be regarded as a category by taking morphisms to be commutative diagrams of Λ-homomorphisms thus

$$\begin{matrix} \mathbf{E} \\ \downarrow h = \\ \mathbf{F} \end{matrix} \begin{pmatrix} 0 \to E_+ \to E_0 \to E_- \to 0 \\ \downarrow h_+ \quad \downarrow h_0 \quad \downarrow h_- \\ 0 \to F_+ \to F_0 \to F_- \to 0 \end{pmatrix}$$

The projection

$$w_- \begin{pmatrix} 0 \to E_+ \to P_0 \to E_- \to 0 \\ \quad \downarrow h_+ \quad \downarrow h_0 \quad \downarrow h_- \\ 0 \to E'_+ \to P'_0 \to E'_- \to 0 \end{pmatrix} = \begin{matrix} E_- \\ \downarrow h_- \\ E'_- \end{matrix}$$

defines a functor $\omega_- : \mathbf{Ext}^1_{\mathcal{C}} \longrightarrow \mathcal{C}$. When $A, B \in \mathcal{C}$, $\mathbf{Ext}^1_{\mathcal{C}}(A, B)$ will denote the full subcategory of $\mathbf{Ext}^1_{\mathcal{C}}$ whose objects $\mathbf{E}$ satisfy $E_- = A$ and $E_+ = B$.

Let $\mathbf{Proj}^1_{\mathcal{C}}$ denote the full subcategory of $\mathbf{Ext}^1_{\mathcal{C}}$ whose objects are *projective covers* in $\mathcal{C}$; that is, sequences of the form

$$\mathcal{P} = (0 \to D \overset{i}{\to} P \overset{\epsilon}{\to} M \to 0)$$

in which D, P, and M are objects in $\mathcal{C}$, and where P is projective. Any $M \in \mathcal{C}$ admits such a projective cover. The following proposition is fundamental:

Proposition 19.7: Let

$$\begin{pmatrix} 0 \to D \overset{i}{\to} P \overset{\epsilon}{\to} M \to 0 \\ \quad \downarrow h_+ \quad \downarrow h \quad \downarrow h_- \\ 0 \to K \overset{j}{\to} X \overset{\eta}{\to} M' \to 0 \end{pmatrix}$$

be a morphism in $\mathbf{Ext}^1_{\mathcal{C}}$, in which P is projective; then

$$h_- \approx 0 \Rightarrow h_+ \approx 0$$

Proof: Consider the special case where $h_- = 0$; then $\mathrm{Im}(h) \subset \mathrm{Im}(j)$ and h_+ admits the factorization $D \overset{i}{\to} P \overset{\lambda}{\to} K$ through the projective module P, where $\lambda = j^{-1}h$. In general, suppose that $h_- = \alpha \circ \beta$

$$M \overset{\beta}{\to} Q \overset{\alpha}{\to} M'$$

where Q is projective. By the universal property for Q, since η is surjective there exists a homomorphism $\hat{\alpha} : Q \to X$ such that $\alpha = \eta \circ \hat{\alpha}$. Put $\hat{h} = \hat{\alpha} \circ \beta \circ \epsilon : P \to X$ so that the following diagram commutes

$$\begin{matrix} 0 \to D \overset{i}{\to} P \overset{\epsilon}{\to} M \to 0 \\ \quad \downarrow 0 \quad \downarrow \hat{h} \quad \downarrow h_- \\ 0 \to K \overset{j}{\to} X \overset{\eta}{\to} M' \to 0 \end{matrix}$$

Consequently, we have a commutative diagram

$$
\begin{array}{ccccccccc}
0 & \to & D & \xrightarrow{i} & P & \xrightarrow{\epsilon} & M & \to & 0 \\
& & \downarrow h_+ & & \downarrow h - \hat{h} & & \downarrow 0 & & \\
0 & \to & K & \xrightarrow{j} & X & \xrightarrow{\eta} & M' & \to & 0
\end{array}
$$

so that by the above special case, $h_+ : D \to K$ factors through a projective. $\square$

We also have the dual statement:

Proposition 19.8: Let

$$
\begin{pmatrix}
0 & \to & K & \xrightarrow{j} & X & \xrightarrow{\eta} & M' & \to & 0 \\
& & \downarrow h_+ & & \downarrow h & & \downarrow h_- & & \\
0 & \to & D & \xrightarrow{i} & P & \xrightarrow{\epsilon} & M & \to & 0
\end{pmatrix}
$$

be a morphism in $\mathbf{Ext}_{\mathcal{C}}^1$, in which P is projective; then

$$
h_+ \approx 0 \implies h_- \approx 0
$$

Proof: Consider the special case where $h_+ = 0$; then, since $h_{|\mathrm{Im}(j)} \equiv 0$, there exists a unique homomorphism $\lambda : M' \to P$ such that $h = \lambda \circ \eta$. Then $h_- \circ \eta = \epsilon \circ h = \epsilon \circ \lambda \circ \eta$ so that $h_- = \epsilon \circ \lambda$ since η is an epimorphism. However, $\epsilon \circ \lambda$ is a factorization through the projective P, which is the desired conclusion.

In general, we may only assume that h_+ factorizes as $h_+ = \mu \circ \lambda$, thus

$$
K \xrightarrow{\lambda} Q \xrightarrow{\mu} D
$$

where Q is projective. Since Q is strongly injective relative to $\mathcal{C}$, there exists a homomorphism $\nu : X \to Q$ such that $\lambda = \nu \circ i$, thus

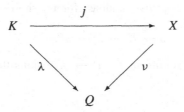

Put $\hat{h} = j \circ \mu \circ \nu : X \to P$; then

$$
\begin{aligned}
\hat{h} \circ i &= j \circ \mu \circ \nu \circ i \\
&= j \circ \mu \circ \lambda \\
&= j \circ h_+
\end{aligned}
$$

Moreover, $\eta \circ j = 0$ implies that $\eta \circ \hat{h} = 0$, so that the following diagram commutes

$$\begin{pmatrix} 0 \to K \xrightarrow{j} & X \xrightarrow{\eta} & M' \to 0 \\ \downarrow 0 \quad \downarrow h - \hat{h} & \downarrow h_- \\ 0 \to D \xrightarrow{i} & P \xrightarrow{\epsilon} & M \to 0 \end{pmatrix}$$

The conclusion $h_- \approx 0$ follows by the above special case.   □

Taking (19.7) and (19.8) together, we obtain:

Corollary 19.9: Let $\mathcal{C}$ be a tame class, and let

$$\begin{pmatrix} 0 \to D \xrightarrow{i} & P \xrightarrow{\epsilon} & M \to 0 \\ \downarrow h_+ \quad \downarrow h & \downarrow h_- \\ 0 \to D' \xrightarrow{j} & P' \xrightarrow{\eta} & M' \to 0 \end{pmatrix}$$

be a morphism in $\mathbf{Proj}_{\mathcal{C}}^1$; then

$$h_- \approx 0 \iff h_+ \approx 0$$

20 Derived functors

When $\mathcal{C}$ is a tame class, its derived category admits a sequence of functors, the so-called 'derived functors'

$$\mathbf{D}_n : \mathcal{D}er(\mathcal{C}) \to \mathcal{D}er(\mathcal{C})$$

which we now describe. In the simplest case, if $0 \to D \xrightarrow{i} P \xrightarrow{\epsilon} M \to 0$ is a projective cover and $f : M \to M'$ is a Λ-homomorphism, then by the universal property for projective modules, for any short exact sequence

$$0 \to K \xrightarrow{j} X \xrightarrow{\eta} M' \to 0$$

there exists a Λ-homomorphism $\tilde{f} : P \to X$ making the following diagram commute

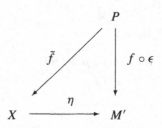

We thus get a commutative diagram

$$
\begin{array}{ccccccccc}
0 & \to & D & \to & P & \overset{\epsilon}{\to} & M & \to & 0 \\
 & & & & \downarrow \omega_+(\tilde{f}) & & \downarrow \tilde{f} & & \downarrow f \\
0 & \to & K & \to & X & \overset{\eta}{\to} & M' & \to & 0
\end{array}
$$

where $\omega_+(\tilde{f})$ is the restriction to D. The existence of projective covers shows that $\omega_- : \mathbf{Proj}^1_{\mathcal{C}} \to \mathcal{C}$ is surjective on objects. Moreover, if $f : M \to M'$ is a morphism in $\mathcal{C}$, and $\mathcal{P}$, $\mathcal{Q}$ are *any* objects in $\mathbf{Proj}^1_{\mathcal{C}}$ such that $\omega_-(\mathcal{P}) = M$ and $\omega_-(\mathcal{Q}) = M'$, then there exists a morphism $\tilde{f} : \mathcal{P} \to \mathcal{Q}$ in $\mathbf{Proj}^1$ such that $\omega_-(\tilde{f}) = f$; that is:

Proposition 20.1: $\omega_- : \mathbf{Proj}^1_{\mathcal{C}} \to \mathcal{C}$ is an epifunctor.

We construct a functor $\mathbf{D}_1 : \mathcal{D}\mathrm{er}(\mathcal{C}) \to \mathcal{D}\mathrm{er}(\mathcal{C})$ in the following way: for each $M \in \mathcal{C}$ we make a definite choice of a projective cover

$$
\mathcal{P}_M = (0 \to \mathbf{D}_1(M) \to \mathbf{P}_M \to M \to 0)
$$

For each morphism $f : M \to N$ in $\mathcal{C}$ there exists a morphism $\tilde{f} : \mathcal{P}_M \to \mathcal{P}_N$ lifting f; that is

$$
\tilde{f} = \begin{pmatrix} 0 \to \mathbf{D}_1(M) \to \mathbf{P}_M \to M \to 0 \\ \quad \downarrow \omega_+(\tilde{f}) \quad \downarrow \tilde{f} \quad \downarrow f \\ 0 \to \mathbf{D}_1(N) \to \mathbf{P}_N \to N \to 0 \end{pmatrix}
$$

Suppose that $\hat{f} : \mathcal{P}_M \to \mathcal{P}_N$ is another morphism lifting f; on considering the difference $\tilde{f} - \hat{f}$, it follows from (19.7) that $\omega_+(\tilde{f}) - \omega_+(\hat{f}) \approx 0$, and that the class of $[\omega_+(\tilde{f})] \in \mathcal{D}\mathrm{er}(\mathcal{C})$ is uniquely determined by that of f. Thus the correspondence $M \to \mathbf{D}_1(M)$; $f \mapsto [\omega_+(\tilde{f})]$ determines a functor $\mathbf{D}_1 : \mathcal{D}\mathrm{er}(\mathcal{C}) \to \mathcal{D}\mathrm{er}(\mathcal{C})$.

Changing the projective cover used to define $\mathbf{D}_1$ does not change the isomorphism type of $\mathbf{D}_1(M)$ within $\mathcal{D}\mathrm{er}(\mathcal{C})$; it does, however, change the isomorphism by which the identification is made. Thus suppose that

$$
\mathcal{P}_M = (0 \to \mathbf{D}_1^P \to \mathbf{P} \to M \to 0)
$$

and

$$
\mathcal{Q}_M = (0 \to \mathbf{D}_1^Q \to \mathbf{Q} \to M \to 0)
$$

are both projective covers of M, then there is a morphism $\tilde{\mathrm{Id}} : \mathcal{P}_M \to \mathcal{Q}_M$

lifting Id_M; that is

$$\tilde{\mathrm{Id}} = \begin{pmatrix} 0 \to \mathbf{D}_1^P \to \mathbf{P} \to M \to 0 \\ \quad\;\; \downarrow \alpha_{PQ} \;\; \downarrow \tilde{\mathrm{Id}} \;\; \downarrow \mathrm{Id} \\ 0 \to \mathbf{D}_1^Q \to \mathbf{Q} \to M \to 0 \end{pmatrix}$$

It is easy to check that, in the particular case where $Q = P$, one has $\alpha_{PP} \approx \mathrm{Id}$. It follows that:

Proposition 20.2: α_{PQ} is an isomorphism in the derived category.

Proof: $\alpha_{PQ} \circ \alpha_{QP} = \alpha_{PP} \approx \mathrm{Id}$ and likewise $\alpha_{QP} \circ \alpha_{PQ} \approx \mathrm{Id}$. □

We also have a functor $\omega_+ : \mathbf{Proj}_{\mathcal{C}}^1 \longrightarrow \mathcal{C}$ given by

$$w_+ \begin{pmatrix} 0 \to D \to P \to M \to 0 \\ \quad\;\; \downarrow h_+ \;\; \downarrow h_0 \;\; \downarrow h_- \\ 0 \to D' \to P' \to M' \to 0 \end{pmatrix} = \begin{matrix} D \\ \downarrow h_+ \\ D' \end{matrix}$$

If M' is a Λ-module then, by property $\mathcal{T}(4)$, there exists an exact sequence in $\mathcal{C}$ of the form

$$0 \to M' \to P' \to Q' \to 0$$

with P' projective. Suppose that M, M' are Λ-modules and that $f : M \to M'$ is a Λ-homomorphism. Let

$$0 \to M' \xrightarrow{j} P' \xrightarrow{\eta} Q' \to 0$$

be an exact sequence in $\mathcal{C}$ with P' projective and let

$$0 \to M \xrightarrow{i} X \xrightarrow{\epsilon} Q \to 0$$

be any short exact sequence in $\mathcal{C}$ begining in M. By the strong relative injectivity of the projective module P', there exists a Λ-homomorphism $\hat{f} : X \to P'$ making the following diagram commute

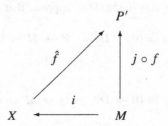

We get a commutative diagram

$$
\begin{array}{ccccccccc}
0 & \to & M & \xrightarrow{i} & X & \xrightarrow{\epsilon} & Q & \to & 0 \\
 & & \downarrow f & & \downarrow \hat{f} & & \downarrow \omega_+(\hat{f}) & & \\
0 & \to & K' & \xrightarrow{j} & P' & \xrightarrow{\eta} & Q' & \to & 0
\end{array}
$$

We see that:

Proposition 20.3: $\omega_+ : \mathbf{Proj}^1(\mathcal{C}) \to \mathcal{C}$ is an epifunctor.

We construct a functor $\mathbf{D}_{-1} : \mathcal{D}er(\mathcal{C}) \to \mathcal{D}er(\mathcal{C})$ in the following way: for each $M \in \mathcal{C}$ choose an exact sequence in $\mathcal{C}$

$$
\mathcal{P}_M = (0 \to M \to P \to Q \to 0)
$$

with P projective, and for each Λ-homomorphism $f : M \to N$ choose a specific morphism $\hat{f} : \mathcal{P}_M \to \mathcal{P}_N$ such that $\omega_+(\hat{f}) = f$; then define

$$
\mathbf{D}_{-1}(M) = [Q]
$$

and

$$
\mathbf{D}_{-1}(f) = [\omega_-(\hat{f})]
$$

It is straightforward to see that on the category $\mathcal{D}er(\mathcal{C})$ we have

$$
\mathbf{D}_1 \circ \mathbf{D}_{-1} \approx \mathbf{D}_{-1} \circ \mathbf{D}_1 \approx \mathrm{Id}
$$

that is:

Proposition 20.4: Let $\mathcal{C}$ be a tame class; then up to a natural equivalence, the functors $\mathbf{D}_1, \mathbf{D}_{-1} : \mathcal{D}er(\mathcal{C}) \to \mathcal{D}er(\mathcal{C})$ are mutually inverse *additive* equivalences of categories.

For any integer $n \geq 2$, we obtain functors $\mathbf{D}_n, \mathbf{D}_{-n} : \mathcal{D}er(\mathcal{C}) \to \mathcal{D}er(\mathcal{C})$ as follows

$$
\mathbf{D}_n = \underbrace{\mathbf{D}_1 \circ \mathbf{D}_1 \circ \cdots \circ \mathbf{D}_1}_{n}
$$

and

$$
\mathbf{D}_{-n} = \underbrace{\mathbf{D}_{-1} \circ \mathbf{D}_{-1} \circ \cdots \circ \mathbf{D}_{-1}}_{n}
$$

Obviously $\mathbf{D}_n$ and $\mathbf{D}_{-n}$ are also additive self-equivalences of the derived category $\mathcal{D}\text{er}(\mathcal{C})$. It follows that for all $M, N \in \mathcal{C}$

$$\text{Hom}_{\mathcal{D}\text{er}}(\mathbf{D}_1(M), \mathbf{D}_1(N)) \cong \text{Hom}_{\mathcal{D}\text{er}}(M, N) \cong \text{Hom}_{\mathcal{D}\text{er}}(\mathbf{D}_{-1}(M), \mathbf{D}_{-1}(N))$$

In particular, we obtain the following *adjointness formula*

(20.5) $$\text{Hom}_{\mathcal{D}\text{er}}(\mathbf{D}_n(M), N) \cong \text{Hom}_{\mathcal{D}\text{er}}(M, \mathbf{D}_{-n}(N))$$

The nth-cohomology group $\mathcal{H}^n(M, N)$ of M with coefficients in N is defined thus

(20.6) $$\mathcal{H}^n(M, N) \cong \text{Hom}_{\mathcal{D}\text{er}}(\mathbf{D}_n(M), N)$$

From (20.5) and (20.6) it is clear that

(20.7) $$\mathcal{H}^n(M, N) \cong \text{Hom}_{\mathcal{D}\text{er}}(M, \mathbf{D}_{-n}(N))$$

Though not essential to our arguments, it is true that, for $n \geq 1$, cohomology defined in the above way for modules in a tame class coincides with the standard Eilenberg–Maclane definition in terms of projective resolutions. The details are given in Section 27 below. With this interpretation, (20.6) above is a *Corepresentation Formula*, and (20.7) is the corresponding *Representation Formula*.

21 The long exact sequence in cohomology

Let $\mathcal{C}$ be a tame class of Λ-modules, and fix a short exact sequence $\mathcal{E}$ within $\mathcal{C}$

$$\mathcal{E} = \left(0 \to A \overset{i}{\to} B \overset{p}{\to} C \to 0\right)$$

It is straightforward to show that, for any Λ-module M, the sequence

$$0 \to \text{Hom}_\Lambda(M, A) \overset{i_*}{\to} \text{Hom}_\Lambda(M, B) \overset{p_*}{\to} \text{Hom}_\Lambda(M, C)$$

is exact. Passage to the derived category yields a more restricted statement:

Proposition 21.1: For any Λ-module M, the sequence

$$\text{Hom}_{\mathcal{D}\text{er}}(M, A) \overset{i_*}{\to} \text{Hom}_{\mathcal{D}\text{er}}(M, B) \overset{p_*}{\to} \text{Hom}_{\mathcal{D}\text{er}}(M, C)$$

is exact.

Proof: Clearly $p_* i_* = 0$. Let $\beta \in \mathrm{Hom}_\Lambda(M, B)$ and suppose that $p_*(\beta) \approx 0$; it suffices to show that there exists $\alpha \in \mathrm{Hom}_\Lambda(M, A)$ such that $\beta \approx i_*(\alpha)$. That is, we have a commutative diagram as follows, where P is projective

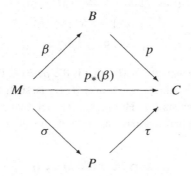

Since p is surjective, there exists $\lambda : P \to B$ such that $p \circ \lambda = \tau$; thus $p \circ \beta = \tau \circ \sigma$, and hence $p \circ \beta = p \circ \lambda \circ \sigma$. It follows that we get a mapping

$$\beta - \lambda \circ \sigma : M \to \mathrm{Ker}(p) = \mathrm{Im}(i)$$

Put $\alpha = i^{-1}(\beta - \lambda \circ \sigma)$ so that $\alpha : M \to A$. Then $\beta - i_*(\alpha) = \lambda \circ \sigma$. However $\lambda \circ \sigma$ is a factorization through the projective P, and, as required, $\beta \approx i_*(\alpha)$ for some $\alpha \in \mathrm{Hom}_\Lambda(M, A)$. $\qquad\square$

It is not in general true that $i_* : \mathrm{Hom}_{\mathcal{D}\mathrm{er}}(M, A) \to \mathrm{Hom}_{\mathcal{D}\mathrm{er}}(M, B)$ is injective. This will become clear (see (21.5) below) as we proceed to extend the above exact sequence to the right. Within $\mathcal{C}$, we fix a module M and a projective cover $\mathcal{P} = (0 \to D \xrightarrow{j} P \xrightarrow{\pi} M \to 0)$; any Λ-homomorphism $\gamma : M \to C$ admits a lifting to a morphism of exact sequences $\mathcal{P} \to \mathcal{E}$ of the following form

$$
\begin{array}{ccccccccc}
0 \to & D & \to & P & \to & M & \to 0 \\
& \downarrow \omega_+(\gamma) & & \downarrow \hat{\gamma} & & \downarrow \gamma \\
0 \to & A & \to & B & \to & C & \to 0
\end{array}
$$

and, as we have already seen, the correspondence $\gamma \mapsto \omega_+(\gamma)$ gives a well-defined additive homomorphism $\partial : \mathrm{Hom}_{\mathcal{D}\mathrm{er}}(M, C) \longrightarrow \mathrm{Hom}_{\mathcal{D}\mathrm{er}}(D, A)$; with this notation:

Proposition 21.2: The sequence

$$\mathrm{Hom}_{\mathcal{D}\mathrm{er}}(M, B) \xrightarrow{p_*} \mathrm{Hom}_{\mathcal{D}\mathrm{er}}(M, C) \xrightarrow{\partial} \mathrm{Hom}_{\mathcal{D}\mathrm{er}}(D, A)$$

is exact.

Proof: Let $b \in \mathrm{Hom}_\Lambda(M, B)$; since $\pi \circ j = 0$ then $b \circ \pi \circ j = 0$ so that $p_*(b)$ fits into the following commutative diagram

$$0 \to D \xrightarrow{j} P \xrightarrow{\pi} M \to 0$$
$$\downarrow 0 \quad \downarrow b \circ \pi \quad \downarrow p_*(b)$$
$$0 \to A \xrightarrow{i} B \xrightarrow{p} C \to 0$$

Thus $0 : D \to A$ represents $\partial \circ p_*(b)$; that is $\partial \circ p_*(b) \approx 0$, and so $\partial \circ p_* = 0$.

Conversely, suppose that $\gamma \in \mathrm{Hom}_\Lambda(M, C)$ is such that $\partial(\gamma) \approx 0$; we must show that there exists $b \in \mathrm{Hom}_\Lambda(M, B)$ such that $p_*(b) \approx \gamma$. Consider first the special case where we have a commutative diagram of the following form

$$0 \to D \xrightarrow{j} P \xrightarrow{\pi} M \to 0$$
$$\downarrow 0 \quad \downarrow \beta \quad \downarrow \gamma$$
$$0 \to A \xrightarrow{i} B \xrightarrow{p} C \to 0$$

that is, where we may actually take the zero map $0 : D \to A$ to represent $\partial(\gamma)$. Since $\beta_{|\mathrm{Im}(j)} \equiv 0$, there exists a unique homomorphism $b : M \to B$ such that $\beta = b \circ \pi$. However $\gamma \circ \pi = p \circ \beta$ so that $\gamma \circ \pi = p \circ b \circ \pi$. Since π is an epimorphism, the desired conclusion $p_*(b) = \gamma$ now follows.

In general, however, we can only assume that we have a commutative diagram

$$0 \to D \xrightarrow{j} P \xrightarrow{\pi} M \to 0$$
$$\downarrow \alpha \quad \downarrow \beta \quad \downarrow \gamma$$
$$0 \to A \xrightarrow{i} B \xrightarrow{p} C \to 0$$

where α factors through a projective module Q; that is $\alpha = \mu \circ \lambda$ where $\lambda : D \to Q$ and $\mu : Q \to A$ for some projective module $Q \in \mathcal{C}$. However, Q is injective relative to $\mathcal{C}$, so there exists a homomorphism $\nu : P \to Q$ making the following diagram commute

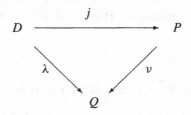

Then $\mu \circ \nu \circ j = \mu \circ \lambda = \alpha$. Put $\beta = \hat{\beta} - i \circ \mu \circ \nu : P \to B$ so that

$$
\begin{aligned}
\beta \circ j &= \hat{\beta} \circ j - i \circ \mu \circ \nu \circ j \\
&= i \circ \alpha - i \circ \mu \circ \lambda \\
&= i \circ \alpha - i \circ \alpha \\
&= 0
\end{aligned}
$$

However, the diagram below commutes

$$
\begin{array}{ccccccccc}
0 & \to & D & \xrightarrow{j} & P & \xrightarrow{\pi} & M & \to & 0 \\
& & \downarrow 0 & & \downarrow \beta & & \downarrow \gamma & & \\
0 & \to & A & \xrightarrow{i} & B & \xrightarrow{p} & C & \to & 0
\end{array}
$$

We are thus in the special case treated above, and the stated conclusion follows. □

Proposition 21.3: The sequence

$$
\operatorname{Hom}_{\mathcal{D}\mathrm{er}}(M, C) \xrightarrow{\partial} \operatorname{Hom}_{\mathcal{D}\mathrm{er}}(D, A) \xrightarrow{i_*} \operatorname{Hom}_{\mathcal{D}\mathrm{er}}(D, B)
$$

is exact.

Proof: If $\gamma : M \to C$ then γ lifts to a commutative diagram

$$
\begin{array}{ccccccccc}
0 & \to & D & \xrightarrow{j} & P & \xrightarrow{\pi} & M & \to & 0 \\
& & \downarrow \alpha & & \downarrow \beta & & \downarrow \gamma & & \\
0 & \to & A & \xrightarrow{i} & B & \xrightarrow{p} & C & \to & 0
\end{array}
$$

Clearly $i_*(\alpha) = \beta \circ j$ is a factorization of $i_*(\alpha)$ through the projective P, so that $i_*(\alpha) \approx 0$. By definition, however, α represents $\partial(\gamma)$ so that $i_*(\partial(\gamma)) \approx 0$, and we have shown that $i_* \circ \partial = 0$ in $\operatorname{Hom}_{\mathcal{D}\mathrm{er}}(D, B)$.

Conversely, suppose that $i_*(\alpha) \approx 0$ where $\alpha : D \to A$, and let $i \circ \alpha = \tau \circ \sigma$ be a factorization of $i \circ \alpha$ through a projective Q

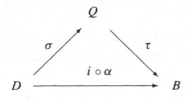

Since Q is injective relative to $\mathcal{C}$, σ factorizes thus

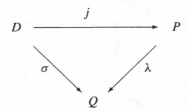

Put $\beta = \tau \circ \lambda : P \to B$; then $\beta \circ j = \tau \circ \lambda \circ j = \tau \circ \sigma = i \circ \alpha$. In particular, since $p \circ i = 0$, then $p \circ \beta \circ j = p \circ i \circ \alpha = 0$; that is, $(p \circ \beta)_{|\mathrm{Im}(j)} \equiv 0$ and $p \circ \beta$ induces a unique homomorphism $\gamma : M \to C$ making the following commute

$$0 \to D \xrightarrow{j} P \xrightarrow{\pi} M \to 0$$
$$\downarrow \alpha \quad \downarrow \beta \quad \downarrow \gamma$$
$$0 \to A \xrightarrow{i} B \xrightarrow{p} C \to 0$$

Thus $\alpha \approx \partial(\gamma)$ for some $\gamma \in \mathrm{Hom}_\Lambda(M, C)$, as claimed. $\square$

We obtain *the basic exact sequence in cohomology*:

Theorem 21.4: Let $\mathcal{E} = (0 \to A \xrightarrow{i} B \xrightarrow{p} C \to 0)$ be an exact sequence in $\mathcal{C}$, where $\mathcal{C}$ is a tame class of Λ-modules; then, for any $M \in \mathcal{C}$, the sequence below is exact

$$\mathcal{H}^0(M, A) \xrightarrow{i_*} \mathcal{H}^0(M, B) \xrightarrow{p_*} \mathcal{H}^0(M, C) \xrightarrow{\partial} \mathcal{H}^1(M, A) \xrightarrow{i_*} \mathcal{H}^1(M, B)$$

The boundary map $\partial : \mathcal{H}^0(M, C) \to \mathcal{H}^1(M, A)$ depends on a specific choice of projective cover $\mathcal{P} = (0 \to D \to P \to M \to 0)$. If $\mathcal{Q} = (0 \to D' \to Q \to M \to 0)$ is another choice of projective cover for M, it is straightforward to see that

$$\partial^{\mathcal{Q}} = (\alpha_{\mathcal{P}\mathcal{Q}})_* \circ \partial^{\mathcal{P}}$$

where $(\alpha_{\mathcal{P}\mathcal{Q}})_* : \mathrm{Hom}_{\mathcal{D}er}(D, A) \to \mathrm{Hom}_{\mathcal{D}er}(D', A)$ is the isomorphism in the derived category which was discussed in Section 20. It is clear from the definition that $\mathcal{H}^n(M, A) = \mathcal{H}^0(\mathbf{D}_n(M), A) = \mathrm{Hom}_{\mathcal{D}er}(\mathbf{D}_n(M), A)$; moreover, it is also evident $\mathbf{D}_{n+1} = \mathbf{D}_1 \circ \mathbf{D}_n$. It follows immediately that:

Theorem 21.5: Let $\mathcal{E} = (0 \to A \xrightarrow{i} B \xrightarrow{p} C \to 0)$ be an exact sequence in the tame class $\mathcal{C}$; then, for any $M \in \mathcal{C}$, the sequence below is exact for all $n \in \mathbf{Z}$.

$$\cdots \to \mathcal{H}^n(M, A) \xrightarrow{i_*} \mathcal{H}^n(M, B) \xrightarrow{p_*} \mathcal{H}^n(M, C) \xrightarrow{\partial} \mathcal{H}^{n+1}(M, A) \xrightarrow{i_*} \mathcal{H}^{n+1}(M, B) \xrightarrow{p_*} \cdots$$

22 Stable modules and the derived category

For a tame class $\mathcal{C}$, the derived module category $\mathcal{D}er(\mathcal{C})$ can be regarded as the quotient category obtained from $\mathcal{C}$ by setting all projective modules equal to zero. It is useful to have some criterion for characterizing projective modules in this context.

Proposition 22.1: If $\mathcal{C}$ is a tame class then the following conditions on a module $L \in \mathcal{C}$ are equivalent:

(i) L is projective;

(ii) $\mathrm{Id}_L \approx 0$;

(iii) $\mathrm{Hom}_{\mathcal{D}er}(M, L) = 0$ for all $M \in \mathcal{C}$;

(iv) $\mathrm{Hom}_{\mathcal{D}er}(L, M) = 0$ for all $M \in \mathcal{C}$;

(v) there exists an integer n such that $\mathcal{H}^n(M, L) = 0$ for all $M \in \mathcal{C}$;

(vi) there exists an integer n such that $\mathcal{H}^n(L, M) = 0$ for all $M \in \mathcal{C}$.

Proof: (i) $\Rightarrow$ (ii) is obvious.

(ii) $\Rightarrow$ (i) Suppose that $\mathrm{Id}_L \approx 0$, then Id_L factorizes. Thus $\mathrm{Id}_L = \alpha \circ \beta$ where $\alpha : P \to L$ and $\beta : L \to P$ where P is projective. In particular, α is surjective and β is injective. Put $K = \mathrm{Ker}(\alpha)$; then there exists a short exact sequence

$$0 \to K \to P \xrightarrow{\alpha} L \to 0$$

which is split on the right by β; that is, $\alpha \circ \beta = \mathrm{Id}_L$. Thus $P \cong K \oplus L$, and L, being a direct summand of a projective module, is itself projective.

(ii) $\Rightarrow$ (iii) Let $f \in \mathrm{Hom}_{\mathcal{D}er}(M, L)$; then $\mathrm{Id}_L \circ f = f$. Thus, if $\mathrm{Id}_L \approx 0$, then $\mathrm{Id}_L \circ f \approx 0$ so that $f \approx 0$.

(iii) $\Rightarrow$ (ii) If $\mathrm{Hom}_{\mathcal{D}er}(M, L) = 0$ for all $L \in \mathcal{C}$, then, in particular, $\mathrm{Hom}_{\mathcal{D}er}(L, L) = 0$. Thus $\mathrm{Id}_L \approx 0$.

(ii) $\Rightarrow$ (iv) This follows by the same proof that (ii) $\Rightarrow$ (iii) on replacing $\mathrm{Id}_L \circ f$ by $f \circ \mathrm{Id}_L$.

(iv) $\Rightarrow$ (ii) This follows by the same proof as (iii) $\Rightarrow$ (ii).

The equivalence of (iii) with (v) is clear, since $\mathcal{H}^n(M, L) \cong \mathrm{Hom}_{\mathcal{D}er}(\mathbf{D}_n(M), L)$. Likewise, (iv) and (vi) are equivalent since $\mathcal{H}^n(L, M) \cong \mathrm{Hom}_{\mathcal{D}er}(L, \mathbf{D}_{-n}(M))$. $\square$

Recall that, in Chapter 3, we introduced a stability relation '$\sim$' on Λ-modules by means of

$$M \sim N \iff M \oplus \Lambda^\mu \cong N \oplus \Lambda^\nu$$

When C is a tame class we put

$$\text{Stab}(C) = C/\sim$$

In addition to the stability relation $\sim$, there is an analogous, but coarser, relation which arises naturally in connection with the derived category. We define the *hyper-stability relation* '$\sim\sim$' on C by writing

$$M_1 \sim\sim M_2 \iff M_1 \oplus P_1 \cong M_2 \oplus P_2$$

for some finitely generated projective modules P_1, P_2. Denote by $\text{Hyp}(C)$ the set of equivalence classes of C under $\sim\sim$

$$\text{Hyp}(C) = C/\sim\sim$$

The objects in $\text{Hyp}(C)$ will be called *hyper-stable modules*.* If $M \in C$, denote by $\langle M \rangle \in \text{Hyp}(C)$ the class of M under $\sim\sim$.

Hyper-stable modules are precisely the same as isomorphism classes in $\mathcal{D}er(C)$:

Proposition 22.2: Let C be a tame class, and let M, $N \in C$; then

$$M \cong_{\text{Der}} N \iff M \sim\sim N$$

Proof: ($\Leftarrow$) Let $P \in C$ be projective. For any $M \in C$, let $\iota_M : M \to M \oplus P$ denote the inclusion $\iota_M(x) = (x, 0)$, and let $\pi_M : M \oplus P \to M$ denote the projection $\pi_M(x, y) = x$. Then $\pi_M \circ \iota_M = \text{Id}_M$ whilst $\iota_M \circ \pi_M - \text{Id}_M \approx 0$, so that ι_M and π_M are mutually inverse isomorphisms in $\mathcal{D}er(C)$.

($\Rightarrow$) Suppose that $f : M \to N$ and $g : N \to M$ are Λ-module homomorphisms such that

$$f \circ g \sim \text{Id}_N; \quad g \circ f \sim \text{Id}_M$$

Let $\varphi : Q \to N$ be a surjective homomorphism from a projective module $Q \in C$, put $M_1 = M \oplus Q$ and let

$$F : M_1 \to N$$

be the homomorphism

$$F(m, q) = f(m) + \varphi(q)$$

---

* We apologize for this neologism but it seems difficult to avoid something like it. If, as we wish to, we follow the the usual convention within Algebraic K-Theory, then 'stabilize' is already reserved to mean 'addition of a free summand'. See the note on terminology at the end of this section.

Put $K = \text{Ker}(F)$, and let $j : K \to M_1$ be the inclusion; then:

(I) $K \in \mathcal{C}$;

(II) F is surjective;

(III) $F : M_1 \to N$ is an isomorphism in $\mathcal{D}er(\mathcal{C})$.

By (III), for any module $L \in \mathcal{C}$, the induced map $F_* : \text{Hom}_{\mathcal{D}er}(L, M_1) \to \text{Hom}_{\mathcal{D}er}(L, N)$ is an isomorphism of abelian groups. In particular, F_* : $\mathcal{H}^0(L, M_1) \to \mathcal{H}^0(L, N)$ is an isomorphism. Moreover, since $\mathcal{H}^0(\mathbf{D}_1(L), -) \cong \mathcal{H}^1(L, -)$, it is also true that the induced map $F_* : \mathcal{H}^1(L, M_1) \to \mathcal{H}^1(L, N)$ is an isomorphism.

From the long exact coefficient sequence

$$\mathcal{H}^0(L; M_1) \overset{F_*}{\to} \mathcal{H}^0(L; N) \overset{\delta}{\to} \mathcal{H}^1(L; K) \overset{j_*}{\to} \mathcal{H}^1(L; M_1) \overset{F_*}{\to} \mathcal{H}^1(L; N)$$

we see that $\mathcal{H}^1(L; K) = 0$ for all modules $L \in \mathcal{C}$. Thus K is projective by (22.1). In particular, K is also injective relative to $\mathcal{C}$, so that the exact sequence $0 \to K \to M_1 \to N \to 0$ splits, and $M_1 \cong N \oplus K$. Hence $M \oplus Q \cong N \oplus K$ and $M \sim\sim N$ as required. $\square$

Thus, when $\mathcal{C}$ is tame, isomorphism classes in $\mathcal{D}er(\mathcal{C})$ are parametrized by hyper-stable modules so that the derived functors $\mathbf{D}_n$ give rise to correspondences

$$\mathbf{D}_n : \mathcal{C} \to \text{Hyp}(\mathcal{C})$$

If $\mathcal{C}$ is a tame class, then $\text{Stab}(\mathcal{C})$ is naturally an abelian monoid under '$\oplus$'. A case of particular interest is the class $\mathcal{C} = \mathbf{P}(\Lambda)$ of finitely generated projective modules. Then $\text{Stab}(\mathbf{P}(\Lambda))$ is simply the reduced projective class group $\widetilde{K}_0(\Lambda)$ of Λ.

In general, for any tame class $\mathcal{C}$, $\text{Hyp}(\mathcal{C})$ is also an abelian monoid under $\oplus$, and the correspondence $[M] \mapsto \langle M \rangle$ gives a monoid homomorphism $\mu :$ $\text{Stab}(\mathcal{C}) \to \text{Hyp}(\mathcal{C})$ given by $\mu([M]) = \langle M \rangle$, whose kernel is easily seen to be $\widetilde{K}_0(\Lambda)$. Thus, μ fails to be injective precisely when $\widetilde{K}_0(\Lambda) \neq 0$.

A projective cover $0 \to \Omega \to S \to A \to 0$ is said to be *stably free* when S is stably free. If $0 \to \Omega \to S \to A \to 0$ and $0 \to \Omega' \to S' \to A \to 0$ are stably free covers, Schanuel's Lemma gives an isomorphism $\Omega \oplus S' \cong \Omega' \oplus S$; hence $\Omega \sim \Omega'$. By restricting to stably free covers, the constructions $\mathbf{D}_n$ are modified to produce correspondences

$$\Omega_n : \mathcal{C} \to \text{Stab}(\mathcal{C})$$

We proceed as follows; to each module $M \in \mathcal{C}$, we associate a sequence $(\Omega_r(M))_{r \geq 0}$ of stable modules defined by the condition that $\Omega_n(M)$ is the stable class $[D]$ of any module $D \in \mathcal{C}$ for which there exists an exact sequence of the form

$$0 \to D \to S_{n-1} \to \cdots \to S_0 \to M \to 0$$

where each S_r is a finitely generated stably free module over Λ. Likewise, to each module $M \in \mathcal{C}$, we associate a sequence $(\Omega_{-r}(M))_{r\geq 0}$ of stable modules defined by the condition that $\Omega_{-n}(M)$ is the stable class $[D]$ of any module $D \in \mathcal{C}$ for which there exists an exact sequence of the form

$$0 \to M \to S_0 \to \cdots \to S_{n-1} \to D \to 0$$

where each S_r is a finitely generated stably free module over Λ. In effect, we produce liftings of $\mathbf{D}_n$ through μ

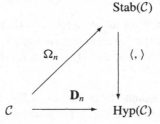

The following is clear:

Proposition 22.3: For any module $M \in \mathcal{C}$ the following relations hold:

(i) $\Omega_m(\Omega_n(M)) = \Omega_{m+n}(M)$;

(ii) $\Omega_n(\Omega_{-n}(M)) = [M]$.

The correspondences $(\Omega_n)_{n\in\mathbf{Z}}$ are no longer functors on the derived category; in particular, we have not defined the action of morphisms under Ω_n.

A note on terminology

In the world of Algebraic K-Theory, 'stabilization' has a reserved meaning, namely 'addition of a free summand'. We have accordingly chosen 'hyper-stable' to connote 'addition by a projective summand'. In Carlson's book [11], what we have called the 'derived module category' is called the 'stable module category'. In that case, there is no confusion since projectives are then necessarily free; 'hyper-stable module category' seemed too much. We have used 'derived' because it objectifies the 'derived functor' construction. Our 'derived module category' should not be confused with 'derived categories' of chain complexes [23], [22], although the relation is close. In any case, as anyone familiar with derived categories will know, there are already so many variations (positive, negative, bounded $\cdots$) that one more will not hurt.

23 Module extensions and $\mathbf{Ext}^1$

Let $\mathcal{C}$ denote a tame class of Λ-modules; associated with $\mathbf{Ext}^1_{\mathcal{C}}$ are a number of natural constructions.

Pullback

Given a Λ-homomorphism $f : A_1 \to A_2$ there is a 'pullback functor'

$$f^* : \mathbf{Ext}^1_{\mathcal{C}}(A_2, B) \to \mathbf{Ext}^1_{\mathcal{C}}(A_1, B)$$

defined as follows; if $\mathbf{E} = (0 \to B \to E_0 \xrightarrow{\eta} A_2 \to 0) \in \mathbf{Ext}^1_{\mathcal{C}}(A_2, B)$, we put $f^*(\mathbf{E}) = (0 \to B \to F_0 \xrightarrow{\epsilon} A_1 \to 0)$, where F_0 is the fibre product

$$F_0 = E_0 \underset{\eta, f}{\times} A_1 = \{(x, y) : \eta(x) = f(y)\}$$

and $\epsilon : F_0 \to A_1$ is the projection $\epsilon(x, y) = y$. If $g : A_2 \to A_3$ is a Λ-homomorphism, it is straightforward to see that $(g \circ f)^*(\mathbf{E}) = f^* \circ g^*(\mathbf{E})$. There is a natural transformation $\mu_f : f^* \to \mathrm{Id}$ defined by

$$
\begin{array}{c}
f^*(\mathbf{E}) \\
\downarrow \mu_f = \\
\mathbf{E}
\end{array}
\begin{pmatrix}
0 \to B \to F_0 \to A_1 \to 0 \\
\quad\; \downarrow \mathrm{Id} \;\; \downarrow \mu_0 \;\; \downarrow f \\
0 \to B \to E_0 \to A_2 \to 0
\end{pmatrix}
$$

where $\mu_0 : F_0 \to E_0$ is the projection $\mu_0(x, y) = x$.

Pushout

Let A, B_1, B_2 be Λ-modules; if $f : B_1 \to B_2$ is a Λ-homomorphism, the 'pushout' functor $f_* : \mathbf{Ext}^1_{\mathcal{C}}(A, B_1) \to \mathbf{Ext}^1_{\mathcal{C}}(A, B_2)$ is defined thus. Let

$$\mathbf{E} = \left(0 \to B_1 \to E_0 \xrightarrow{\eta} A \to 0\right) \in \mathbf{Ext}^1_{\mathcal{C}}(A, B_1)$$

and put

$$f_*(\mathbf{E}) = \left(0 \to B_2 \xrightarrow{j} F_0 \xrightarrow{\epsilon} A \to 0\right)$$

where F_0 is the colimit

$$F_0 = \underset{\to}{\lim}(f, i) = (B_2 \oplus E_0)/\mathrm{Im}(f \times -i)$$

and $j : B_2 \to F_0$ is the injection $j(x) = [x, 0]$. It is straightforward to see that, if $g : B_2 \to B_3$, then

$$(g \circ f)_*(\mathbf{E}) = g_* \circ f_*(\mathbf{E})$$

There is a natural transformation $\nu_f : \mathrm{Id} \to f_*$ obtained as follows

$$
\begin{array}{cc}
\mathbf{E} & \begin{pmatrix} 0 \to B_1 \to E_0 \to A \to 0 \\[4pt] \quad\;\; \downarrow f \;\; \downarrow \nu_0 \;\; \downarrow \mathrm{Id} \\[4pt] 0 \to B_2 \to F_0 \to A \to 0 \end{pmatrix} \\
\downarrow \nu_f = & \\
f_*(\mathbf{E}) &
\end{array}
$$

where $\nu_0 : E_0 \to F_0$ is the inclusion $\nu_0(x) = [0, x]$.

Direct product

Let A_1, A_2, B_1, B_2 be Λ-modules, and let $\mathbf{E}(r) \in \mathrm{Ext}_C^n(A_r, B_r)$ for $r = 1, 2$

$$\mathbf{E}(r) = (0 \to B_r \to E(r)_0 \to A_r \to 0)$$

The direct product $\mathbf{E}(1) \times \mathbf{E}(2)$ is the extension

$$(0 \to B_1 \times B_2 \to E(1)_0 \times E(2)_0 \to A_1 \times A_2 \to 0)$$

The sequence $\mathbf{E}(1) \times \mathbf{E}(2)$ is exact, and we get a functorial pairing

$$\times : \mathbf{Ext}_C^1(A_1, B_1) \times \mathbf{Ext}_C^1(A_2, B_2) \to \mathbf{Ext}_C^1(A_1 \oplus A_2, B_1 \oplus B_2)$$

If $\mathbf{E}, \mathbf{F} \in \mathbf{Ext}_C^1(A, B)$, a morphism $\varphi : \mathbf{E} \to \mathbf{F}$ is said to be a *congruence* when it induces the identity at both ends thus

$$
\begin{array}{cc}
\mathbf{E} & \begin{pmatrix} 0 \to B \to E_0 \to A \to 0 \\[4pt] \quad\;\; \downarrow \mathrm{Id} \;\; \downarrow \varphi_0 \;\; \downarrow \mathrm{Id} \\[4pt] 0 \to B \to F_0 \to A \to 0 \end{pmatrix} \\
\downarrow \varphi = & \\
\mathbf{F} &
\end{array}
$$

By the Five Lemma, congruence is an equivalence relation on $\mathbf{Ext}_C^1(A, B)$. We denote by $\mathrm{Ext}_C^1(A, B)$ the collection of equivalence classes in $\mathbf{Ext}_C^1(A, B)$ under the relation of congruence. Elementary considerations show that $\mathbf{Ext}_C^1(A, B)$ is equivalent to a small category, so that $\mathrm{Ext}_C^1(A, B)$ is actually a set. There is a natural group structure on $\mathrm{Ext}_C^1(A, B)$ obtained as follows; direct product gives a functorial pairing

$$\times : \mathbf{Ext}_C^1(A_1, B_1) \times \mathbf{Ext}_C^1(A_2, B_2) \to \mathbf{Ext}_C^1(A_1 \oplus A_2, B_1 \oplus B_2)$$

Let A, B_1, B_2 be Λ-modules; there is a functorial pairing, *external sum*

$$\oplus : \mathbf{Ext}_C^1(A, B_1) \times \mathbf{Ext}_C^1(A, B_2) \to \mathbf{Ext}_C^1(A, B_1 \oplus B_2)$$

given by

$$\mathcal{E}_1 \oplus \mathcal{E}_2 = \Delta^*(\mathcal{E}_1 \times \mathcal{E}_2)$$

where $\Delta : A \to A \times A$ is the diagonal. The addition map $+ : B \times B \to B$ can also be regarded as a Λ-homomorphism

$$\alpha : B \oplus B \to B; \quad \alpha(b_1, b_2) = b_1 + b_2$$

Combining external sum with pushout, we obtain the so-called 'Baer sum'; let $\mathcal{E}_r \in \mathbf{Ext}_C^1(A, B)$ for $r = 1, 2$, and define the *Baer sum* $\mathcal{E}_1 + \mathcal{E}_2$ by

$$\mathcal{E}_1 + \mathcal{E}_2 = \alpha_*(\mathcal{E}_1 \oplus \mathcal{E}_2) \ (= \alpha_* \Delta^*(\mathcal{E}_1 \times \mathcal{E}_2))$$

This gives a functorial pairing

$$+ : \mathbf{Ext}_C^1(A, B) \times \mathbf{Ext}_C^1(A, B) \to \mathbf{Ext}_C^1(A, B)$$

It is straightforward to see that congruence in $\mathbf{Ext}_C^1$ is compatible with Baer sum, and that:

Proposition 23.1: $\mathrm{Ext}_C^1(A, B)$ is an abelian group with respect to Baer sum.

Proof: The identity is given by the trivial extension

$$\mathcal{T} = (0 \to B \to B \oplus A \to A \to 0)$$

If $\mathcal{E} = (0 \to B \overset{i}{\to} X \overset{p}{\to} A \to 0)$, then the congruence class $-\mathcal{E}$ is represented by $(0 \to B \overset{i}{\to} X \overset{-p}{\to} A \to 0)$. The required congruence $\Psi_* : \mathcal{E} + (-\mathcal{E}) \to \mathcal{T}$ is induced from the morphism

$$
\begin{array}{ccccccccc}
0 & \to & B \oplus B & \to & X \underset{p,-p}{\times} X & \to & A & \to & 0 \\
 & & \downarrow + & & \downarrow \Psi & & \downarrow \mathrm{Id} & & \\
0 & \to & B & \to & B \oplus A & \to & A & \to & 0
\end{array}
$$

where $\Psi(x_1, x_2) = (x_1 + x_2, p(x))$. $\qquad\square$

Projective covers have some notable invariance properties under isomorphisms, inclusions and projections. For the purposes of the discussion, fix a projective cover

$$\mathbf{P} = (0 \to D \to P \to M \to 0) \in \mathbf{Proj}_C^1(M, D)$$

Let $\alpha : D \to D'$ be a morphism in $\mathcal{C}$; then we have a natural map

$$
\begin{array}{c} \mathbf{P} \\ \downarrow \nu(\alpha) = \\ \alpha_*(\mathbf{P}) \end{array}
\begin{pmatrix}
0 \to & D & \overset{i}{\to} & P \to & M \to 0 \\
& \downarrow \alpha & & \downarrow \alpha_0 & \downarrow \mathrm{Id} \\
0 \to & D' & \to & \underset{\to}{\lim}(\alpha, i) \to & M \to 0
\end{pmatrix}
$$

We claim:

Proposition 23.2: With the above notation, $\alpha_*(\mathbf{P})$ is a projective cover if and only if $[\alpha]$ is an isomorphism in $\mathcal{D}\mathrm{er}(\mathcal{C})$.

Proof: ($\Leftarrow$). Suppose that $\alpha : D \to D'$ gives an isomorphism in $\mathcal{D}\mathrm{er}(\mathcal{C})$. We must show that the colimit $L = \underset{\to}{\lim}(\alpha, i)$ is projective. For any $X \in \mathcal{C}$, there is a commutative ladder of coefficient sequences

$$
\begin{array}{ccccccccc}
\mathcal{H}^0(X, M) & \to & \mathcal{H}^1(X, D) & \to & \mathcal{H}^1(X, P) & \to & \mathcal{H}^1(X, M) & \to & \mathcal{H}^2(X, D) \\
\downarrow \mathrm{Id} & & \downarrow \alpha_* & & \downarrow (\alpha_0)_* & & \downarrow \mathrm{Id} & & \downarrow \alpha_* \\
\mathcal{H}^0(X, M) & \to & \mathcal{H}^1(X, D') & \to & \mathcal{H}^1(X, L) & \to & \mathcal{H}^1(X, M) & \to & \mathcal{H}^2(X, D')
\end{array}
$$

in which Id and α_* are isomorphisms; then $(\alpha_0)_* : \mathcal{H}^1(X, P) \to \mathcal{H}^1(X, L)$ is also an isomorphism. However, P is projective so that we have $\mathcal{H}^1(X, L) \cong \mathcal{H}^1(X, P) = 0$ for all $X \in \mathcal{C}$, and L is projective by (22.1). This completes the proof of ($\Leftarrow$).

($\Rightarrow$) To show the converse, note that, if $\mathbf{P} = (0 \to D \to P \to A \to 0)$ is a projective cover and $\mu : \mathbf{P} \to \mathbf{P}$ is a lifting of the zero map $0 : A \to A$, then $\mu_+ : D \to D$ factors through the projective P. Consequently, given a lifting of Id_A

$$
\begin{array}{c} \mathbf{P} \\ \downarrow = \\ \mathbf{P} \end{array}
\begin{pmatrix}
0 \to D \to P \to A \to 0 \\
\downarrow \omega \quad \downarrow \omega_0 \quad \downarrow \mathrm{Id} \\
0 \to D \to P \to A \to 0
\end{pmatrix}
$$

then $\omega - \mathrm{Id}$ factors through P; that is, $[\omega] = [\mathrm{Id}]$ in $\mathrm{End}_{\mathcal{D}\mathrm{er}}(D)$.

Now suppose that $\alpha : D \to D'$ is a Λ-homomorphism such that $L = \underset{\to}{\lim}(\alpha, i)$ is projective, then there exists a morphism $\alpha_*(\mathbf{P}) \to \mathbf{P}$ lifting Id

$$
\begin{array}{c} \alpha_*(\mathbf{P}) \\ \downarrow \quad = \\ \mathbf{P} \end{array}
\begin{pmatrix}
0 \to & D' & \to & \underset{\to}{\lim}(\alpha, i) \to & A \to 0 \\
& \downarrow \beta & & \downarrow \beta_0 & \downarrow \mathrm{Id} \\
0 \to & D & \to & P \to & A \to 0
\end{pmatrix}
$$

By the argument above, both $\alpha \circ \beta - \mathrm{Id}_{D'}$ and $\beta \circ \alpha - \mathrm{Id}_D$ factor through projectives, so that $[\alpha]$ is an isomorphism in $\mathcal{D}er(\mathcal{C})$, with inverse $[\beta]$. This proves ($\Rightarrow$) and completes the proof. $\qquad \square$

The proof of the corresponding statement for pullbacks is dual to the above; we leave the details to the reader:

Proposition 23.3: If $\beta : M' \to M$ is a homomorphism in $\mathcal{C}$, then $\beta^*(\mathbf{P})$ is a projective cover if and only if $[\beta]$ is an isomorphism in $\mathcal{D}er(\mathcal{C})$.

Let $Q \in \mathcal{C}$ also be projective, let $i_D : D \to D \oplus Q$ be the inclusion $i(d) = (d, 0)$, and let $\pi_M : M \oplus Q \to M$ be the projection $\pi(m, q) = m$; then we have

(23.4) $$(i_D)_*(\mathbf{P}) \in \mathbf{Proj}^1_{\mathcal{C}}(M, D \oplus Q)$$

(23.5) $$\pi_M^*(\mathbf{P}) \in \mathbf{Proj}^1_{\mathcal{C}}(M \oplus Q, D)$$

Now suppose that $\mathbf{P}' = (0 \to D \oplus Q \to P \to M \to 0)$ is an extension in $\mathbf{Proj}^1_{\mathcal{C}}$ in which Q is also projective, then

(23.6) $$(\pi_D)_*(\mathbf{P}') \in \mathbf{Proj}^1_{\mathcal{C}}(M, D)$$

Finally, if $\mathbf{P}'' = (0 \to D \to P \to M \oplus Q \to 0)$ is an extension in $\mathbf{Proj}^1_{\mathcal{C}}$ in which Q is also projective; then

(23.7) $$i_M^*(\mathbf{P}'') \in \mathbf{Proj}^1_{\mathcal{C}}(M, D)$$

The proofs of (23.4)–(23.5) are completely straightforward. Those of the dual statements (23.6), (23.7) are only slightly less obvious, but *do* require specific appeal to the properties $\mathcal{T}(2)$ and $\mathcal{T}(3)$ of the tame class $\mathcal{C}$. We leave them to the reader.

For any module $M \in \mathcal{C}$, $\mathbf{D}_1(M)$ represents all modules $D' \in \mathcal{C}$ which are isomorphic in $\mathcal{D}er(\mathcal{C})$ to some particular module $D \in \mathcal{C}$ which occurs in a projective cover $\mathbf{P} = (0 \to D \to P \to M \to 0)$. By (22.2), we can identify $\mathbf{D}_1(M)$ with the class $\langle D \rangle$ under the hyper-stability relation '$\sim\sim$'. These identifications are completely compatible with the construction of projective covers, since by (22.2), (23.2), (23.4), (23.6) it follows easily that:

Proposition 23.8: Let M, D be modules in $\mathcal{C}$ which occur in a projective cover $\mathbf{P} = (0 \to D \to P \to M \to 0)$, and let $D' \in \mathcal{C}$. Then the following statements are equivalent:

(i) $D' \sim\sim D$;
(ii) $D' \in \mathbf{D}_1(M)$;
(iii) there exists a projective cover of the form $\mathbf{P}' = (0 \to D' \to P' \to M \to 0)$.

The dual statements likewise follow easily from (22.2), (23.3), (23.5), (23.7):

Proposition 23.9: Let M, D be modules in C which occur in a projective cover $\mathbf{P} = (0 \to D \to P \to M \to 0)$, and let $M' \in C$; then the following statements are equivalent:

(i) $M' \sim\sim M$;

(ii) $M' \in \mathbf{D}_{-1}(D)$;

(iii) there exists a projective cover of the form $\mathbf{P}' = (0 \to D \to P' \to M' \to 0)$.

Fix $A, B \in C$, and make a specific choice of projective cover

$$\mathbf{P} = (0 \to D \to P \to A \to 0)$$

Then D is a representative of $\mathbf{D}_1(A)$, and there is a mapping

$$\mathrm{ext}_\mathbf{P} : \mathrm{Hom}_\Lambda(D, B) \to \mathrm{Ext}^1_C(A, B)$$

defined by means of pushout, thus: $\mathrm{ext}_\mathbf{P}(f) = f_*(\mathbf{P})$. The definition of the addition

$$+ : \mathrm{Hom}_\Lambda(D, B) \times \mathrm{Hom}_\Lambda(D, B) \to \mathrm{Hom}_\Lambda(D, B)$$

exactly mirrors that of Baer sum, namely $f + g = \alpha_* \Delta^*(f, g)$, where $\Delta : A \to A \times A$ is the diagonal and $\alpha : B \times B \to B$ is the addition map. It follows easily that:

Proposition 23.10: The mapping $\mathrm{ext}_\mathbf{P} : \mathrm{Hom}_\Lambda(D, B) \to \mathrm{Ext}^1_C(A, B)$ is a group homomorphism.

There is also a function $c_\mathbf{P} : \mathrm{Ext}^1_C(A, B) \to \mathrm{Hom}_{\mathcal{D}\mathrm{er}}(D, B) = \mathcal{H}^1(A, B)$ obtained, thus, for each $\mathbf{E} \in \mathrm{Ext}^1_C(A, B)$, choose a morphism $\varphi : \mathbf{P} \to \mathbf{E}$ lifting Id_A

$$\begin{matrix} \mathbf{P} \\ \downarrow \varphi = \\ \mathbf{E} \end{matrix} \begin{pmatrix} 0 \to D \to P \to A \to 0 \\ \downarrow \varphi_+ \quad \downarrow \varphi \quad \downarrow \mathrm{Id} \\ 0 \to B \to E \to A \to 0 \end{pmatrix}$$

It follows from (19.7) that, if ψ is any other lifting of Id_A, the difference $\varphi_+ - \psi_+ : D \to B$ factors through a projective module. We define

$$c_\mathbf{P}(\mathbf{E}) = [\varphi_+] \in \mathcal{H}^1(A, B)$$

for any such lifting $\varphi : \mathbf{P} \to \mathbf{E}$ of Id_A, where $[\ ] : \mathrm{Hom}_\Lambda(D, B) \to \mathcal{H}^1(A, B)$ is the quotient map.

Proposition 23.11: The following diagram commutes

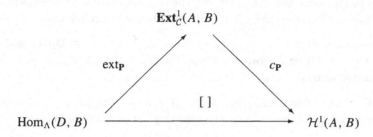

The proof of the following is straightforward:

Proposition 23.12: Let $\mathbf{E}, \mathbf{F} \in \mathbf{Ext}_{\mathcal{C}}^1(A, B)$; if $\mathbf{E} \approx \mathbf{F}$ then $c_{\mathbf{P}}(\mathbf{E}) = c_{\mathbf{P}}(\mathbf{F})$.

It follows that $c_{\mathbf{P}}$ induces a mapping, denoted by the same symbol

$$c_{\mathbf{P}} : \mathrm{Ext}_{\mathcal{C}}^1(A, B) \to \mathrm{Hom}_{\mathcal{D}\mathrm{er}}(D, B)$$

Composing $\mathrm{ext}_{\mathbf{P}} : \mathrm{Hom}_{\Lambda}(D, B) \to \mathbf{Ext}_{\mathcal{C}}^1(A, B)$ with the quotient map

$$\mathbf{Ext}_{\mathcal{C}}^1(A, B) \to \mathbf{Ext}_{\mathcal{C}}^1(A, B)/ \approx \; = \; \mathrm{Ext}_{\mathcal{C}}^1(A, B)$$

we obtain a function $\mathrm{Hom}_{\Lambda}(D, B) \to \mathrm{Ext}_{\mathcal{C}}^1(A, B)$, also denoted by '$\mathrm{ext}_{\mathbf{P}}$'. The following is the main technical result of Yoneda Theory:

Theorem 23.13: The mapping $\mathrm{ext}_{\mathbf{P}} : \mathrm{Hom}_{\Lambda}(D, B) \to \mathrm{Ext}_{\mathcal{C}}^1(A, B)$ factors through $\mathrm{Hom}_{\mathcal{D}\mathrm{er}}(D, B)$; that is, there is a mapping

$$e_{\mathbf{P}} : \mathrm{Hom}_{\mathcal{D}\mathrm{er}}(D, B) \to \mathrm{Ext}_{\mathcal{C}}^1(A, B)$$

such that $\mathrm{ext}_{\mathbf{P}} = e_{\mathbf{P}} \circ [\ ]$.

Proof: We have to show that $\mathrm{ext}_{\mathbf{P}} : \mathrm{Hom}_{\Lambda}(D, B) \to \mathrm{Ext}_{\mathcal{C}}^1(A, B)$ factors through $\mathrm{Hom}_{\mathcal{D}\mathrm{er}}(D, B)$. Suppose that $\varphi : D \to B$ factors as $\varphi = \lambda \circ \psi$ where $\psi : D \to Q, \lambda : Q \to B$, and Q is projective. Consider the extensions $\psi_*(\mathbf{P}) = (0 \to Q \to X_0 \to A \to 0)$ and $\varphi_*(\mathbf{P}) = (0 \to B \to E_0 \to A \to 0)$ where $X_0 = \varinjlim(\psi, i) = (Q \oplus P_0)/\mathrm{Im}(\psi \times -i)$ and $E_0 = \varinjlim(\varphi, i) = (B \oplus P_0)/\mathrm{Im}(\varphi \times -i)$. There is a morphism $\tilde{\lambda} : \psi_*(\mathbf{P}) \to \varphi_*(\mathbf{P})$ given by

$$
\begin{matrix}
\psi_*(\mathbf{P}) \\
\downarrow \tilde{\lambda} \\
\varphi_*(\mathbf{P})
\end{matrix}
\; = \;
\begin{pmatrix}
0 \to Q \to X_0 \to A \to 0 \\
\quad \downarrow \lambda \;\; \downarrow \lambda_0 \;\; \downarrow \mathrm{Id} \\
0 \to B \to E_0 \to A \to 0
\end{pmatrix}
$$

where $\lambda_0[q, p] = [\lambda(q), p]$. Q, being projective, is injective relative to $\mathcal{C}$, by property $\mathcal{T}(3)$, so that $\psi_*(\mathbf{P})$ splits. Let $s : A \to X_0$ be a right splitting of $\psi_*(\mathbf{P})$; then $\lambda_0 \circ s : A \to E_0$ is a right splitting of $\varphi_*(\mathbf{P})$ which therefore represents the zero class in $\text{Ext}^1_{\mathcal{C}}(A, B)$. Thus $\text{ext}_{\mathbf{P}}$ descends to a homomorphism $e_{\mathbf{P}} : \text{Hom}_{\mathcal{D}\text{er}}(D, B) \to \text{Ext}^1_{\mathcal{C}}(A, B)$, which is the desired conclusion. $\qquad\square$

Theorem 23.14: Let A, B, D be modules in $\mathcal{C}$ with $D \in \mathbf{D}_1(A)$, and let $\mathbf{P} \in \text{Ext}^1_{\mathcal{C}}(A, D)$ be a projective cover; then $c_{\mathbf{P}} : \text{Ext}^1_{\mathcal{C}}(A, B) \to \mathcal{H}^1(A, B)$ is bijective with $c_{\mathbf{P}}^{-1} = e_{\mathbf{P}}$.

Proof: Observe that the map $e_{\mathbf{P}} : \text{Hom}_{\mathcal{D}\text{er}}(D, B) \to \text{Ext}^1_{\mathcal{C}}(A, B)$ is surjective. If $\mathbf{E} \in \text{Ext}^1_{\mathcal{C}}(A, B)$, then choose a morphism $\varphi : \mathbf{P} \to \mathbf{E}$ lifting Id_A

$$
\begin{array}{cc}
\mathbf{P} \\
\downarrow \varphi =
\end{array}
\left(
\begin{array}{ccccccc}
0 \to & D & \xrightarrow{j} & P & \to & A \to & 0 \\
 & \downarrow \varphi_+ & & \downarrow \varphi & & \downarrow \text{Id}_A & \\
0 \to & B & \xrightarrow{i} & E & \to & A \to & 0
\end{array}
\right)
$$
$$
\mathbf{E}
$$

Then there is a congruence $\psi : (\varphi_+)_*(\mathbf{P}) \to \mathbf{E}$ as can be seen by the following

$$
\begin{array}{cc}
(\varphi_+)_*(\mathbf{P}) \\
\downarrow \psi \; =
\end{array}
\left(
\begin{array}{ccccccc}
0 \to & B & \to & \underset{\to}{\lim} & \to & A \to & 0 \\
 & \downarrow \text{Id}_B & & \downarrow \psi & & \downarrow \text{Id}_A & \\
0 \to & B & \xrightarrow{i} & E & \to & A \to & 0
\end{array}
\right)
$$
$$
\mathbf{E}
$$

Here $\lim = \lim(\varphi_+, j)$ and $\psi[b, x] = i(b) + \varphi(x)$. Hence $\text{ext}_{\mathbf{P}}(\varphi_+) \approx \mathbf{E}$ in $\text{Ext}^1_{\mathcal{C}}(\vec{A}, B)$, so that $e_{\mathbf{P}}[\varphi_+] = [\mathbf{E}]$ in $\text{Ext}^1_{\mathcal{C}}(A, B)$, and $e_{\mathbf{P}}$ is surjective as claimed.

It follows from the relation $c_{\mathbf{P}} \circ \text{ext}_{\mathbf{P}} = [\;]$ that $c_{\mathbf{P}} \circ e_{\mathbf{P}} = \text{Id}$; in particular, $e_{\mathbf{P}} = \text{Id}$ is injective, and, hence, by the above, bijective. Since $c_{\mathbf{P}}$ is a left inverse for the bijective map $e_{\mathbf{P}}$, it follows that $c_{\mathbf{P}}$ is two-sided inverse for $e_{\mathbf{P}}$, which is the desired result. $\qquad\square$

24 General module extensions and Ext^n

As above, $\mathcal{C}$ is a tame class of Λ-modules; for $n \geq 2$, $\mathbf{Ext}^n_{\mathcal{C}}$ will denote the category whose objects are exact sequences in $\mathcal{C}$ of the form

$$
\mathbf{E} = (0 \to E_+ \to E_{n-1} \to \cdots \to E_0 \to E_- \to 0)
$$

and whose morphisms are commutative diagrams of Λ-homomorphisms

$$
\begin{array}{cc}
\mathbf{E} \\
\downarrow h =
\end{array}
\left(
\begin{array}{ccccccccc}
0 \to & E_+ & \to & E_{n-1} & \to & \cdots & \to & E_0 & \to & E_- \to & 0 \\
 & \downarrow h_+ & & \downarrow h_{n-1} & & & & \downarrow h_0 & & \downarrow h_- & \\
0 \to & F_+ & \to & F_{n-1} & \to & \cdots & \to & F_0 & \to & F_- \to & 0
\end{array}
\right)
$$
$$
\mathbf{F}
$$

As in the case of $\mathbf{Ext}^1$, there are functors $\omega_-, \omega_+ : \mathbf{Ext}^n_{\mathcal{C}} \longrightarrow \mathcal{C}$ given by

$$
\omega_- \begin{pmatrix} 0 \to K \to P_{n-1} \to \cdots \to P_0 \to M \to 0 \\ \quad\ \downarrow h_+ \ \downarrow h_{n-1} \qquad\qquad \downarrow h_0 \ \downarrow h_- \\ 0 \to K' \to P'_{n-1} \to \cdots \to P'_0 \to M' \to 0 \end{pmatrix} \begin{matrix} M \\ = \downarrow h_- \\ M' \end{matrix}
$$

and

$$
\omega_+ \begin{pmatrix} 0 \to K \to P_{n-1} \to \cdots \to P_0 \to M \to 0 \\ \quad\ \downarrow h_+ \ \downarrow h_{n-1} \qquad\qquad \downarrow h_0 \ \downarrow h_- \\ 0 \to K' \to P'_{n-1} \to \cdots \to P'_0 \to M' \to 0 \end{pmatrix} \begin{matrix} K \\ = \downarrow h_+ \\ K' \end{matrix}
$$

If A, B are Λ-modules we denote by $\mathbf{Ext}^n_{\mathcal{C}}(A, B)$ the full subcategory of $\mathbf{Ext}^n_{\mathcal{C}}$ whose objects $\mathbf{E}$ satisfy $E_- = A$ and $E_+ = B$. The constructions introduced for $\mathbf{Ext}^1$ all have analogues for $\mathbf{Ext}^n$ ($n \geq 2$).

Pullback

If $f : A_1 \to A_2$ is a Λ-homomorphism, we obtain a functor $f^* : \mathbf{Ext}^n_{\mathcal{C}}(A_2, B) \to \mathbf{Ext}^n_{\mathcal{C}}(A_1, B)$; let $\mathbf{E} \in \mathbf{Ext}^n_{\mathcal{C}}(A_2, B)$

$$
\mathbf{E} = \left(0 \to B \to E_{n-1} \overset{\partial_{n-1}}{\to} \cdots \overset{\partial_1}{\to} E_0 \overset{\eta}{\to} A_2 \to 0 \right)
$$

and put

$$
f^*(\mathbf{E}) = \left(0 \to B \to F_{n-1} \overset{\delta_{n-1}}{\to} \cdots \overset{\delta_1}{\to} F_0 \overset{\epsilon}{\to} A_1 \to 0 \right)
$$

where F_0 is the fibre product $F_0 = E_0 \underset{\eta, f}{\times} A_1 = \{(x, y) : \eta(x) = f(y)\}$, $\delta_1 : F_1 \to F_0$ is the map $\delta_1(x) = (\partial_1(x), 0)$; $\epsilon : F_0 \to A_1$ is the projection $\epsilon(x.y) = y$, $F_r = E_r$ for $r \geq 1$; $F_r = E_r$ for $r \geq 1$, and, finally, $\delta_r = \partial_r$ for $r \geq 2$. It is straightforward to check that, if $g : A_2 \to A_3$, then

$$
(g \circ f)^* = f^* \circ g^*
$$

Observe also that there is a natural transformation $\mu_f : f^* \to \mathrm{Id}$ obtained as

$$
\begin{matrix} f^*(\mathbf{E}) \\ \downarrow \mu_f = \\ \mathbf{E} \end{matrix} \begin{pmatrix} 0 \to B \to F_{n-1} \to \cdots \to F_0 \to A_1 \to 0 \\ \quad\ \downarrow \mathrm{Id} \ \downarrow \mu_{n-1} \qquad\quad \downarrow \mu_0 \ \downarrow f \\ 0 \to B \to E_{n-1} \to \cdots \to E_0 \to A_2 \to 0 \end{pmatrix}
$$

where $\mu_r = \mathrm{Id}$ for $r \geq 1$, and $\mu_0 : F_0 \to E_0$ is the projection $\mu_0(x, y) = x$.

Pushout

If A, B_1, B_2 are Λ-modules and $f : B_1 \to B_2$ is a Λ-homomorphism, we obtain a functor $f_* : \mathbf{Ext}^n_{\mathcal{C}}(A, B_1) \to \mathbf{Ext}^n_{\mathcal{C}}(A, B_2)$ as follows; let $\mathbf{E} \in \mathbf{Ext}^n_{\mathcal{C}}(A, B_1)$

$$\mathbf{E} = \left(0 \to B_1 \xrightarrow{i} E_{n-1} \xrightarrow{\partial_{n-1}} \cdots \xrightarrow{\partial_1} E_0 \xrightarrow{\eta} A \to 0 \right)$$

and put

$$f_*(\mathbf{E}) = \left(0 \to B_2 \xrightarrow{j} F_{n-1} \xrightarrow{\delta_{n-1}} \cdots \xrightarrow{\delta_1} F_0 \xrightarrow{\epsilon} A_1 \to 0 \right)$$

where $F_r = E_r$ and $\delta_r = \partial_r$ for $r \leq n - 2$; F_{n-1} is the colimit

$$F_{n-1} = \varinjlim (f, i) = (B_2 \oplus E_{n-1})/\mathrm{Im}(f \times -i)$$

$j : B_2 \to F_{n-1}$ is the injection $j(x) = [x, 0]$, and $\delta_{n-1} : F_{n-1} \to F_{n-2}$ is the map $\delta_{n-1}[x, y] = \partial_{n-1}(y)$. It is straightforward to see that, if $g : B_2 \to B_3$, then

$$(g \circ f)_* = g_* \circ f_*$$

In this case there is a natural transformation $\nu_f : \mathrm{Id} \to f_*$ obtained as follows

$$
\begin{array}{cc}
\mathbf{E} & \\
\downarrow \nu_f = & \left(\begin{array}{ccccccc} 0 \to & B_1 \to & E_{n-1} \to & \cdots \to & E_0 \to & A \to 0 \\ & \downarrow f & \downarrow \nu_{n-1} & & \downarrow \nu_0 & \downarrow \mathrm{Id} \\ 0 \to & B_2 \to & F_{n-1} \to & \cdots \to & F_0 \to & A \to 0 \end{array} \right) \\
f_*(\mathbf{E}) &
\end{array}
$$

where $\nu_r = \mathrm{Id}$ for $r \leq n - 2$, and $\nu_{n-1} : E_0 \to F_0$ is the inclusion $\nu_{n-1}(x) = [0, x]$.

Direct product

Let A_1, A_2, B_1, B_2 be Λ-modules, and let $\mathbf{E}(r) \in \mathbf{Ext}^n_{\mathcal{C}}(A_r, B_r)$ for $r = 1, 2$

$$\mathbf{E}(r) = (0 \to B_r \to E(r)_{n-1} \to \cdots \to E(r)_0 \to A_r \to 0)$$

The direct product $\mathbf{E}(1) \times \mathbf{E}(2)$ is the sequence

$$(0 \to B_1 \times B_2 \to E(1)_{n-1} \times E(2)_{n-1} \to \cdots \to E(1)_0 \times E(2)_0 \to A_1 \times A_2 \to 0)$$

which is easily shown to be exact. We obtain a functorial pairing

$$\times : \mathbf{Ext}^n_{\mathcal{C}}(A_1, B_1) \times \mathbf{Ext}^n_{\mathcal{C}}(A_2, B_2) \to \mathbf{Ext}^n_{\mathcal{C}}(A_1 \oplus A_2, B_1 \oplus B_2)$$

If $\mathbf{E}, \mathbf{F} \in \mathbf{Ext}_C^n(A, B)$, a morphism $\varphi : \mathbf{E} \to \mathbf{F}$ is said to be an *elementary congruence* when it induces the identity at both ends, thus

$$
\begin{array}{c}
\mathbf{E} \\
\downarrow \varphi \ = \\
\mathbf{F}
\end{array}
\left(
\begin{array}{ccccccc}
0 \to & B \to & E_{n-1} \to & \cdots & \to E_0 \to & A \to 0 \\
 & \downarrow \text{Id} & \downarrow \varphi_{n-1} & & \downarrow \varphi_0 & \downarrow \text{Id} \\
0 \to & B \to & F_{n-1} \to & \cdots & \to F_0 \to & A \to 0
\end{array}
\right)
$$

In the case $n = 1$ we observed that elementary congruence is an equivalence relation. This fails to be true however when $n \geq 2$. We write $\mathbf{E} \rightsquigarrow \mathbf{F}$ when there exists an elementary congruence $\varphi : \mathbf{E} \to \mathbf{F}$, and we denote by '$\approx$' the equivalence relation generated by '$\rightsquigarrow$'; that is $\mathbf{E}, \mathbf{F} \in \mathbf{Ext}_C^n(A, B)$ are said to be *congruent*, written $\mathbf{E} \approx \mathbf{F}$, when there exists a sequence $\mathbf{E}_r, (0 \leq r \leq m) \in \mathbf{Ext}_C^n(A, B)$ such that $\mathbf{E} = \mathbf{E}_0, \mathbf{F} = \mathbf{E}_m$, and, for each $r \leq m - 1$, either $\mathbf{E}_r \rightsquigarrow \mathbf{E}_{r+1}$ or $\mathbf{E}_{r+1} \rightsquigarrow \mathbf{E}_r$. We denote by $\text{Ext}_C^n(A, B)$ the quotient set

$$
\text{Ext}_C^n(A, B) = \mathbf{Ext}_C^n(A, B)/ \approx
$$

In addition, there is a natural construction, Yoneda product, involving all $\mathbf{Ext}^n$.

Yoneda product

If A, B, C are Λ-modules we define a pairing

$$
\circ : \mathbf{Ext}_C^m(B, C) \times \mathbf{Ext}_C^n(A, B) \to \mathbf{Ext}_C^{m+n}(A, C)
$$

as follows. Let

$$
\mathbf{E}_1 = \left(0 \to C \overset{i}{\to} Y_{m-1} \overset{\partial^1_{m-1}}{\to} \cdots \overset{\partial^1_1}{\to} Y_0 \to B \to 0 \right) \in \mathbf{Ext}_C^m(A, B)
$$

$$
\mathbf{E}_2 = \left(0 \to B \to X_{n-1} \overset{\partial^2_{n-1}}{\to} \cdots \overset{\partial^2_1}{\to} X_0 \overset{\eta}{\to} A \to 0 \right) \in \mathbf{Ext}_C^n(B, C)
$$

and let

$$
\mathbf{E}_1 \circ \mathbf{E}_2 = \left(0 \to C \to Z_{n-1} \overset{\delta_{m+n-1}}{\to} \cdots \overset{\delta_1}{\to} Z_0 \to A \to 0 \right) \in \mathbf{Ext}_C^{m+n}(A, C)
$$

be the extension defined by $(Z_r, \delta_r) = (X_r, \partial^2_r)$ for $r \leq n - 1$, $Z_r = Y_{r-n}$ for $n - 1 < r \leq m + n - 1$, $\delta_n = \eta \circ i$, and $\delta_r = \partial^1_{r-n}$ for $n < r \leq m + n - 1$.

The Yoneda product is compatible with congruence, and descends to give a pairing of set valued functors

$$
\circ : \text{Ext}_C^m(B, C) \times \text{Ext}_C^n(A, B) \to \text{Ext}_C^{m+n}(A, C)
$$

We shall eventually impose a group structure on any $\text{Ext}^n_C(A, B)$, with respect to which these pairings are bi-additive. This point will not be pursued here, but will emerge later, as a consequence of Yoneda's classification of extensions.

25 Classification of general module extensions

If M is a Λ-module, by a *projective* (resp. *free*) *n-stem* of M we mean an exact sequence of Λ-homomorphisms of the form

$$\mathbf{P} = (0 \to K \to P_{n-1} \to \cdots \to P_0 \to M \to 0)$$

where P_r is projective (resp. free) for $0 \le r \le n - 1$. A projective 1-stem is the same as a projective cover. We denote by $\mathbf{Proj}^n_C$ the full subcategory of $\mathbf{Ext}^n_C$ whose objects are *projective n-stems*, and we denote by $\text{Proj}^n_C(M, D)$ the corresponding subset of $\text{Ext}^n_C(M, D)$. Projective n-stems exist; in fact we have:

Proposition 25.1: Let C be a tame class of Λ-modules. Then any module $A \in C$ has a free n-stem for each $n \ge 1$.

Proof: This is true for $n = 1$, since any $A \in C$ is finitely generated, and there exists an epimorphism $\eta : F_n \to A$ where $F(0)$ is a finitely generated free Λ-module. Putting $K = \text{Ker}(\eta)$, the exact sequence $\mathbf{P}(1) = (0 \to K \to F(0) \to A \to 0)$ is a free 1-stem, and, since C contains all finitely generated projectives, it contains $F(0)$, so that $K \in C$ by property $\mathcal{T}(2)$. By induction, there is a free $(n - 1)$-stem

$$\mathbf{P}(n - 1) = (0 \to D \to F(n - 1) \to \cdots \to F(1) \to K \to 0)$$

over K, and the Yoneda product $\mathbf{P}(n - 1) \circ \mathbf{P}(1)$ is a free n-stem over A. $\quad\square$

In dealing with projective n-stems, there is a simple technique which facilitates arguments by induction, namely 'cutting and splicing'. Given a projective n-stem

$$\mathbf{P} = (0 \to N \to P_{n-1} \to \cdots \to P_0 \to M \to 0)$$

by cutting at an intermediate point we form two shorter projective stems

$$\mathbf{P}_1 = (0 \to N \to P_{n-1} \to \cdots \to P_m \to K \to 0)$$
$$\mathbf{P}_2 = (0 \to K \to P_{m-1} \to \cdots \to P_0 \to M \to 0)$$

where $K = \text{Im}(P_m \to P_{m-1}) = \text{Ker}(P_{m-1} \to P_{m-2})$. The reverse process, 'splicing', re-combines the two shorter sequences by means of the Yoneda product

$$\mathbf{P} = \mathbf{P}_1 \circ \mathbf{P}_2$$

Let A, A', B be Λ-modules, let $\mathbf{P}$ be a projective n-stem over A, and let $\mathbf{E} \in \mathrm{Ext}_C^n(A', B)$; if $\varphi : A \to A'$ is a Λ-homomorphism, then a morphism $\hat{\varphi} : \mathbf{P} \to \mathbf{E}$ is said to be a *lifting of* φ when $\hat{\varphi}_- = \varphi$; that is, when the following diagram commutes

$$\begin{array}{c} \mathbf{P} \\ \downarrow \hat{\varphi} = \\ \mathbf{E} \end{array} \begin{pmatrix} 0 \to D \to P_{n-1} \to \cdots \to P_0 \to A \to 0 \\ \qquad \downarrow \varphi_+ \ \downarrow \varphi_{n-1} \qquad\qquad \downarrow \varphi_0 \ \downarrow \varphi \\ 0 \to B \to E_{n-1} \to \cdots \to E_0 \to A' \to 0 \end{pmatrix}$$

The following is easily proved by induction, using cutting and splicing, starting from the corresponding result for projective covers (19.7):

Proposition 25.2: Let $\mathbf{P} = (0 \to D \to P_{n-1} \to \cdots \to P_0 \to L \to 0)$ be a projective n-stem over A, and let $\mathbf{E} \in \mathrm{Ext}_C^n(A', B)$; if $\varphi : A \to A'$ is a Λ-homomorphism, then there exists a morphism $\hat{\varphi} : \mathbf{P} \to \mathbf{E}$ lifting φ; if $\tilde{\varphi}$ is any other lifting of φ, the difference $\hat{\varphi}_+ - \tilde{\varphi}_+ : D \to B$ factors through a projective module.

Likewise, the dual result follows by induction from (19.8):

Proposition 25.3: Let $\mathbf{P} = (0 \to D \to P_{n-1} \to \cdots \to P_0 \to L \to 0)$ be a projective n-stem over A, and let $\mathbf{E} \in \mathrm{Ext}_C^n(A', B)$; if $\varphi : B \to D$ is a Λ-homomorphism, then there exists a morphism $\hat{\varphi} : \mathbf{E} \to \mathbf{P}$ extending φ; if $\tilde{\varphi}$ is any other extension of φ, then the difference $\hat{\varphi}_- - \tilde{\varphi}_- : A' \to A$ factors through a projective module.

We make a specific choice of projective n-stem $\mathbf{P}$ over A

$$\mathbf{P} = (0 \to D \to P_{n-1} \to \cdots \to P_0 \to A \to 0)$$

so that $D \in C$ is a representative of $\mathbf{D}_n(A)$. Pushout defines a mapping $\mathrm{ext}_{\mathbf{P}} : \mathrm{Hom}_\Lambda(D, B) \to \mathrm{Ext}_C^n(A, B)$ by means of

$$\mathrm{ext}_{\mathbf{P}}(f) = f_*(\mathbf{P})$$

There is also a function $c_{\mathbf{P}} : \mathrm{Ext}_C^n(A, B) \to \mathrm{Hom}_{\mathcal{D}\mathrm{er}}(D, B) = \mathcal{H}^n(A, B)$ obtained as follows: for each $\mathbf{E} \in \mathrm{Ext}_C^n(A, B)$, we may, by (25.2) above, choose a morphism $\varphi : \mathbf{P} \to \mathbf{E}$ lifting Id_A

$$\begin{array}{c} \mathbf{P} \\ \downarrow \varphi = \\ \mathbf{E} \end{array} \begin{pmatrix} 0 \to D \to P_{n-1} \to \cdots \to P_0 \to A \to 0 \\ \qquad \downarrow \varphi_+ \ \downarrow \varphi_{n-1} \qquad\qquad \downarrow \varphi_0 \ \downarrow \mathrm{Id} \\ 0 \to B \to E_{n-1} \to \cdots \to E_0 \to A \to 0 \end{pmatrix}$$

If ψ is any other lifting of Id_A, the difference $\varphi_+ - \psi_+ : D \to B$ factors through a projective module. We define

$$c_\mathbf{P}(\mathbf{E}) = [\varphi_+] \in \mathcal{H}^n(A, B)$$

for any such lifting $\varphi : \mathbf{P} \to \mathbf{E}$ of Id_A, where $[\ ] : \mathrm{Hom}_\Lambda(D, B) \to \mathcal{H}^n(A, B)$ is the quotient map.

Theorem 25.4: The following diagram commutes

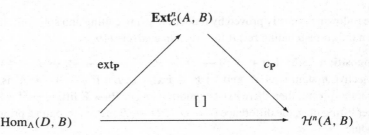

The proof of the following is entirely straightforward:

Proposition 25.5: Let $\mathbf{E}, \mathbf{F} \in \mathbf{Ext}^n_{\mathcal{C}}(A, B)$; if $\mathbf{E} \approx \mathbf{F}$ then $c_\mathbf{P}(\mathbf{E}) = c_\mathbf{P}(\mathbf{F})$.

It follows that $c_\mathbf{P}$ induces a mapping, denoted by the same symbol

$$c_\mathbf{P} : \mathrm{Ext}^n_{\mathcal{C}}(A, B) \to \mathrm{Hom}_{\mathcal{D}\mathrm{er}}(D, B)$$

Composing $\mathrm{ext}_\mathbf{P} : \mathrm{Hom}_\Lambda(D, B) \to \mathbf{Ext}^n_{\mathcal{C}}(A, B)$ with the quotient map

$$\mathbf{Ext}^n_{\mathcal{C}}(A, B) \to \mathrm{Ext}^n_{\mathcal{C}}(A, B) = \mathbf{Ext}^n_{\mathcal{C}}(A, B)/\approx$$

we obtain a function $\mathrm{Hom}_\Lambda(D, B) \to \mathrm{Ext}^n_{\mathcal{C}}(A, B)$, also denoted by '$\mathrm{ext}_\mathbf{P}$'. We have the following extension of (23.13):

Theorem 25.6: For any $n \geq 1$, $\mathrm{ext}_\mathbf{P} : \mathrm{Hom}_\Lambda(D, B) \to \mathrm{Ext}^n_{\mathcal{C}}(A, B)$ factors through $\mathrm{Hom}_{\mathcal{D}\mathrm{er}}(D, B)$; that is, there is a mapping

$$e_\mathbf{P} : \mathrm{Hom}_{\mathcal{D}\mathrm{er}}(D, B) \to \mathrm{Ext}^n_{\mathcal{C}}(A, B)$$

such that $\mathrm{ext}_\mathbf{P} = e_\mathbf{P} \circ [\ ]$.

Proof: The statement for $n = 1$ is true by (23.13). The case of n-fold extensions reduces to (23.13) using an explicit form of dimension shifting. Let $A, B \in \mathcal{C}$ and let

$$\mathbf{P} = \left(0 \to D_n \to P_{n-1} \overset{\partial_{n-1}}{\to} \cdots \to P_0 \to A \to 0 \right)$$

be a specific choice of projective n-stem over A. We split this at the $(n - 1)$ stem to get a projective $(n - 1)$-stem over A

$$\widehat{\mathbf{P}} = (0 \to D_{n-1} \to P_{n-2} \to \cdots \to P_0 \to A \to 0)$$

where $D_{n-1} = \text{Im}(\partial_{n-1}) = \text{Ker}(\partial_{n-2})$, and a projective cover

$$\mathbf{P}(1) = (0 \to D_n \to P_{n-1} \to D_{n-1} \to 0)$$

which are reassembled into the original by using the Yoneda product

$$\mathbf{P} = \mathbf{P}(1) \circ \widehat{\mathbf{P}}$$

If $\varphi : D_n \to B$ is a Λ-homomorphism, then

$$\varphi_*(\mathbf{P}) = \varphi_*(\mathbf{P}(1)) \circ \widehat{\mathbf{P}}$$

By the Main Theorem for $n = 1$, if $\varphi, \psi : D_n \to B$ are Λ-homomorphisms and $\varphi - \psi$ factors through a projective, then $\varphi_*(\mathbf{P}(1)) \approx \psi_*(\mathbf{P}(1))$ so that $\varphi_*(\mathbf{P}(1)) \circ \widehat{\mathbf{P}} \approx \psi_*(\mathbf{P}(1)) \circ \widehat{\mathbf{P}}$ and hence

$$\varphi_*(\mathbf{P}) \approx \psi_*(\mathbf{P})$$

Thus the construction $\varphi \mapsto \text{ext}_{\mathbf{P}}(\varphi) = \varphi_*(\mathbf{P})$ factors through $\text{Hom}_{\mathcal{D}er}(D, B)$. In particular, there is a mapping

$$e_{\mathbf{P}} : \text{Hom}_{\mathcal{D}er}(D, B) \to \text{Ext}^n_{\mathcal{C}}(A, B)$$

such that $\text{ext}_{\mathbf{P}} = e_{\mathbf{P}} \circ [\ ]$. This completes the proof of (25.6). $\qquad \square$

Using (25.6) we now prove:

Theorem 25.7: Let A, B, D be modules in $\mathcal{C}$ with $D \in \mathbf{D}_n(A)$, and let $\mathbf{P} \in \text{Ext}^n_{\mathcal{C}}(A, D)$ be a projective n-stem; then $c_{\mathbf{P}} : \text{Ext}^n_{\mathcal{C}}(A, B) \to \mathcal{H}^n(A, B)$ is bijective with $c_{\mathbf{P}}^{-1} = e_{\mathbf{P}}$.

Proof: The map $e_{\mathbf{P}} : \text{Hom}_{\mathcal{D}er}(D, B) \to \text{Ext}^n_{\mathcal{C}}(A, B)$ is surjective. If $\mathbf{E} \in \text{Ext}^n_{\mathcal{C}}(A, B)$, then, by (25.2), we choose a morphism $\varphi : \mathbf{P} \to \mathbf{E}$ lifting Id_A

$$
\begin{array}{cc}
\mathbf{P} \\
\downarrow \varphi \ = \\
\mathbf{E}
\end{array}
\left(
\begin{array}{ccccccc}
0 \to D \xrightarrow{j} P_{n-1} \to & \cdots & \to P_0 \to & A \to 0 \\
\downarrow \varphi_+ \ \downarrow \varphi_{n-1} & & \downarrow \varphi_0 \ \downarrow \text{Id}_A \\
0 \to B \xrightarrow{i} E_{n-1} \to & \cdots & \to E_0 \to & A \to 0
\end{array}
\right)
$$

Then there is an elementary congruence $\psi : (\varphi_+)_*(\mathbf{P}) \to \mathbf{E}$, as can be seen by the following

$$
\begin{array}{c}
(\varphi_+)_*(\mathbf{P}) \\
\downarrow \psi \\
\mathbf{E}
\end{array}
\;=\;
\begin{pmatrix}
0 \to B \to & \overset{\lim}{\underset{\to}{}} \to & P_{n-2} \to \cdots \to P_0 \to A \to 0 \\
\downarrow \mathrm{Id}_B & \downarrow \psi_{n-1} \; \downarrow \psi_{n-2} & \downarrow \psi_0 \; \downarrow \mathrm{Id}_A \\
0 \to B \overset{i}{\to} & E_{n-1} \to & E_{n-2} \to \cdots \to E_0 \to A \to 0
\end{pmatrix}
$$

Here $\overset{\lim}{\to} = \overset{\lim}{\to}(\varphi_+, j)$, $\psi_{n-1}[b, x] = i(b) + \varphi_{n-1}(x)$, and $\psi_r = \varphi_r$ for $0 \le r \le n - 2$. Hence $\mathrm{ext}_{\mathbf{P}}(\varphi_+) \approx \mathbf{E}$ in $\mathrm{Ext}_C^n(A, B)$, so that $e_{\mathbf{P}}[\varphi_+] = [\mathbf{E}]$ in $\mathrm{Ext}_C^n(A, B)$, and $e_{\mathbf{P}}$ is surjective as claimed.

It follows from the relation $c_{\mathbf{P}} \circ \mathrm{ext}_{\mathbf{P}} = [\;]$ that $c_{\mathbf{P}} \circ e_{\mathbf{P}} = \mathrm{Id}$; in particular, $e_{\mathbf{P}} = \mathrm{Id}$ is injective, and hence, by the above, bijective. Since $c_{\mathbf{P}}$ is a left inverse for the bijective map $e_{\mathbf{P}}$, it follows that $c_{\mathbf{P}}$ is two-sided inverse for $e_{\mathbf{P}}$, which is the desired result. $\qquad\qquad\square$

If A is fixed and $\mathbf{P}$ is a specific choice of projective n-stem over A, then $\mathrm{Ext}_C^n(A, -)$ is naturally equivalent to $\mathcal{H}^n(A, -)$ by (25.7). Suppose, furthermore, that $f : A \to A'$ is a Λ-homomorphism, and let

$$
\mathbf{P} = (0 \to D \to P_{n-1} \to \cdots \to P_0 \to A \to 0)
$$

and

$$
\mathbf{P}' = (0 \to D' \to P'_{n-1} \to \cdots \to P'_0 \to A' \to 0)
$$

be specific choices of projective n-stems over A, A'. Then f lifts to a commutative diagram

$$
\begin{array}{c}
\mathbf{P} \\
\downarrow \tilde{f} \\
\mathbf{P}'
\end{array}
\;=\;
\begin{pmatrix}
0 \to D \to & P_{n-1} \to \cdots \to P_0 \to A \to 0 \\
\downarrow \omega_+(f) & \downarrow \tilde{f}_{n-1} \quad\quad \downarrow \tilde{f}_0 \; \downarrow f \\
0 \to D' \to & P'_{n-1} \to \cdots \to P'_0 \to A' \to 0
\end{pmatrix}
$$

Note that the following diagram commutes

$$
\begin{array}{ccc}
\mathrm{Ext}^n(A', B) & \overset{c'_{\mathbf{P}}}{\longrightarrow} & \mathrm{Hom}_{\mathcal{D}\mathrm{er}}(D', B) \\
f^* \downarrow & & \omega_+(f)^* \downarrow \\
\mathrm{Ext}^n(A, B) & \overset{c_{\mathbf{P}}}{\longrightarrow} & \mathrm{Hom}_{\mathcal{D}\mathrm{er}}(D, B)
\end{array}
$$

and that:

Theorem 25.8: (Yoneda [83]) If for each $A \in \mathcal{C}$, $\mathbf{P}_A$ is a projective n-stem over A, then $c = (c_{\mathbf{P}_A})_{A \in \mathcal{C}}$ defines a natural equivalence

$$c : \mathrm{Ext}^n(A, -) \to \mathcal{H}^n(A, -)$$

Since the functor $\mathcal{H}^n(-, -)$ is group valued, we obtain a natural group structure on $\mathrm{Ext}_\mathcal{C}^n(A, B)$, by requiring that $c_{\mathbf{P}}$ be a group isomorphism; it is now clear also that:

Proposition 25.9: The pairing $\circ : \mathrm{Ext}_\mathcal{C}^m(B, C) \times \mathrm{Ext}_\mathcal{C}^n(A, B) \to \mathrm{Ext}_\mathcal{C}^{m+n}(A, C)$ given by the Yoneda product is bi-additive with respect to the natural group structures.

An extension $\mathbf{E} \in \mathbf{Ext}_\mathcal{C}^n(A, B)$ is said to be *almost free* (resp. *almost projective*) when E_j is free (resp. projective) for $j \neq n - 1$. When $n = 1$, every object in $\mathbf{Ext}_\mathcal{C}^1(A, B)$ is almost free. If we repeat the proof of (25.7) taking a free n-stem $\mathbf{P}$, then the extension $\hat{\mathbf{E}} = (\varphi_+)_*(\mathbf{P})$ is almost free. Thus we have:

Proposition 25.10: If $\mathbf{E} \in \mathbf{Ext}_\mathcal{C}^n(A, B)$ then there exists an almost free extension $\hat{\mathbf{E}} \in \mathbf{Ext}_\mathcal{C}^n(A, B)$ and an elementary congruence $\psi : \hat{\mathbf{E}} \to \mathbf{E}$.

For projective n-stems, we can improve on (25.10) by a different construction. We show that a projective n-stem is congruent to an extension of the form

$$0 \to B \to P \to F_{n-2} \to \cdots \to F_0 \to A \to 0$$

where $F_0, \ldots, F_{n-2}$ are free and P is projective; that is:

Proposition 25.11: If $\mathbf{E} \in \mathbf{Proj}_\mathcal{C}^n(A, B)$ is a projective n-stem, then there exists an almost free extension $\hat{\mathbf{E}}$ also belonging to $\mathbf{Proj}_\mathcal{C}^n(A, B)$ such that $\hat{\mathbf{E}} \approx \mathbf{E}$.

Proof: When $n = 1$, the statement is empty and we can take $\hat{\mathbf{E}} = \mathbf{E}$ and $\psi = \mathrm{Id}$. When $n = 2$, write

$$\mathbf{E} = \left(0 \to B \xrightarrow{i} E_1 \xrightarrow{\partial} E_0 \xrightarrow{p} A \to 0\right)$$

Since E_0 is projective, we can choose a complementary module D such that $E_0 \oplus D$ is free. Put

$$\hat{\mathbf{E}} = \left(0 \to B \xrightarrow{\lambda(i)} E_1 \oplus D \xrightarrow{\sigma(\partial)} E_0 \oplus D \xrightarrow{\rho(p)} A \to 0\right)$$

where

$$\lambda(i) = \begin{pmatrix} i \\ 0 \end{pmatrix}; \quad \sigma(\partial) = \begin{pmatrix} \partial & 0 \\ 0 & \mathrm{Id} \end{pmatrix}; \quad \rho(p) = (\, p, \ 0\,)$$

Then $\hat{\mathbf{E}}$ is almost free, and the projections $E_i \oplus D \to E_i$ induces an elementary congruence $\hat{\mathbf{E}} \to \mathbf{E}$.

Suppose proved for $n - 1$, and let $n \geq 3$. Decompose $\mathbf{E}$ into Yoneda factors, thus

$$\mathbf{E} = \mathbf{E}_+ \circ \mathbf{E}_-$$

where $\mathbf{E}_- \in \mathbf{Proj}^2(A, C)$, and $\mathbf{E}_+ \in \mathbf{Proj}^{n-2}(C, B)$. Let $\hat{\mathbf{E}}_- \in \mathbf{Proj}^2(A, C)$ be an almost free extension congruent to $\mathbf{E}_-$, and put $\mathbf{F} = \mathbf{E}_+ \circ \hat{\mathbf{E}}_- \in \mathbf{Proj}^n(A, B)$. We may now decompose $\mathbf{F}$ thus

$$\mathbf{F} = \mathbf{F}_+ \circ \mathbf{F}_-$$

where $\mathbf{F}_- \in \mathbf{Free}^1(A, D)$ is a free extension, and $\mathbf{F}_+ \in \mathbf{Proj}^{n-1}(D, B)$. By induction, there exists an almost free extension $\mathbf{L} \in \mathbf{Proj}^{n-1}(D, B)$, which is congruent to $\mathbf{F}_+$. Putting $\hat{\mathbf{E}} = \mathbf{L} \circ \mathbf{F}_- \in \mathbf{Proj}^n(A, B)$ gives the result. $\square$

26 Classification of projective n-stems

Fix a specific projective n-stem

$$\mathbf{P} = (0 \to D \to P_{n-1} \to \cdots \to P_0 \to M \to 0)$$

the set of endomorphisms of D in the derived category $\mathcal{D}\mathrm{er}(\mathcal{C})$ is then naturally a ring. We put

$$\mathcal{A}(D) = \mathrm{Aut}_{\mathcal{D}\mathrm{er}}(D)$$

that is, $\mathcal{A}(D)$ is the unit group of the ring $\mathrm{End}_{\mathcal{D}\mathrm{er}}(D)$. Then Yoneda's Theorem (25.7), shows that the correspondence $\alpha \mapsto \alpha_*(\mathbf{P})$ defines a bijection $e_{\mathbf{P}} : \mathrm{End}_{\mathcal{D}\mathrm{er}}(D) \to \mathrm{Ext}^n(M, D)$. The main result proved in this section ((26.5) below) is that $e_{\mathbf{P}}$ restricts to give a bijection

$$e_{\mathbf{P}} : \mathcal{A}(D) \to \mathrm{Proj}_{\mathcal{C}}^n(M, D)$$

allowing us to parametrize projective n-stems by the elements of $\mathcal{A}(D)$.

Note that (23.8), (23.9) have analogues for general projective n-stems:

Proposition 26.1: Let M, D be modules in $\mathcal{C}$ which occur in a projective n-stem $\mathbf{P} = (0 \to D \to P_{n-1} \to \cdots P_0 \to M \to 0)$ and let $D' \in \mathcal{C}$; then the following statements are equivalent:

(i) $D' \sim\sim D$;

(ii) $D' \in \mathbf{D}_n(M)$;

(iii) there exists a projective n-stem of the form

$$\mathbf{P}' = (0 \to D' \to P'_{n-1} \to \cdots P'_0 \to M \to 0)$$

Again there is a dual statement:

Proposition 26.2: Let M, D be modules in C which occur in a projective n-stem $\mathbf{P} = (0 \to D \to P_{n-1} \to \cdots P_0 \to M \to 0)$ and let $M' \in C$; then the following statements are equivalent:

(i) $M' \sim\sim M$;
(ii) $M' \in \mathbf{D}_{-n}(D)$;
(iii) there exists a projective n-stem of the form

$$\mathbf{P}' = (0 \to D \to P'_{n-1} \to \cdots P'_0 \to M' \to 0)$$

The proofs of (26.1), (26.2) follow easily by induction from those of (23.8), (23.9) by decomposing the projective n-stem into the Yoneda product of a projective cover and a projective $(n-1)$-stem.

Morphisms $\alpha : D \to D'$, $\beta : M' \to M$ in C give functors

$$\alpha_* : \mathbf{Ext}^n(M, D) \to \mathbf{Ext}^n(M, D'); \qquad \beta^* : \mathbf{Ext}^n(M, D) \to \mathbf{Ext}^n(M', D)$$

It is not generally true that that $\mathbf{Proj}^n(-, -)$ is stable under all these functors. However, generalizing (23.2) and (23.3), $\mathbf{Proj}^n(-, -)$ is stable under functors α_*, β^* where α, β are *isomorphisms in the derived category*.

When $n \geq 2$, the statement for $\mathbf{Proj}^n(-, -)$ corresponding to (23.2) reduces immediately to the case where $n = 1$, and hence to (23.2), on decomposing an n-fold extension $\mathbf{P}$ as a Yoneda product $\mathbf{P} = \mathbf{P}_+ \circ \mathbf{P}_-$ where $\mathbf{P}_+$ is a 1-fold extension and $\mathbf{P}_-$ is an $(n-1)$-fold extension. We leave the details to the reader, and note the result.

(26.3) Let $\mathbf{P} \in \mathbf{Proj}^n(M, D)$ and let $\alpha : D \to D'$ be a Λ-homomorphism; then $\alpha_*(\mathbf{P}) \in \mathbf{Proj}^n(M, D')$ if and only if $[\alpha]$ is an isomorphism in $\mathcal{D}\mathrm{er}(C)$.

The above result is fundamental in the detailed classification of projective n-stems both now and later. Though we shall not need to use it, the corresponding statement for pullbacks is also true, and is easily reduced to (23.3). We leave the details to the reader.

(26.4) Let $\mathbf{P} \in \mathbf{Proj}^n(M, D)$ and let $\beta : M' \to M$ be a Λ-homomorphism; then $\beta^*(\mathbf{P}) \in \mathbf{Proj}^n(M', D)$ if and only if $[\beta]$ is an isomorphism in $\mathcal{D}\mathrm{er}(C)$.

We can express (26.3) in terms of the following (left) module structure of $\mathrm{Ext}_C^n(M, D)$ over the ring $\mathrm{End}_{\mathcal{D}\mathrm{er}}(D)$

$$\mathrm{End}_{\mathcal{D}\mathrm{er}}(D) \times \mathrm{Ext}_C^n(M, D) \to \mathrm{Ext}_C^n(M, D)$$
$$[\alpha] * [\mathcal{E}] = [\alpha_*(\mathcal{E})]$$

e_P then takes the form

$$e_P([\alpha]) = [\alpha] * [\mathbf{P}]$$

We will show that $\mathrm{Proj}^n_C(M, D)$ forms a single orbit under the group action

$$\mathcal{A}(D) \times \mathrm{Ext}^n_C(M, D) \to \mathrm{Ext}^n_C(M, D)$$

obtained by restriction.

Theorem 26.5: Let $\mathbf{P} \in \mathbf{Proj}^n_C(M, D)$ be a projective n-stem; then e_P restricts to give a bijection

$$e_P : \mathcal{A}(D) \to \mathrm{Proj}^n_C(M, D)$$

Proof: We first treat the case $n = 1$. If $\mathbf{Q}$ also belongs to $\mathbf{Proj}^1_C(M, D)$, then there is a morphism $\omega : \mathbf{P} \to \mathbf{Q}$ which lifts Id_M. Let $\alpha = \omega_+ : D \to D$ be the Λ-homomorphism so obtained. Then by (23.2), $[\alpha]$ is an isomorphism in $\mathcal{D}\mathrm{er}(C)$. Furthermore, $\mathbf{Q}$ is congruent to $\alpha_*(\mathbf{P})$, so that $\mathrm{Proj}^1_C(M, D)$ consists of a single orbit under the action of $\mathcal{A}(D)$ on $\mathrm{Ext}^1(M, D)$. Thus e_P gives a surjection

$$e_P : \mathcal{A}(D) \to \mathrm{Proj}^1_C(M, D)$$

As in the proof of (25.7), e_P is injective on $\mathrm{End}_{\mathcal{D}\mathrm{er}}(D)$, and so gives a bijection

$$e_P : \mathcal{A}(D) \to \mathrm{Proj}^1_C(M, D)$$

as claimed. This completes the proof for $n = 1$. As in the proof of Yoneda's Theorem, the proof for $n > 1$ is reduced to the case $n = 1$ by dimension shifting. In effect, the proof for the action of $\mathcal{A}(D_n)$ on $\mathrm{Proj}^n_C(M, D_n)$ is the same as that for the action of $\mathcal{A}(D_n)$ on $\mathrm{Proj}^1_C(D_{n-1}, D_n)$. □

Let $M, D \in \mathcal{F}(\mathbf{Z}[G])$ and suppose that

$$\mathbf{P} = (0 \to D \to P_{n-1} \to \cdots \to P_0 \to M \to 0)$$

is a projective n-stem. Since the functor $\mathbf{D}_n$ is an equivalence of categories on $\mathcal{D}\mathrm{er}(C)$

$$\mathrm{Hom}_{\mathcal{D}\mathrm{er}}(D, D) \cong \mathrm{Hom}_{\mathcal{D}\mathrm{er}}(M, M)$$

so that we have both an isomorphism of rings

$$\mathrm{End}_{\mathcal{D}\mathrm{er}}(D) \cong \mathrm{End}_{\mathcal{D}\mathrm{er}}(M)$$

and of groups

$$\mathcal{A}(D) \cong \mathcal{A}(M)$$

An isomorphism $\mathrm{End}_{\mathcal{D}\mathrm{er}}(D) \cong \mathrm{End}_{\mathcal{D}\mathrm{er}}(M)$ may be calculated explicitly in the following way.

If $\alpha : D \to D$ is a $\mathbf{Z}[G]$-homomorphism we may extend α to a morphism in $\mathbf{Ext}^n(M, D)$ thus

$$
\begin{array}{c}
\mathbf{P} \\
\downarrow \widetilde{\alpha} = \\
\mathbf{P}
\end{array}
\begin{pmatrix}
0 \to D \to P_{n-1} \to \cdots \to P_0 \to M \to 0 \\
\quad\quad \downarrow \alpha \;\; \downarrow \alpha_{n-1} \quad\quad\quad \downarrow \alpha_0 \;\; \downarrow \widetilde{\alpha} \\
0 \to D \to P_{n-1} \to \cdots \to P_0 \to M \to 0
\end{pmatrix}
$$

We write

$$
\kappa_{\mathbf{P}}(\alpha) = [\widetilde{\alpha}]
$$

Proposition 26.6: $\kappa_{\mathbf{P}}$ is an isomorphism of rings

$$
\kappa_{\mathbf{P}} : \mathrm{End}_{\mathcal{D}\mathrm{er}}(D) \cong \mathrm{End}_{\mathcal{D}\mathrm{er}}(M)
$$

Proof: Note that $\kappa_{\mathbf{P}}$ is automatically bijective, since $\mathbf{D}_n$ is an equivalence of categories. For $\mathbf{Z}[G]$-homomorphisms $\alpha, \beta : D \to D$, we obtain commutative diagrams thus

$$
\begin{array}{c}
\mathbf{P} \\
\downarrow \widetilde{\alpha} = \\
\mathbf{P}
\end{array}
\begin{pmatrix}
0 \to D \to P_{n-1} \to \cdots \to P_0 \to M \to 0 \\
\quad\quad \downarrow \alpha \;\; \downarrow \alpha_{n-1} \quad\quad\quad \downarrow \alpha_0 \;\; \downarrow \widetilde{\alpha} \\
0 \to D \to P_{n-1} \to \cdots \to P_0 \to M \to 0
\end{pmatrix}
$$

and

$$
\begin{array}{c}
\mathbf{P} \\
\downarrow \widetilde{\beta} = \\
\mathbf{P}
\end{array}
\begin{pmatrix}
0 \to D \to P_{n-1} \to \cdots \to P_0 \to M \to 0 \\
\quad\quad \downarrow \beta \;\; \downarrow \beta_{n-1} \quad\quad\quad \downarrow \beta_0 \;\; \downarrow \widetilde{\beta} \\
0 \to D \to P_{n-1} \to \cdots \to P_0 \to M \to 0
\end{pmatrix}
$$

From the commutative diagrams

$$
\begin{array}{c}
\mathbf{P} \\
\downarrow \widetilde{\alpha + \beta} = \\
\mathbf{P}
\end{array}
\begin{pmatrix}
0 \to D \to \quad P_{n-1} \to \cdots \to P_0 \to \quad M \to \quad 0 \\
\quad \downarrow \alpha + \beta \;\; \downarrow \alpha_{n-1} + \beta_{n-1} \;\; \downarrow \alpha_0 + \beta_0 \;\; \downarrow \widetilde{\alpha} + \widetilde{\beta} \\
0 \to D \to \quad P_{n-1} \to \cdots \to P_0 \to \quad M \to \quad 0
\end{pmatrix}
$$

and

$$
\begin{array}{c}
\mathbf{P} \\
\downarrow \widetilde{\alpha \circ \beta} = \\
\mathbf{P}
\end{array}
\begin{pmatrix}
0 \to D \to \quad P_{n-1} \to \cdots \to P_0 \to \quad M \to \quad 0 \\
\quad \downarrow \alpha \circ \beta \;\; \downarrow \alpha_{n-1} \circ \beta_{n-1} \;\; \downarrow \alpha_0 \circ \beta_0 \;\; \downarrow \widetilde{\alpha} \circ \widetilde{\beta} \\
0 \to D \to \quad P_{n-1} \to \cdots \to P_0 \to \quad M \to \quad 0
\end{pmatrix}
$$

we see that $\kappa_{\mathbf{P}}(\alpha + \beta) = \kappa_{\mathbf{P}}(\alpha) + \kappa_{\mathbf{P}}(\beta)$ and $\kappa_{\mathbf{P}}(\alpha \circ \beta) = \kappa_{\mathbf{P}}(\alpha) \circ \kappa_{\mathbf{P}}(\beta)$. Finally, note that the commutative diagram

$$
\begin{array}{l}
\mathbf{P} \\
\downarrow \mathrm{Id} = \\
\mathbf{P}
\end{array}
\left(
\begin{array}{ccccccc}
0 \to D \to P_{n-1} \to & \cdots & \to P_0 \to M \to 0 \\
\downarrow \mathrm{Id} & \downarrow \mathrm{Id} & & \downarrow \mathrm{Id} & \downarrow \mathrm{Id} \\
0 \to D \to P_{n-1} \to & \cdots & \to P_0 \to M \to 0
\end{array}
\right)
$$

shows that $\kappa_{\mathbf{P}}(\mathrm{Id}) = \mathrm{Id}$, which is the final requirement for $\kappa_{\mathbf{P}}$ to be a ring isomorphism. $\square$

Recall also, that if $\mathbf{Q} = (0 \to D \to Q_{n-1} \to \cdots \to Q_0 \to M \to 0)$ is another element of $\mathbf{Proj}^n(M, D)$ then there is a unique automorphism $\lambda_{\mathbf{PQ}}$ of M in the derived category making the following diagram commute

$$
\begin{array}{ccccccc}
0 \to D \to Q_{n-1} \to & \cdots & \to Q_0 \to M \to & 0 \\
\downarrow \mathrm{Id} & \downarrow \lambda_{n-1} & & \downarrow \lambda_0 & \downarrow \lambda_{\mathbf{PQ}} \\
0 \to D \to P_{n-1} \to & \cdots & \to P_0 \to M \to & 0
\end{array}
$$

Clearly $\lambda_{\mathbf{QP}} = \lambda_{\mathbf{PQ}}^{-1}$ and $\kappa_{\mathbf{Q}}$ is related to $\kappa_{\mathbf{P}}$ by means of

(26.7) $\kappa_{\mathbf{P}}(\alpha) = \lambda_{\mathbf{PQ}} \kappa_{\mathbf{Q}}(\alpha) \lambda_{\mathbf{PQ}}^{-1}$

27 The standard cohomology theory of modules

We recall briefly the basics of the Eilenberg–Maclane cohomology theory of modules [12], [34]. Let M be a Λ-module; a *resolution* of M is an exact sequence of Λ-homomorphisms

$$
\mathcal{A} = \left(\cdots \to X_{n+1}^{\mathcal{A}} \xrightarrow{\partial_{n+1}^{\mathcal{A}}} X_n^{\mathcal{A}} \xrightarrow{\partial_n^{\mathcal{A}}} X_{n-1}^{\mathcal{A}} \xrightarrow{\partial_{n-1}^{\mathcal{A}}} \cdots \xrightarrow{\partial_1^{\mathcal{A}}} X_0^{\mathcal{A}} \xrightarrow{\epsilon^{\mathcal{A}}} M^{\mathcal{A}} \to 0 \right)
$$

in which $M^{\mathcal{A}} = M$. We abbreviate this to $\mathcal{A} = (X_*^{\mathcal{A}} \to M)$. By a morphism of resolutions, $\varphi : \mathcal{A} \to \mathcal{B}$, we mean a collection (φ_r) of Λ-homomorphisms completing a commutative diagram

$$
\begin{array}{l}
\mathcal{A} \\
\downarrow \varphi = \\
\mathcal{B}
\end{array}
\left(
\begin{array}{ccccccc}
\cdots \to X_n^{\mathcal{A}} \to X_{n-1}^{\mathcal{A}} \to & \cdots & \to X_0^{\mathcal{A}} \to M^{\mathcal{A}} \to 0 \\
\downarrow \varphi_n & \downarrow \varphi_{n-1} & & \downarrow \varphi_0 & \downarrow \varphi_M \\
\cdots \to X_n^{\mathcal{B}} \to X_{n-1}^{\mathcal{B}} \to & \cdots & \to X_0^{\mathcal{B}} \to M^{\mathcal{B}} \to 0
\end{array}
\right)
$$

We say that a resolution $\mathcal{A} = (X_*^{\mathcal{A}} \to M)$ is *projective* (resp. *free*) when each $X_r^{\mathcal{A}}$ is a projective (resp. free) Λ-module. The following is standard:

Proposition 27.1: Every Λ-module has a free (and hence a projective) resolution.

Projective resolutions will be denoted by $\mathcal{A} = (P_*^{\mathcal{A}} \to M)$ or $(P_* \to M)$ if $\mathcal{A}$ is clear from context, and free resolutions will be denoted $\mathcal{A} = (F_*^{\mathcal{A}} \to M)$ or $(F_* \to M)$.

Suppose that $\mathcal{A} = (X_*^{\mathcal{A}} \to M^{\mathcal{A}})$ and $\mathcal{B} = (X_*^{\mathcal{B}} \to M^{\mathcal{B}})$ are resolutions, and $\varphi, \psi : \mathcal{A} \to \mathcal{B}$ are morphisms such that $\varphi_M = \psi_M$; φ, ψ are said to be *homotopic* (written $\varphi \simeq \psi$) when there exists a collection $H = (H_r)_{r \geq 0}$ of Λ-homomorphisms $H_r : X_r^{\mathcal{A}} \to X_{r+1}^{\mathcal{B}}$ such that

(i) $\varphi_0 - \psi_0 = \partial_{r+1}^{\mathcal{B}} H_r$, and
(ii) $\varphi_r - \psi_r = \partial_{r+1}^{\mathcal{B}} H_r + H_{r-1} \partial_r^{\mathcal{A}}$ for all $r \geq 1$.

If $f : M^{\mathcal{A}} \to M^{\mathcal{B}}$ is a Λ-homomorphism, a morphism

$$\varphi : (X_*^{\mathcal{A}} \to M^{\mathcal{A}}) \to (X_*^{\mathcal{B}} \to M^{\mathcal{B}})$$

is said to be a *lifting* of f when $\varphi_M = f$. We recall the following:

Proposition 27.2: Let $f : M \to N$ be a Λ-homomorphism; if $\mathcal{B} = (X_* \to N)$ is a resolution and $\mathcal{A} = (P_* \to M)$ is a projective resolution, then:

(i) there exists a lifting $\tilde{f} : \mathcal{A} \to \mathcal{B}$ of f; moreover,
(ii) any two liftings of the same homomorphism are homotopic.

If M, C are Λ-modules, the cohomology groups $H^n(M, C)$ are defined as follows: choose a projective resolution $\mathcal{P} = (P_* \to M)$ of M, and define

$$P_r^C = \operatorname{Hom}_\Lambda(P_r, C)$$

Denote by $\partial_r^C : P_r^C \to P_{r+1}^C$ the induced map

$$\partial_r^C(\alpha) = \alpha \circ \partial_r$$

Put

$$H^0(M, C) = \operatorname{Ker}(\partial_1^C)$$

and, for $n \geq 1$

$$H^n(M, C) = \operatorname{Ker}(\partial_{n+1}^C) / \operatorname{Im}(\partial_n^C)$$

Proposition 27.3: The isomorphism type of $H^n(M, C)$ is independent of the particular projective resolution of M.

If P is projective, then as a projective resolution of P we may take

$$\mathcal{P} = \left(\cdots \to 0 \to 0 \to 0 \to \cdots \to P \xrightarrow{\epsilon} P \to 0 \right)$$

where $\epsilon = \mathrm{Id}_P$. It follows that:

Proposition 27.4: If P is projective, then for any module C, $H^n(P, C) = 0$ when $n \geq 1$.

Furthermore, when P is projective, the functor $B \mapsto \mathrm{Hom}_\Lambda(B, P)$ is *exact*, from which we see:

Proposition 27.5: If P is projective, then for any module N, $H^n(N, P) = 0$ when $n \geq 1$.

We now relate cohomology in this definition to that of our earlier definition in terms of the derived category. Thus suppose that C is a tame class, and $f : M_1 \to M_2$ is a morphism in C which factors through the projective module P, then the induced map on cohomology $f^* : H^n(M_2, N) \to H^n(M_1, N)$ factors through $H^n(P, N) = 0$. Hence, for $n \geq 1$, H^n considered as a functor of the first (contravariant) variable factors through the derived category; that is, if $f, g : M \to N$ are morphisms in $\mathcal{M}od_\Lambda$ such that $f \approx g$, then we have equality of induced maps $f^* = g^* : H^*(N, B) \to H^*(M, B)$ for any fixed module B. This extends to the following:

Theorem 27.6: Let C be a tame class; for any Λ-modules M, N, there is a (canonical) surjective homomorphism

$$\nu : H^n(M, N) \to \mathcal{H}^n(M, N)$$

Proof: Since C is tame and $M \in C$, we may suppose, without loss of generality, that M has a projective resolution $\mathcal{P} = (P_* \to M)$ in which each P_r is finitely generated

$$\mathcal{P} = \left(\cdots \to P_{n+1} \xrightarrow{\partial_{n+1}} P_n \xrightarrow{\partial_n} P_{n-1} \xrightarrow{\partial_{n-1}} \cdots \xrightarrow{\partial_1} P_0 \xrightarrow{\epsilon} M \to 0 \right)$$

If $f : A \to B$ is a Λ-homomorphism we denote by $f^N : \mathrm{Hom}_\Lambda(B, N) \to \mathrm{Hom}_\Lambda(A, N)$ the induced map

$$f^N(\alpha) = \alpha \circ f$$

Put

$$\mathrm{Ker}\left(\partial_{n+1}^N\right) = \mathcal{Z}^n; \quad \mathrm{Im}\left(\partial_n^N\right) = \mathcal{B}^n$$

Then we have

$$H^0(M, N) = \mathcal{Z}^1; \quad H^n(M, N) = \mathcal{Z}^n/\mathcal{B}^n \ (n \geq 1)$$

Put $D_n = \operatorname{Im}(\partial_n : P_n \to P_{n-1})$, and factorize $\partial_n : P_n \to P_{n-1}$ as $\partial_n = \iota_n \circ \pi_n$ where $\iota_n : D_n \subset P_{n-1}$ is the inclusion, and $\pi_n : P_n \to D_n$ is the canonical surjection. From $\partial_n \circ \partial_{n+1} = 0$ it is straightforward to see that $\operatorname{Im}(\pi_n^N) \subset \mathcal{Z}^n$.

This inclusion is actually equality; to observe the inclusion in the opposite direction, note that $z \in \mathcal{Z}^n$ is a map $z : P_n \to N$ satisfying $z \circ \partial_{n+1} = 0$, and so induces a homomorphism $\check{z} : P_n/\operatorname{Im}(\partial_{n+1}) \to N$, and so also $\check{z} : P_n/\operatorname{Ker}(\partial_n) \to N$ since $\operatorname{Im}(\partial_{n+1}) = \operatorname{Ker}(\partial_n)$. Via π_n, we may identify $P_n/\operatorname{Ker}(\partial_n)$ with $\operatorname{Im}(\partial_n) = D_n$, under which identification $\pi_n^*(\check{z}) = z$, so that $\mathcal{Z}^n \subset \operatorname{Im}(\pi_n^*)$, and so, as claimed, we have

$$\operatorname{Im}(\pi_n^N) = \mathcal{Z}^n.$$

In particular, we have a surjective homomorphism

$$\pi_n^N : \operatorname{Hom}_\Lambda(D_n, N) \to \mathcal{Z}^n$$

Put

$$\mathcal{P}(D_n, N) = \{\lambda \in \operatorname{Hom}_\Lambda(D_n, N) : \lambda \approx 0\}$$

so that

$$\operatorname{Hom}_{\mathcal{D}er}(D_n, N) = \operatorname{Hom}_\Lambda(D_n, N)/\mathcal{P}(D_n, N).$$

We claim that

$$\left(\pi_n^N\right)^{-1}(\mathcal{B}^n) \subset \mathcal{P}(D_n, N)$$

For, if $\beta \in (\pi_n^N)^{-1}(\mathcal{B}^n)$, then

$$\beta \circ \pi_n = \partial_n^N(\alpha)$$

for some $\alpha \in \operatorname{Hom}(P_{n-1}, N)$; that is

$$\beta \circ \pi_n = \alpha \circ \partial_n = \alpha \circ \iota_n \circ \pi_n$$

But π_n is surjective, so that

$$\beta = \alpha \circ \iota_n$$

Hence β factors through the P_{n-1}, and $(\pi_n^N)^{-1}(\mathcal{B}^n) \subset \mathcal{P}(D_n, N)$ as claimed. Recalling that

$$\operatorname{Hom}(D_n, N)/\mathcal{P}(D_n, N) = \operatorname{Hom}_{\mathcal{D}er}(\mathbf{D}_n(M), N)$$

it follows that there is a short exact sequence

$$0 \to \mathcal{P}(D_n, N)/\big(\pi_n^N\big)^{-1}(\mathcal{B}^n) \to \operatorname{Hom}(D_n, N)/\big(\pi_n^N\big)^{-1}(\mathcal{B}^n)$$
$$\to \operatorname{Hom}_{\mathcal{D}er}(\mathbf{D}_n(M), N) \to 0$$

But $\operatorname{Hom}(D_n, N) = (\pi_n^N)^{-1}(\mathcal{Z}^n)$ and π_n^N induces an isomorphism

$$\big(\pi_n^N\big)^{-1}(\mathcal{Z}^n)/\big(\pi_n^N\big)^{-1}(\mathcal{B}^n) \cong \mathcal{Z}^n/\mathcal{B}^n \cong H^n(M, N)$$

so that the exact sequence becomes

$$0 \to \mathcal{P}(D_n, N)/\big(\pi_n^N\big)^{-1}(\mathcal{B}^n) \to H^n(M, N) \to \operatorname{Hom}_{\mathcal{D}er}(\mathbf{D}_n(M), N) \to 0$$

This completes the proof. $\square$

When $\mathcal{C}$ is a tame class and $n \geq 1$, the functor $M \mapsto H^n(M, -)$ is co-representable on the derived category, with co-representing object $\mathbf{D}_n(M)$; that is:

Theorem 27.7: Let $\mathcal{C}$ be a tame class; if $n \geq 1$ then

$$\nu : H^n(M, N) \to \operatorname{Hom}_{\mathcal{D}er}(\mathbf{D}_n(M), B) = \mathcal{H}^n(M, N)$$

is an isomorphism for any $M, N \in \mathcal{C}$.

Proof: As before, let $\mathcal{P} = (P_* \to M)$ be a resolution of M by finitely generated projectives. Put $D_n = \operatorname{Im}(\partial_n : P_n \to P_{n-1})$, and factorize $\partial_n : P_n \to P_{n-1}$ as $\partial_n = \iota_n \circ \pi_n$. The surjectivity of ν follows from the fact, observed above, that $(\pi_n^*)^{-1}(\mathcal{B}^n) \subset \mathcal{P}(D_n, N)$. To show ν is an isomorphism, it now suffices to show that $(\pi_n)^*(\mathcal{P}(D_n, N)) \subset \mathcal{B}^n$.

Thus suppose that $f : D_n \to N$ factors through a projective, say $f = j \circ \varphi$, where $\varphi : D_n \to Q$ and $j : Q \to N$, with Q projective. Without loss, we may suppose that Q is finitely generated, and so $Q \in \mathcal{C}$. Since $n \geq 1$, we have an exact sequence in $\mathcal{C}$ thus

$$0 \to D_n \xrightarrow{i_n} P_{n-1} \xrightarrow{\pi_{n-1}} D_{n-1} \to 0$$

Q is projective, so that, by $\mathcal{T}(3)$ and (19.5), Q is strongly injective relative to $\mathcal{C}$, and there exists a homomorphism $\Phi : P_{n-1} \to Q$ such that $\varphi = \Phi \circ i_n$. Thus

$$f = j \circ \varphi = j \circ \Phi \circ i_n$$

so that

$$(\pi_n)^*(f) = f \circ \pi_n = j \circ \Phi \circ i_n \circ \pi_n = j \circ \Phi \circ \partial_n$$

and, consequently

$$(\pi_n)^*(f) = (\partial_n)^*(j \circ \Phi)$$

so that $(\pi_n)^*(\mathcal{P}(D_n, N)) \subset \mathcal{B}^n$ as desired. □

We note that the conclusion of (27.7) fails for $n = 0$; in general, $\mathcal{H}^0(M, N)$ is a proper quotient of $H^0(M, N)$. In particular, as we shall see in Chapter 6, when $\Lambda = \mathbf{Z}[G]$ is the integral group ring of a finite group G and $\mathbf{Z}$ is the trivial $\mathbf{Z}[G]$-module structure on the group of integers, then $\mathcal{H}^0(\mathbf{Z}, \mathbf{Z}) \cong \mathbf{Z}/|G|$ is finite whilst $H^0(\mathbf{Z}, \mathbf{Z}) \cong \mathbf{Z}$ is infinite. In connection with the translation functors $\mathbf{D}_n$ we have the phenomenon of *dimension shifting* which holds independently of any proof that ν is an isomorphism.

Proposition 27.8: For any modules $M, N \in \mathcal{C}$, we have

$$H^n(\mathbf{D}_m(M), N) \cong H^{n+m}(M, N)$$

Chapter 5

The derived category of a finite group

We specialize the theory developed in Chapter 4 to the case where $\Lambda = \mathbf{Z}[G]$, the integral group ring of a finite group G. We begin by showing that the class $\mathcal{F}(\mathbf{Z}[G])$ of $\mathbf{Z}[G]$-lattices is tame. It follows that for coefficients $N \in \mathcal{F}(\mathbf{Z}[G])$, the group cohomology $H^n(G, N) = \mathcal{H}^n(\mathbf{Z}, N)$ is both representable, $H^n(G, N) \cong \text{Hom}_{\mathcal{D}er}(\mathbf{Z}, \Omega_{-n}(N))$, and co-representable, $H^n(G, N) \cong \text{Hom}_{\mathcal{D}er}(\Omega_n(\mathbf{Z}), N)$, in the derived category, by means of the stable module construction $\Omega_n(-)$.

The stable modules $(\Omega_n(\mathbf{Z}))_{n \in \mathbf{Z}}$ then become our primary object of study; in particular, we analyse the tree structure introduced in Chapter 3. Since $\mathbf{Z}[G]$ is an order in the semisimple algebra $\mathbf{Q}[G]$, the cancellation theory developed in Chapter 3 also becomes available, and we deduce fron the Swan–Jacobinski Theorem that the stable modules $\Omega_{2n+1}(\mathbf{Z})$ always have the structure of a *fork*. Though less crucial for our study, the structure of the even modules $\Omega_{2n}(\mathbf{Z})$ is also transparent.

Finally, in Sections 31 and 32 we consider the relations which hold between the derived category of the finite group G and those of its subgroups.

28 Lattices over a finite group

Throughout, G will denote a *finite* group, $\mathbf{A}$ will denote a commutative ring, and $\mathbf{A}[G]$ will denote the group algebra of G over $\mathbf{A}$. We denote by $\mathcal{F}(\mathbf{A}[G])$ the category of $\mathbf{A}[G]$-lattices; that is, right $\mathbf{A}[G]$-modules which are finitely generated and projective as modules over $\mathbf{A}$. A fortiori, they are finitely generated over $\mathbf{A}[G]$. More generally, if Λ is an algebra over $\mathbf{A}$, which is finitely generated as an $\mathbf{A}$-module, we denote by $\mathcal{F}(\Lambda)$ the category of right Λ-modules which are finitely generated and projective as modules over $\mathbf{A}$. When the algebra structure of Λ over $\mathbf{A}$ is understood we shall refer to elements of $\mathcal{F}(\Lambda)$ simply as Λ-lattices.

114

By an *involution* τ on a ring Λ, we mean an isomorphism $\tau : \Lambda \to \Lambda^{opp}$ of Λ with its opposite ring such that $\tau \circ \tau = \text{Id}$. If Λ is any ring, and M is a right Λ-module, we denote by $M^{\wedge}$ the abelian group $\text{Hom}_{\Lambda}(M, \Lambda)$ with a *left* Λ structure given by

$$(\lambda * f)(m) = \lambda . f(m)$$

When Λ admits an involution τ, we can convert this to a right action '$\bullet$' by means of

$$f \bullet \lambda = {}^{\tau}\lambda * f$$

If G is a finite group, the group ring $A[G]$ admits a natural involution $\lambda \mapsto \overline{\lambda}$ given by

$$\overline{\left(\sum_{g \in G} a_g g\right)} = \left(\sum_{g \in G} a_g g^{-1}\right)$$

We begin by showing that $\mathcal{F}(A[G])$ is a tame class. In this we are aided by the fact that the natural involution on $A[G]$ allows us to introduce a duality into $\mathcal{F}(A[G])$; let $M \in \mathcal{F}(A[G])$, and convert the left module $M^{\wedge}$ into a right module by means of the canonical involution; we get the following explicit model for $M^{\wedge}$; elements of $M^{\wedge}$ are sequences $\alpha = (\alpha_g)_{g \in G}$, with each $\alpha_g \in \text{Hom}_A(M, A)$, satisfying the condition

$$\alpha_{gx}(m) = \alpha_x(mg^{-1})$$

Moreover, the right G-action on $M^{\wedge}$ is given by

$$(\alpha \bullet h)_x = \alpha_{hx}$$

Within $\mathcal{F}(A[G])$ however, there is another possible model for the dual module; we denote by M^* the right $A[G]$-module whose underlying A-module is $\text{Hom}_A(M, A)$ and whose G-action is given by

$$(f \bullet g)(m) = f(mg^{-1})$$

The two constructions are naturally equivalent. This special case of 'Shapiro's Lemma' is effected explicitly by means of the isomorphism

$$L : M^{\wedge} \to M^*; \quad L((\alpha_g)_{g \in G}) = \alpha_1$$

Since it is a matter of taste which model to employ, we use $M \mapsto M^*$ for simplicity.

There is a natural transformation $\nu : M \to M^{**}$ given by $\nu(x)(\alpha) = \alpha(x)$. Since M is free over A, ν is easily seen to be an isomorphism over A; moreover,

ν is easily shown to be equivariant with respect to the actions of G. It follows that:

Proposition 28.1: For all $M \in \mathcal{F}(\mathbf{A}[G])$, $\nu : M \to M^{**}$ is an natural isomorphism of $\mathbf{A}[G]$-modules.

Consider first the case where a module $M \in \mathcal{F}(\mathbf{A}[G])$ is free, rather than merely projective, over $\mathbf{A}$. In matrix terms, M determines, and is determined by, a representation $\rho : G \to GL_m(\mathbf{A})$, where $m = \mathrm{rk}_{\mathbf{A}}(M)$. The dual module M^* then corresponds to the representation $\rho^* : G \to GL_m(\mathbf{A})$ defined by $\rho^*(g) = \rho(g^{-1})^T$ where μ^T is the *transpose* of the matrix μ. (28.1) is simply a reflection of the fact that $\rho^{**} = \rho$.

Although $M \cong M^{**}$, it is *not true in general* that $M^* \cong M$. This fact does not appear to be as well known as perhaps it should be; the point is perhaps best illustrated by reference to the most important examples with which we shall have to deal:

Example 1: The regular representation
The regular representation is simply the matrix description of the free module of rank 1; that is, the group ring $\mathbf{A}[G]$ considered as a module over itself. If G is a finite group with $|G| = N$, (left) translation in the group ring gives rise to the (left) regular representation $\lambda : G \to GL_N(\mathbf{A})$; that is: $\lambda(g)(\mathbf{x}) = g\mathbf{x}$ for $g \in G$ and $\mathbf{x} \in \mathbf{A}[G]$. Each $\lambda(g)$ is a permutation matrix and so satisfies the orthogonality condition $\lambda(g)^{-1} = \lambda(g)^T$, and so $\lambda \equiv \lambda^*$. Expressed in terms of modules, we have that $\mathbf{A}[G] \cong_{\mathbf{A}[G]} \mathbf{A}[G]^*$. In particular, we may legitimately identify $\mathbf{A}[G]$ with $\mathbf{A}[G]^*$ by confusing the canonical basis $\{g\}_{g \in G}$ with its dual basis $\{g^*\}_{g \in G}$.

Example 2: The trivial representation
The coefficient ring $\mathbf{A}$ can be considered as a module over $\mathbf{A}[G]$ in which each group element $g \in G$ acts as the identity. In matrix terms, the trivial representation is given by the trivial homomorphism $\tau : G \to GL_1(\mathbf{A})$. Since we have $\tau(g) = \tau^*(g) = \mathrm{Id}$ for all $g \in G$, we see also that $A \cong_{\mathbf{A}[G]} A^*$.

The fact that the two most obvious examples are self-dual is perhaps deceptive. More typical are:

Example 3: The augmentation ideal and its dual
The regular representation and the trivial representation are related by the *augmentation homomorphism*, which is the ring homomorphism $\epsilon : \mathbf{A}[G] \to A$ defined by $\epsilon(\sum_{g \in G} a_g g) = \sum_{g \in G} a_g$. The kernel of ϵ is called the *augmentation ideal* of G and is denoted by $\mathbf{I}_{\mathbf{A}}(G)$. Clearly, $\mathbf{I}_{\mathbf{A}}(G) \in \mathcal{F}(\mathbf{A}[G])$.

In general, the augmentation ideal and its dual $\mathbf{I}_{\mathbf{A}}(G)^*$ are not isomorphic. For us, the principal case of interest is when $\mathbf{A} = \mathbf{Z}$; then we write the integral augmentation ideal simply as $\mathbf{I}(G)$. To anticipate matters slightly, it can be shown that the following statements (I)–(IV) are equivalent:

(I) $\mathbf{I}(G) \cong_{\mathbf{Z}[G]} \mathbf{I}(G)^*$;

(II) $\mathbf{I}(G) \sim \mathbf{I}(G)^*$;

(III) G has cohomological period 2;

(IV) G is cyclic.

Here (I) $\Longrightarrow$ (II) is obvious; when the definition of cohomological period is known (see Chapter 7), (II) $\Longrightarrow$ (III) should be clear from the Representation and Co-Representation Formulae, (20.7) and (20.6), once it is appreciated that $\langle \mathbf{I}(G) \rangle = \mathbf{D}_1(\mathbf{Z})$; the implication (III) $\Longrightarrow$ (IV) follows from a result of Swan ([60] Lemma (5.2)) that cyclic groups are the only groups of period 2; finally, the implication (IV) $\Longrightarrow$ (I) is a straightforward, though not unappealing, exercise in elementary basis change.

Non-isomorphism of M and M^* is related to the non-invertibility of the order of G in $\mathbf{A}$. When $\mathbf{A}$ is a field of characteristic coprime to G, M and M^* are isomorphic. The self-duality of the regular representation generalizes easily to give:

Proposition 28.2: If $M \in \mathcal{F}(\mathbf{A}[G])$, then

$$M \text{ is free} \Longleftrightarrow M^* \text{ is free}$$

If $M \in \mathcal{F}(\mathbf{A}[G])$ is projective, then, for some Q, $M \oplus Q \cong F^n$. Since $(F^n)^* \cong F^n$, we dualize to get $M^* \oplus Q^* \cong F^n$, so that M^* is projective. Conversely, if M^* is projective, then $M^{**} \cong M$ is projective. (28.2) extends to Projectives, thus:

Proposition 28.3: If $M \in \mathcal{F}(\mathbf{A}[G])$, then

$$M \text{ is projective} \Longleftrightarrow M^* \text{ is projective.}$$

One point which must be treated circumspectly is the relation which holds, under duality, between injective and surjective morphisms. Over a field, the dual of a surjective map is injective, and, conversely, the dual of an injective map is surjective. Whilst the former continues to hold in $\mathcal{F}(\mathbf{A}[G])$, the latter is no longer true in general; for example, the dual of the morphism $M \to M$; $x \mapsto 2x$ fails to be surjective whenever M is a nonzero lattice in $\mathcal{F}(\mathbf{Z}[G])$. In this case, of course, $M/2M$ is no longer a lattice.

The most useful statement we can make is in terms of exact sequences in $\mathcal{F}(\mathbf{A}[G])$; that is by exact sequences

$$0 \to E_0 \to E_1 \to \cdots \to E_n \to 0$$

in which all objects are in $\mathcal{F}(\mathbf{A}[G])$; then everything we shall need is covered by:

Proposition 28.4: If $0 \to E_0 \to E_1 \to \cdots \to E_n \to 0$ is an exact sequence in $\mathcal{F}(\mathbf{A}[G])$, then the dual sequence $0 \to E_n^* \to \cdots \to E_1^* \to E_0^* \to 0$ is also an exact sequence in $\mathcal{F}(\mathbf{A}[G])$.

Theorem 28.5: For any commutative ring $\mathbf{A}$ the class $\mathcal{F}(\mathbf{A}[G])$ of $\mathbf{A}$-lattices over $\mathbf{A}[G]$ is tame.

Proof: Evidently, $\mathcal{F}(\mathbf{A}[G])$ is closed under isomorphism, and property $\mathcal{T}(0)$ is true directly from the definition. It remains to show that $\mathcal{T}(1)$-$\mathcal{T}(4)$ hold also:

$\mathcal{T}(1)$: If P is a finitely generated projective module over $\mathbf{A}[G]$, then, *a fortiori*, P is finitely generated projective over $\mathbf{A}$. Hence $P \in \mathcal{F}(\mathbf{A}[G])$.

$\mathcal{T}(2)$: Let $0 \to K \to M \to Q \to 0$ be a short exact sequence of $\mathbf{A}[G]$-modules and $Q \in \mathcal{F}(\mathbf{A}[G])$. Then Q is projective over $\mathbf{A}$, so that the sequence splits over $\mathbf{A}$, and we have an isomorphism of $\mathbf{A}$-modules

$$M \cong_A K \oplus Q$$

If M is a direct summand of a free $\mathbf{A}$-module so also is K. Conversely, since Q is a direct summand of a free $\mathbf{A}$-module, $A^m \cong Q \oplus Q'$ say, then

$$M \oplus Q' \cong_A K \oplus A^m$$

Hence if K is a direct summand of a free $\mathbf{A}$-module, so also is M. Thus $\mathcal{F}(\mathbf{A}[G])$ satisfies $\mathcal{T}(2)$.

$\mathcal{T}(3)$: If $\mathcal{E} = (0 \to P \to M \to Q \to 0)$ is an exact sequence in $\mathcal{F}(\mathbf{A}[G])$ with P projective, then the dual $\mathcal{E}^* = (0 \to Q^* \to M^* \to P^* \to 0)$ is also exact, and P^* is projective. If $s : P^* \to M^*$ splits $\mathcal{E}^*$ on the right, then identifying $M^{**} \cong M$, $P^{**} \cong P$, the dual map $s^* : M \to P$ splits $\mathcal{E}$ on the left. Hence P is injective relative to $\mathcal{F}(\mathbf{A}[G])$.

Conversely, suppose that P is injective relative to $\mathcal{F}(\mathbf{A}[G])$. Since P^* is finitely generated, there is an exact sequence of $\mathbf{A}[G]$-modules

$$0 \to K \to F^n \to P^* \to 0$$

in which F^n is free. By property $\mathcal{T}(2)$, $K \in \mathcal{F}(\mathbf{A}[G])$, so that this is an exact sequence in $\mathcal{F}(\mathbf{A}[G])$. Making the identification $P^{**} \cong P$, the dual exact sequence

$$0 \to P \to F^n \to K^* \to 0$$

is also in $\mathcal{F}(\mathbf{A}[G])$, and P is injective relative to $\mathcal{F}(\mathbf{A}[G])$. Thus the dual sequence splits, $P \oplus K^* \cong F^n$, so that P is projective.

$\mathcal{T}(4)$: Let $M \in \mathcal{F}(\mathbf{A}[G])$; then $M^* \in \mathcal{F}(\mathbf{A}[G])$. Since M^* is finitely generated over $\mathbf{A}[G]$ there exists an exact sequence in $\mathcal{F}(\mathbf{A}[G])$ of the form

$$0 \to K \to F^n \to M^* \to 0$$

where F^n is the free module of rank n over $\mathbf{A}[G]$. By $\mathcal{T}(2)$, we see that $K \in \mathcal{F}(\mathbf{A}[G])$. Dualizing gives an exact sequence

$$0 \to M^{**} \to F^n \to K^* \to 0$$

in $\mathcal{F}(\mathbf{A}[G])$ in which the middle term is free. However, $M^{**} \cong M$, so that we have an exact sequence

$$0 \to M \to F^n \to K^* \to 0$$

in $\mathcal{F}(\mathbf{A}[G])$ in which the middle term is free. This completes the proof. □

Since $\mathcal{F}(\mathbf{A}[G])$ is a tame class, then as in Section 20, for each module $M \in \mathcal{F}(\mathbf{A}[G])$ we construct a sequence of stable modules $(\Omega_n(M))_{n \in \mathbf{Z}}$: for $n \geq 1$, we require $\Omega_n(M)$ to be the stable class $[D]$ of any module $D \in \mathcal{F}(\mathbf{A}[G])$ for which there exists an exact sequence of the form

$$0 \to D \to F_{n-1} \to \cdots \to F_0 \to M \to 0$$

where each F_r is a finitely generated free module over $\mathbf{A}[G]$. Likewise, we require $\Omega_{-n}(M)$ to be the stable class $[D]$ of any module $D \in \mathcal{F}(\mathbf{Z}[G])$ for which there exists an exact sequence of the form

$$0 \to M \to F_0 \to \cdots \to F_{n-1} \to D \to 0$$

where each F_r is a finitely generated free module over $\mathbf{A}[G]$. Duality is introduced into $\mathrm{Stab}(\mathbf{A}[G])$ by writing

$$[M]^* = [M^*]$$

Observe that for any $M \in \mathcal{F}(\mathbf{Z}[G])$, the tree structures on $[M]$ and $[M^*]$ are isomorphic, even though, in general, the stable modules themselves are not.

Proposition 28.6: For any module $M \in \mathcal{F}(\mathbf{A}[G])$ we have the following relations:

(i) $\Omega_m(\Omega_n(M)) = \Omega_{m+n}(M)$;

(ii) $\Omega_n(M^*) = \Omega_{-n}(M)^*$.

The case of primary interest is $M = \mathbf{Z}$, the trivial module $\mathbf{Z}$ over $\mathbf{Z}[G]$. Note that for any finite group G it follows from the augmentation exact sequence $0 \to \mathbf{I}(G) \to \mathbf{Z}[G] \overset{\epsilon}{\to} \mathbf{Z} \to 0$ that we have

(28.7) $\Omega_1(\mathbf{Z}) = [\mathbf{I}(G)]$

Moreover, the trivial module $\mathbf{Z}$ is self-dual, so that:

(28.8) $\Omega_{-1}(\mathbf{Z}) = [\mathbf{I}(G)^*]$

29 The stable modules $\Omega_n(\mathbf{Z})$

When $\mathbf{F}$ is a field whose characteristic does not divide $|G|$, the $\mathbf{F}$-*isomorphism type* of the stable module $\Omega_n(\mathbf{Z})$ is easily calculated. Since $\mathbf{F}[G]$ is semisimple, the augmentation exact sequence

$$0 \to \mathbf{I}_\mathbf{F}(G) \to \mathbf{F}[G] \overset{\epsilon}{\to} \mathbf{F} \to 0$$

splits so that we have a direct sum $\mathbf{F}[G] \cong_{\mathbf{F}[G]} \mathbf{F} \oplus \mathbf{I}_\mathbf{F}(G)$. In particular, by considering, in sequence, the extensions

$$0 \to \mathbf{I}_\mathbf{F}(G) \to \mathbf{F}[G] \to \mathbf{F} \to 0$$

and

$$0 \to \mathbf{F} \to \mathbf{F}[G] \to \mathbf{I}_\mathbf{F}(G) \to 0$$

we see that the $\mathbf{F}$-isomorphism type of the stable modules $\Omega_n(\mathbf{Z})$ is determined as follows

(29.1): $\Omega_n(\mathbf{Z}) \otimes \mathbf{F} = \Omega_n(\mathbf{F}) = \begin{cases} [\mathbf{F}] & \text{if } n \text{ is even} \\ [\mathbf{I}_\mathbf{F}(G)] & \text{if } n \text{ is odd} \end{cases}$

Proposition 29.2: For each $n \geq 0$, $\Omega_{2n+1}(\mathbf{Z})$ is a fork.

Proof: Wedderburn's Theorem combined with (29.1) shows that $[\mathbf{I}_\mathbf{R}(G)]$ is the unique minimal representative of $\Omega_{2n+1}(\mathbf{Z}) \otimes \mathbf{R}$. If M_0 is a minimal representative of $\Omega_{2n+1}(\mathbf{Z})$, it follows that $M_0 \otimes \mathbf{R} \cong \mathbf{I}_\mathbf{R}(G) \oplus \mathbf{R}[G]^m$ for some $m \geq 0$. Moreover, Wedderburn's Theorem, together with the isomorphism $\mathbf{R}[G] \cong_{\mathbf{R}[G]} \mathbf{R} \oplus \mathbf{I}_\mathbf{R}(G)$, shows that every non-trivial simple module over $\mathbf{R}[G]$, a fortiori, every quaternionic simple module, occurs with multiplicity ≥ 1 in $\mathbf{I}_\mathbf{R}(G)$. Hence every quaternionic simple module over $\mathbf{R}[G]$ occurs with multiplicity $\geq m + 1 \geq 1$ in $M_0 \otimes \mathbf{R}$. The conclusion that $\Omega_{2n+1}(\mathbf{Z})$ is a fork now follows from (15.6) and (14.6). $\square$

Typically, $\Omega_{2n+1}(\mathbf{Z})$ can be pictured, thus

(29.3)

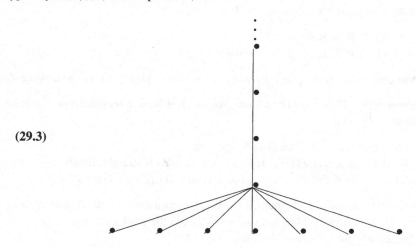

The above argument breaks down in the case of $\Omega_{2n}(\mathbf{Z})$. Nevertheless, the situation is still amenable to analysis, and, although the detailed structure of the even derived modules is not a crucial consideration in what follows, we sketch the results for completeness. In fact, there is only one extra complication; in addition to a fork, there is another possibility, which we call a *crow's foot*

(29.4)

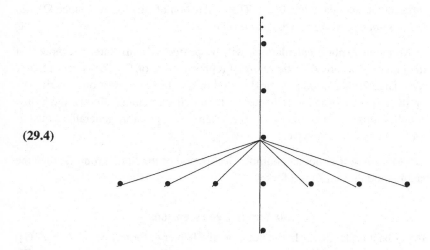

This indicates that there is just one module at the minimal level, a finite number ≥ 1 at level 1, and again just one module at every level ≥ 2.

If M_0 is a minimal representative of $\Omega_{2n}(\mathbf{Z})$, then, by (29.1), there are essentially two possibilities for $M_0 \otimes \mathbf{R}$: *either*

(*) $M_0 \otimes \mathbf{R} \cong \mathbf{R}$ *or*
(**) $M_0 \otimes \mathbf{R} \cong \mathbf{R} \oplus \mathbf{R}[G]^m$ for some $m \geq 1$.

Moreover, (*) occurs precisely when M_0 is isomorphic to the trivial module $\mathbf{Z}$.

Proposition 29.5: Let G be a finite group and let M_0 be a minimal representative of $\Omega_{2n}(\mathbf{Z})$; then:

  (i) if $\mathrm{rk}_{\mathbf{Z}}(M_0) > 1$, then $\Omega_{2n}(\mathbf{Z})$ is a fork;
 (ii) if $M_0 \cong \mathbf{Z}$ and $\mathbf{Z}[G]$ is Eichler, then $\Omega_{2n}(\mathbf{Z})$ is straight; finally
(iii) if $M_0 \cong \mathbf{Z}$ and $\mathbf{Z}[G]$ is not Eichler, then $\Omega_{2n}(\mathbf{Z})$ is a crow's foot.

Proof: If $\mathrm{rk}_{\mathbf{Z}}(M_0) > 1$, then $M_0 \otimes \mathbf{R}$ contains a copy of $\mathbf{R}[G]$, and thus, by (15.1), M_0 is pre-Eichler, and so, by (15.2), has the weak cancellation property. The conclusion of (i) now follows from (14.6).

If $\mathrm{rk}_{\mathbf{Z}}(M_0) = 1$, then $M_0 \cong \mathbf{Z}$, and so $\mathbf{Z}$ is the *unique minimal representative* of $\Omega_{2n}(\mathbf{Z})$. In the case where $\mathbf{Z}[G]$ is Eichler, then $\mathbf{Z} \oplus \mathbf{Z}[G]$ is Eichler. Hence $\mathbf{Z}$ is pre-Eichler, and so has the weak cancellation property. Thus $\Omega_{2n}(\mathbf{Z})$ is a fork, by (14.6), and the conclusion of (ii) follows, since $\Omega_{2n}(\mathbf{Z})$ has a unique minimal representative.

Finally, if $\mathrm{rk}_{\mathbf{Z}}(M_0) = 1$ and $\mathbf{Z}[G]$ is not Eichler, then nevertheless $\mathbf{Z} \oplus \mathbf{Z}[G] \oplus \mathbf{Z}[G]$ is Eichler. Hence $\mathbf{Z} \oplus \mathbf{Z}[G]$ is pre-Eichler and so $\Omega_{2n}(\mathbf{Z})$ has a unique representative at each level ≥ 2. The conclusion of (iii) follows, since $\Omega_{2n}(\mathbf{Z})$ has a unique representative, $\mathbf{Z}$, at level 0. $\qquad\square$

Although explicit calculations will be postponed until later, we note that the situation where $\mathbf{Z}$ is the minimal representative of $\Omega_{2n}(\mathbf{Z})$ is very restrictive. Then $2n$ is a cohomological period of G, placing very strong restrictions on the structure of G: see Chapter 7. To anticipate matters slightly, the Sylow p-subgroups of G are either cyclic or, when $p = 2$, possibly generalized quaternion. In any case, we have:

(29.6) If $m \geq 1$ is not a cohomological period of the finite group G, then the stable module $\Omega_m(\mathbf{Z})$ is a fork.

30 Stably free extensions

Let G be a finite group. In the notation of Chapter 4, for any $M, N \in \mathcal{F}(\mathbf{Z}[G])$ and any $n \geq 1$, $\mathbf{Ext}^n(M, N)$ is the class of all exact sequences of $\mathbf{Z}[G]$-homomorphisms of the form

$$0 \to N \to A_{n-1} \to \cdots \to A_1 \to A_0 \to M \to 0$$

with $A_i \in \mathcal{F}(\mathbf{Z}[G])$ and $\mathbf{Proj}^n(M, N)$ is the subclass of $\mathbf{Ext}^n(M, N)$ consisting of extensions in which each A_i is projective. Let $M, N \in \mathcal{F}(\mathbf{Z}[G])$. It follows from (24.1) that, for each $n \geq 1$, the following conditions are equivalent:

(i) $\mathbf{Proj}^n(M, N) \neq \varnothing$;
(ii) $N \in \mathbf{D}_n(M)$;
(iii) $M \in \mathbf{D}_{-n}(N)$.

Likewise $\mathbf{Free}^n(M, N)$ (resp. $\mathbf{Stab}^n(M, N)$) will denote the subclass of $\mathbf{Ext}^n(M, N)$ where each A_i is free (resp. stably free). We have inclusions

$$\mathbf{Free}^n(M, N) \subset \mathbf{Stab}^n(M, N) \subset \mathbf{Proj}^n(M, N) \subset \mathbf{Ext}^n(M, N)$$

Slightly more delicate than the equivalence of the three conditions above is:

Theorem 30.1: Let $M, N \in \mathcal{F}(\mathbf{Z}[G])$; then for each $n \geq 1$, the following conditions are equivalent:

(i) $\mathbf{Stab}^n(M, N) \neq \varnothing$;
(ii) $N \in \Omega_n(M)$;
(iii) $M \in \Omega_{-n}(N)$.
Moreover, these conditions are equivalent to:
(iv) $\mathbf{Free}^n(M, N) \neq \varnothing$ provided that *either*:
    (a) $n \geq 2$ *or*
    (b) $n = 1$ and G has the free cancellation property.

Proof: We consider first the case where $n \geq 2$; then the equivalence of (ii) and (iii) is obvious, as are the implications (iv) $\Longrightarrow$ (i) and (iv) $\Longrightarrow$ (ii). Thus it suffices to show that (ii) $\Longrightarrow$ (i), and (i) $\Longrightarrow$ (iv).

(ii) $\Longrightarrow$ (i): If F is a finitely generated free module over $\mathbf{Z}[G]$, and $F_1 \subset F$ is a free $\mathbf{Z}[G]$ submodule, then an easy argument using the injectivity of F_1 relative to $\mathcal{F}(\mathbf{Z}[G])$ or, what is the same, by double dualization using the universal property of free modules, shows that F/F_1 is stably free over $\mathbf{Z}[G]$ if and only if F/F_1 is torsion free over $\mathbf{Z}$. Suppose that $N \in \Omega_n(M)$; by definition, there exists a module D which occurs in an exact sequence

$$0 \to D \to \mathbf{Z}[G]^\gamma \to F_{n-2} \to \cdots \to F_0 \to M \to 0$$

with $\gamma > 0$, each F_r finitely generated free over $\mathbf{Z}[G]$, such that

$$N \oplus \mathbf{Z}[G]^a \cong D \oplus \mathbf{Z}[G]^b$$

for some $a, b \geq 0$. Now $D \oplus \mathbf{Z}[G]^b$ occurs in the exact sequence

$$0 \to D \oplus \mathbf{Z}[G]^b \to \mathbf{Z}[G]^{\gamma+b} \to F_{n-2} \to \cdots \to F_0 \to M \to 0$$

so that $N \oplus \mathbf{Z}[G]^a$ occurs in the exact sequence

$$0 \to N \oplus \mathbf{Z}[G]^a \to \mathbf{Z}[G]^{\gamma+b} \to F_{n-2} \to \cdots \to F_0 \to M \to 0$$

We split this last sequence as a pair thus:

(I) $\quad 0 \to N \oplus \mathbf{Z}[G]^a \to \mathbf{Z}[G]^{\gamma+b} \to K \to 0$

(II) $\quad 0 \to K \to F_{n-2} \to \cdots \to F_0 \to M \to 0$

Dividing through (I) by $\mathbf{Z}[G]^m$ gives an extension:

(III) $\quad 0 \to N \to \mathbf{Z}[G]^{\gamma+b}/\mathbf{Z}[G]^a \to K \to 0$

in which $N, K \in \mathcal{F}(\mathbf{Z}[G])$. Thus, $\mathbf{Z}[G]^{\gamma+b}/\mathbf{Z}[G]^a$ is torsion free over $\mathbf{Z}$, so that, by the remark above, $\mathbf{Z}[G]^{\gamma+b}/\mathbf{Z}[G]^a = S$ is stably free over $\mathbf{Z}[G]$. Splicing (II) and (III) back together gives an element of $\mathbf{Stab}^n(M, N)$

$$0 \to N \to S \to F_{n-2} \to \cdots \to F_0 \to M \to 0$$

as required. This completes the proof that (ii) $\Longrightarrow$ (i).

(i) $\Longrightarrow$ (iv) Suppose that $(0 \to N \to S_{n-1} \to \cdots \to S_0 \to M \to 0)$ is an element of $\mathbf{Stab}^n(M, N)$ where each S_i is finitely generated stably free. If $n = 2$ (this is only place where we need the hypothesis that $n \geq 2$), then there exists $a > 0$ so that $S_i \oplus \mathbf{Z}[G]^a$ is free for $i = 1, 2$. Adding a summand of $\mathbf{Z}[G]^a$ to the middle terms gives an extension

$$0 \to N \to S_1 \oplus \mathbf{Z}[G]^a \to S_0 \oplus \mathbf{Z}[G]^a \to M \to 0$$

which defines an element of $\mathbf{Free}^2(M, N)$. If $n > 2$, we may first add a suitable free summand to S_1 and S_0 to make S_0 free; the result follows by induction after an easy splicing argument. This completes the proof that (ii) $\Longrightarrow$ (i), and the proof of the Theorem when $n \geq 2$.

In the case where $n = 1$, the equivalence of (ii) and (iii) is again obvious, and the proof that (ii) $\Longrightarrow$ (i) given above goes through with a minor change of indexing. To show (i) $\Longrightarrow$ (ii), suppose that

$$0 \to N \to S \to M \to 0$$

is an extension in which S is a finitely generated stably free module. Then for some $a \geq 0$, $S \oplus \mathbf{Z}[G]^a$ is free, and from the exact sequence

$$0 \to N \oplus \mathbf{Z}[G]^a \to S \oplus \mathbf{Z}[G]^a \to M \to 0$$

we see that $N \oplus \mathbf{Z}[G]^a \in \Omega_1(M)$. However, this easily implies that $N \in \Omega_1(M)$, showing that (i) $\Longrightarrow$ (ii). This completes the proof that (i), (ii) and (iii) are all equivalent in the case $n = 1$. Finally, if G has the free cancellation property, then (i) is equivalent to (iv). $\qquad\square$

We denote by $\text{Proj}^n(M, N)$ (resp. $\text{Stab}^n(M, N)$, resp. $\text{Free}^n(M, N)$) the image of $\mathbf{Proj}^n(M, N)$ (resp. $\mathbf{Stab}^n(M, N)$, resp. $\mathbf{Free}^n(M, N)$) under the canonical map

$$\mathbf{Ext}^n(M, N) \to \text{Ext}^n(M, N)$$

Evidently $\text{Free}^n(M, N) \subset \text{Stab}^n(M, N)$; in fact, the argument of (30.1) also shows that for $n \geq 2$, $\mathbf{Stab}^n(M, N)$ and $\mathbf{Free}^n(M, N)$ represent the same congruence classes; that is:

Proposition 30.2: For all $n \geq 2$, $\text{Free}^n(M, N) = \text{Stab}^n(M, N)$.

31 Subgroup relations in the derived category

If G is a finite group and H is a subgroup, there is an obvious 'restriction of scalars' functor

$$\mathcal{R}_H^G : \mathcal{F}(\mathbf{Z}[G]) \to \mathcal{F}(\mathbf{Z}[H])$$

which, as the name implies, simply restricts scalars from $\mathbf{Z}[G]$ to $\mathbf{Z}[H]$. There is also an 'extension of scalars' functor

$$\mathcal{E}_H^G : \mathcal{F}(\mathbf{Z}[H]) \to \mathcal{F}(\mathbf{Z}[G])$$

given by

$$\mathcal{E}_H^G(M) = M \otimes_{\mathbf{Z}[H]} \mathbf{Z}[G]$$

Here the tensor factor $\mathbf{Z}[G]$ is simultaneously a left $\mathbf{Z}[H]$-module and a right $\mathbf{Z}[G]$-module; in particular, we have the following identity

$$m \otimes h = mh \otimes 1$$

for $m \in M$ and $h \in H$. The right $\mathbf{Z}[G]$-module structure on $\mathcal{E}_H^G(M)$ is inherited from $\mathbf{Z}[G]$

$$M \otimes_{\mathbf{Z}[H]} \mathbf{Z}[G] \times \mathbf{Z}[G] \to M \otimes_{\mathbf{Z}[H]} \mathbf{Z}[G]$$
$$(m \otimes g_1, g_2) \mapsto m \otimes g_1 g_2$$

When G and H are unambiguous, we abbreviate to $\mathcal{E} = \mathcal{E}_H^G$ and $\mathcal{R} = \mathcal{R}_H^G$.

For any right module $\mathbf{Z}[H]$-module M, the $\mathbf{Z}[H]$-dual $\mathrm{Hom}_{\mathbf{Z}[H]}(M, \mathbf{Z}[H])$ admits a natural left $\mathbf{Z}[H]$-module structure, given by

$$(h \cdot \varphi)(m) = h \cdot \varphi(m)$$

However, in the special case where $M = \mathbf{Z}[G]$, $\mathrm{Hom}_{\mathbf{Z}[H]}(M, \mathbf{Z}[H])$ is naturally a $(\mathbf{Z}[H] - \mathbf{Z}[G])$ bimodule, with natural right $\mathbf{Z}[G]$-module structure given by

$$(\varphi \cdot g)(x) = \varphi(g \cdot x)$$

Evidently $\mathbf{Z}[G]$ also has a natural $(\mathbf{Z}[H] - \mathbf{Z}[G])$ bimodule structure, given by translation on either side. In fact, we have:

Lemma 31.1: There is an isomorphism of $(\mathbf{Z}[H] - \mathbf{Z}[G])$ bimodules

$$\mathbf{Z}[G] \xrightarrow{\simeq} \mathrm{Hom}_{\mathbf{Z}[H]}(\mathbf{Z}[G], \mathbf{Z}[H])$$

Proof: Let $\rho = \{\rho_1, \ldots, \rho_n\}$ be a complete set of representatives for the quotient set $H \backslash G$; that is

$$G = \cup_{i=1}^n H\rho_i \quad \text{where} \quad H\rho_i \cap H\rho_j = \emptyset \text{ if } i \neq j$$

Put $\lambda_i = \rho_i^{-1}$. Then $\lambda = \{\lambda_1, \ldots, \lambda_n\}$ is a complete set of representatives for G/H. As a left $\mathbf{Z}[H]$-module, $\mathbf{Z}[G]$ is free on $\{\rho_1, \ldots, \rho_n\}$. Moreover, as a left $\mathbf{Z}[H]$-module, $\mathrm{Hom}_{\mathbf{Z}[H]}(\mathbf{Z}[G], \mathbf{Z}[H])$ is free on $\{\hat{\lambda}_1, \ldots, \hat{\lambda}_n\}$, where $\hat{\lambda}_i : \mathbf{Z}[G] \to \mathbf{Z}[H]$ is the right $\mathbf{Z}[H]$-homomorphism given by

$$\hat{\lambda}_i \left(\sum_{j=1}^n \lambda_j h_j \right) = h_i$$

It follows that there is an isomorphism of left $\mathbf{Z}[H]$-modules

$$\nu : \mathbf{Z}[G] \to \mathrm{Hom}_{\mathbf{Z}[H]}(\mathbf{Z}[G], \mathbf{Z}[H])$$

given by

$$\nu(\rho_i) = \hat{\lambda}_i$$

and straightforward computation shows that ν is equivariant with respect to the right G action. $\qquad \square$

The functors $\mathcal{E}$, $\mathcal{R}$ have a number of easily verified properties; they are additive, exact, and take free modules to free modules. It follows that they also take projective modules to projective modules. In addition, they have the unusual property that they are simultaneously mutual left and right adjoints without being mutually inverse; that is:

Proposition 31.2: There are natural isomorphisms

$$\operatorname{Hom}_{\mathbf{Z}[G]}(\mathcal{E}(M), N) \cong \operatorname{Hom}_{\mathbf{Z}[H]}(M, \mathcal{R}(N))$$

$$\operatorname{Hom}_{\mathbf{Z}[H]}(\mathcal{R}(N), M) \cong \operatorname{Hom}_{\mathbf{Z}[G]}(N, \mathcal{E}(M))$$

Proof: The first of these is entirely straightforward. In the second, it is useful to make use of a different model for $\mathcal{E}$. Denote by $\mathcal{E}'(M) = \operatorname{Hom}_{\mathbf{Z}[H]}(\mathbf{Z}[G], M)$ where the right $\mathbf{Z}[G]$ structure on $\mathcal{E}'(M)$ is given by

$$(\alpha \cdot g)(x) = \alpha(xg^{-1})$$

There is a preliminary natural equivalence

$$\nu_1 : \mathcal{E}'(M) \to M \otimes_{\mathbf{Z}[H]} \operatorname{Hom}_{\mathbf{Z}[H]}(\mathbf{Z}[G], \mathbf{Z}[H])$$

By (31.1) above, there is now an equivalence

$$\nu_2 : M \otimes_{\mathbf{Z}[H]} \operatorname{Hom}_{\mathbf{Z}[H]}(\mathbf{Z}[G], \mathbf{Z}[H]) \to M \otimes_{\mathbf{Z}[H]} \mathbf{Z}[G] = \mathcal{E}(M)$$

which is natural in M and with respect to the right $\mathbf{Z}[G]$ action. Then

$$\nu = \nu_2 \circ \nu_1 : \mathcal{E}'(M) \to \mathcal{E}(M)$$

is the desired natural equivalence. The adjunction isomorphism

$$\psi : \operatorname{Hom}_{\mathbf{Z}[G]}(N, \mathcal{E}'(M)) \to \operatorname{Hom}_{\mathbf{Z}[H]}(\mathcal{R}(N), M)$$

is given by $[\psi(\alpha)](n) = \alpha(n)(1)$ $\qquad \square$

We need to distinguish between the derived categories and associated constructions corresponding to the different groups G, H. For an arbitrary group Γ we denote by:

  (i) $\mathcal{D}er(\Gamma)$ the derived category of $\mathcal{F}(\mathbf{Z}[\Gamma])$;
 (ii) $\mathbf{D}_n^{\Gamma}(M)$ the nth derived object of the module $M \in \mathcal{F}(\mathbf{Z}[\Gamma])$;
(iii) $\Omega_n^{\Gamma}(M)$ the nth stable module of $M \in \mathcal{F}(\mathbf{Z}[\Gamma])$.

Since the functors $\mathcal{E}, \mathcal{R}$ are exact and preserve projectives they are definable at the level of the derived category; that is we have

$$\mathcal{R}_H^G : \mathcal{D}er(G) \to \mathcal{D}er(H)$$

and

$$\mathcal{E}_H^G : \mathcal{D}er(H) \to \mathcal{D}er(G)$$

Moreover, they each commute with $\mathbf{D}_n$,

$$\mathcal{R}(\mathbf{D}_n^G(N)) = \mathbf{D}_n^H(\mathcal{R}(N))$$

and

$$\mathcal{E}\big(\mathbf{D}_n^H(M)\big) = \mathbf{D}_n^G(\mathcal{E}(M))$$

Because they preserve free modules, they also commute with Ω_n

$$\mathcal{R}\big(\Omega_n^G(N)\big) = \Omega_n^H(\mathcal{R}(N))$$

and

$$\mathcal{E}\big(\Omega_n^H(M)\big) = \Omega_n^G(\mathcal{E}(M))$$

Proposition 31.3: There are natural isomorphisms

$$\mathrm{Hom}_{\mathcal{D}\mathrm{er}(G)}(\mathcal{E}(M), N) \cong \mathrm{Hom}_{\mathcal{D}\mathrm{er}(H)}(M, \mathcal{R}(N))$$

$$\mathrm{Hom}_{\mathcal{D}\mathrm{er}(H)}(\mathcal{R}(N), M) \cong \mathrm{Hom}_{\mathcal{D}\mathrm{er}(G)}(N, \mathcal{E}(M))$$

Proof: If Γ is a finite group, and $M, N \in \mathcal{F}(\mathbf{Z}[\Gamma])$, then

$$\mathrm{Hom}_{\mathcal{D}\mathrm{er}(\Gamma)}(M, N) = \mathrm{Hom}_{\mathbf{Z}[\Gamma]}(M, N)/\mathrm{Hom}_{\mathbf{Z}[\Gamma]}^0(M, N)$$

where $\mathrm{Hom}_{\mathbf{Z}[\Gamma]}^0(M, N)$ denotes the subset of $\mathrm{Hom}_{\mathbf{Z}[\Gamma]}(M, N)$ consisting of homomorphisms which factor through a projective. It suffices to show that the adjunction isomorphisms induce bijections

$$\mathrm{Hom}_{\mathbf{Z}[H]}^0(M, \mathcal{R}(N)) \longleftrightarrow \mathrm{Hom}_{\mathbf{Z}[G]}^0(\mathcal{E}(M), N)$$

and

$$\mathrm{Hom}_{\mathbf{Z}[H]}^0(\mathcal{R}(N), M) \longleftrightarrow \mathrm{Hom}_{\mathbf{Z}[G]}^0(N, \mathcal{E}(M))$$

Suppose $f : M \to \mathcal{R}(N)$ admits a factorization $f = \varphi \circ i$, where $i : M \to P$, $\varphi : P \to \mathcal{R}(N)$ and P is projective. Then the adjoint morphism $\hat{f} : \mathcal{E}(M) \to N$ factorizes as $\hat{f} = \hat{\varphi} \circ \mathcal{E}(i)$ where $\hat{\varphi} : \mathcal{E}(P) \to N$, $\mathcal{E}(i) : \mathcal{M} \to \mathcal{E}(P)$. However $\mathcal{E}(P)$ is projective so that $f \mapsto \hat{f}$ maps $\mathrm{Hom}_{\mathbf{Z}[H]}^0(M, \mathcal{R}(N))$ into $\mathrm{Hom}_{\mathbf{Z}[G]}^0(\mathcal{E}(M), N)$. It suffices to show that

$$\mathrm{Hom}_{\mathbf{Z}[H]}^0(M, \mathcal{R}(N)) \to \mathrm{Hom}_{\mathbf{Z}[G]}^0(\mathcal{E}(M), N); \quad f \mapsto \hat{f}$$

is surjective.

Thus suppose the $\mathbf{Z}[G]$-homomorphism $g : \mathcal{E}(M) \to N$ factorizes as $g = \psi \circ j$ where $j : \mathcal{E}(M) \to Q$, $\psi : Q \to N$ and where Q is projective over $\mathbf{Z}[G]$. Let $i : M \to \mathcal{R}(Q)$ be the composite of j with the canonical inclusion $M \to \mathcal{E}(M); x \mapsto x \otimes 1$. Put $f = \mathcal{R}(\psi) \circ i$; then f factorizes through the $\mathbf{Z}[H]$

projective $\mathcal{R}(Q)$. It is straightforward to see that $g = \hat{f}$, and the correspondence is surjective as claimed. Thus the adjunction map $f \mapsto \hat{f}$ induces a bijection

$$\text{Hom}^0_{\mathbf{Z}[H]}(M, \mathcal{R}(N)) \longleftrightarrow \text{Hom}^0_{\mathbf{Z}[G]}(\mathcal{E}(M), N)$$

as claimed. The bijectivity of $\text{Hom}^0_{\mathbf{Z}[H]}(\mathcal{R}(N), M) \longleftrightarrow \text{Hom}^0_{\mathbf{Z}[G]}(N, \mathcal{E}(M))$ follows by a similar argument. $\square$

Combined with (25.7) and with the representation and co-representation formulae (20.6), (20.7), adjointness gives rise to the following Eckmann–Shapiro relations in cohomology

(31.4) $$\text{Ext}^n_{\mathbf{Z}[H]}(\mathcal{R}(M), N) \cong \text{Ext}^n_{\mathbf{Z}[G]}(M, \mathcal{E}(N))$$

and

(31.5) $$\text{Ext}^n_{\mathbf{Z}[G]}(\mathcal{E}(A), B) \cong \text{Ext}^n_{\mathbf{Z}[H]}(A, \mathcal{R}(B))$$

When M is the trivial $\mathbf{Z}[G]$-module $\mathbf{Z}$, $\mathcal{R}(M)$ is the trivial $\mathbf{Z}[H]$-module $\mathbf{Z}$, and the first of these specializes to the classical 'Shapiro's Lemma'

(31.6) $$\text{H}^n(H, N) \cong \text{H}^n(G, \mathcal{E}(N))$$

32 Restriction and transfer

Adjointness gives rise to some natural transformations. Thus fix a $\mathbf{Z}[G]$-module A; since

$$\text{Hom}_{\mathbf{Z}[H]}(\mathcal{R}(A), \mathcal{R}(A)) \cong \text{Hom}_{\mathbf{Z}[G]}(A, \mathcal{E}\mathcal{R}(A))$$

then, corresponding to $\text{Id}_{\mathcal{R}(A)}$ we obtain a $\mathbf{Z}[G]$-homomorphism

$$\delta : A \to \mathcal{E}\mathcal{R}(A)$$

Let $d = [G : H]$ be the index of H in G, and let $\{\rho_1, \ldots, \rho_d\}$ be a complete set of coset representatives for the quotient set $H \backslash G$; δ takes the following explicit form which is easily shown to be independent of the particular set of coset representatives chosen

$$\delta(a) = \sum_{i=1}^{d} a\rho_i^{-1} \otimes \rho_i$$

In addition, $\text{Hom}_{\mathbf{Z}[H]}(\mathcal{R}(A), \mathcal{R}(A)) \cong \text{Hom}_{\mathbf{Z}[G]}(\mathcal{E}\mathcal{R}(A), A)$, so that, again corresponding to $\text{Id}_{\mathcal{R}(A)}$, we also obtain a $\mathbf{Z}[G]$-homomorphism

$$\epsilon : \mathcal{E}\mathcal{R}(A) \to A$$

ϵ is the canonical 'contraction map' given on primitive tensors by

$$\epsilon(a \otimes g) = ag$$

The composite $\epsilon \circ \delta : A \to A$ is multiplication by the index d; that is

(32.1) $$\epsilon \circ \delta(a) = da \quad \text{for all } a \in A$$

The composite $\delta \circ \epsilon : \mathcal{ER}(A) \to \mathcal{ER}(A)$ is computed as follows; the general element in $\mathcal{ER}(A)$ can be represented canonically as

$$\mathbf{z} = \sum_{i=1}^{d} z_i \rho_i^{-1} \otimes \rho_i$$

where $z_i \in A$. Then $\epsilon(\mathbf{z}) = \sum_{i=1}^{d} z_i$, and

$$\delta\epsilon(\mathbf{z}) = \sum_{i=1}^{d} \epsilon(\mathbf{z})\rho_i^{-1} \otimes \rho_i$$

When A is a (right) lattice over $\mathbf{Z}[G]$, we can give a 'coordinatized' description of $\mathcal{ER}(A)$ as a form of *wreath product* as follows: let $\{\rho_1, \ldots, \rho_d\}$ be a complete set of coset representatives for $H \backslash G$. Let Σ_d denote the group of permutations of $\{1, \ldots, d\}$. The natural action of G on $H \backslash G$ gives a permutation representation $\sigma : G \to \Sigma_d$ by means of

$$H\rho_i g^{-1} = H\rho_{\sigma(g)(i)}$$

Then $\mathcal{ER}(A)$ can be described as

$$A^{(d)} = \underbrace{A \oplus \cdots \oplus A}_{d}$$

with typical element written as $\mathbf{a} = (a_1, a_2, \ldots, a_d)$ and G-action given by

$$\mathbf{a}.g^{-1} = \left(a_{\sigma(g)(1)}g^{-1}, a_{\sigma(g)(2)}g^{-1}, \ldots, a_{\sigma(g)(d)}g^{-1}\right)$$

The mapping

$$\Psi(a_1, a_2, \ldots, a_d) = \sum a_i \rho_i^{-1} \otimes \rho_i$$

is the desired isomorphism of right $\mathbf{Z}[G]$-modules $\Psi : A^{(d)} \to \mathcal{ER}(A)$.

$\mathcal{ER}(A)$ admits another G-action, conjugacy, as follows: if $g \in G$, we define the conjugacy map $c^g : \mathcal{ER}(A) \to \mathcal{ER}(A)$ by

$$c^g \left(\sum_{i=1}^{d} a_i \rho_i^{-1} \otimes \rho_i \right) = \sum_{i=1}^{d} a_i g\rho_i^{-1} \otimes \rho_i g^{-1}$$

The elements of $\mathcal{ER}(A))$ which are fixed by each c^g are those of the form $\delta(a)$ for some $a \in A$. Moreover, c^g induces an automorphism c_*^g on any $\text{Hom}_{\mathcal{D}er(G)}(M, \mathcal{ER}(A))$ and hence on any $H^n(H, \mathcal{R}(A))$, via the Eckmann–Shapiro relation, $H^n(H, \mathcal{R}(A)) \cong H^n(G, \mathcal{ER}(A))$. An element $z \in H^n(H, \mathcal{R}(A))$ is said to be *stable* when $c_*^g(z) = z$ for all $g \in G$. Observe that, for a stable element z, $i^*t(z) = dz$ where d is the index $[G : H]$.

For any $\mathbf{Z}[G]$-lattice N, there is both a canonical 'Shapiro isomorphism'

$$s : H^n(G, \mathcal{ER}(N)) \cong H^n(H, \mathcal{R}(N))$$

and a transformation $\delta_* : H^n(G, N) \to H^n(G, \mathcal{ER}(N))$ induced from the homomorphism of coefficients $\delta : N \to \mathcal{ER}(N)$. If i denotes the inclusion $H \subset G$, it is usual to write

$$i^* = s \circ \delta_* : H^n(G, N) \to H^n(H, \mathcal{R}(N))$$

i^* is the *restriction homomorphism*.

Likewise, we have a transformation $\epsilon_* : H^n(G, \mathcal{ER}(N)) \to H^n(G, N)$ induced from the homomorphism of coefficients $\epsilon : \mathcal{ER}(N) \to N$. It is usual to write

$$t = \epsilon_* \circ s^{-1} : H^n(H, \mathcal{R}(N)) \to H^n(G, N)$$

t is the *transfer homomorphism*. The following is clear:

Proposition 32.2: Let $i : H \subset G$ be the inclusion of a subgroup H in the finite group G; then, for any $\mathbf{Z}[G]$-lattice N, the composite $t \circ i^* : H^n(G, N) \to H^n(G, N)$ is multiplication by $d = [G : H]$.

From this follows the well-known fact that, for $n > 0$, $H^n(G, N)$ consists entirely of torsion of order dividing $|G|$. In fact, taking $H = \{1\}$ to be the trivial subgroup, we see that $\epsilon_* \circ \delta_* : H^n(G, N) \to H^n(G, N)$ factors through $0 = H^n(H, \mathcal{R}(N))$ and has the effect of multiplication by the index, $|G|$, of $\{1\}$ in G.

Corollary 32.3: If G is a finite group and $n > 0$, then for each $\mathbf{Z}[G]$-lattice N, $H^n(G, N)$ is a module over $\mathbf{Z}/|G|$.

Chapter 6

k-invariants

Having established the general structure of $\Omega_n(\mathbf{Z})$ in Chapter 5, we now turn to the problem of classifying extensions

$$0 \to J \to A_{n-1} \to \cdots \to A_0 \to \mathbf{Z} \to 0$$

where $J \in \Omega_n(\mathbf{Z})$ and $A_r \in \mathcal{F}(\mathbf{Z}[G])$. In Chapter 4, this problem was reduced, in principle, to that of computing the group $\mathrm{Ext}^n(\mathbf{Z}, J)$. Here we compute

$$\mathrm{Ext}^n(\mathbf{Z}, J) \cong \mathcal{H}^n(\mathbf{Z}, \Omega_n(\mathbf{Z})) \cong \mathbf{Z}/|G|$$

where $\mathbf{Z}/|G|$ arises as the endomorphism ring, in the derived category, of the trivial module $\mathbf{Z}$. In Section 34 we show how the method of k-invariants, due originally to Maclane and Whitehead [35], allows a classification of projective n-stems by the group of units $(\mathbf{Z}/|G|)^*$.

In Sections 35–37 we give our version of Swan's explicit description of one-dimensional extensions over $\mathbf{Z}$ using the k-invariant method. This is extended to general extensions in Section 38. Finally, in Section 39 we give Tate's generalization of the action of $\mathrm{End}_{\mathcal{D}er}(\mathbf{Z})$ on extensions. We introduce the Tate cohomology groups $\hat{H}^k(G, A)$ for all integer indices k, positive and negative, and show how $\hat{H}^*(G, A)$ becomes a graded module over the graded ring $\hat{H}^*(G, \mathbf{Z})$.

33 Endomorphisms in the derived category

Let G be a finite group, and let $\mathbf{Z}$ denote the trivial module over $\mathbf{Z}[G]$. Note that there are $\mathbf{Z}[G]$-homomorphisms $\Sigma : \mathbf{Z} \to \mathbf{Z}[G]$ and $\epsilon : \mathbf{Z}[G] \to \mathbf{Z}$ given by

$$\Sigma(n) = n \left(\sum_{g \in G} g \right); \quad \epsilon \left(\sum_{g \in G} a_g g \right) = \sum_{g \in G} a_g$$

The following two propositions are easily verified.

(33.1) If $\varphi : \mathbf{Z} \to \mathbf{Z}[G]$ is a $\mathbf{Z}[G]$-homomorphism, then $\varphi = n\Sigma$ for some $n \in \mathbf{Z}$.

(33.2) If $\psi : \mathbf{Z}[G] \to \mathbf{Z}$ is a $\mathbf{Z}[G]$-homomorphism, then $\psi = n\epsilon$ for some $n \in \mathbf{Z}$.

As in Chapter 4, we write $\alpha \approx 0$ when the $\mathbf{Z}[G]$-homomorphism α factors through a projective. We compute the endomorphism ring $\mathrm{End}_{\mathcal{D}\mathrm{er}}(\mathbf{Z})$ of the trivial module $\mathbf{Z}$ in the derived category. Observe that $\mathrm{End}_{\mathbf{Z}[G]}(\mathbf{Z}) \cong \mathbf{Z}$; by contrast we have:

Proposition 33.3: Let $\mathbf{Z}$ denote the trivial module over $\mathbf{Z}[G]$ where G is a finite group; then there is an isomorphism of rings

$$\mathrm{End}_{\mathcal{D}\mathrm{er}}(\mathbf{Z}) \cong \mathbf{Z}/|G|$$

Proof: We may identify $\alpha \in \mathrm{End}_{\mathbf{Z}[G]}(\mathbf{Z})$ with the integer $\alpha(1)$; under this identification we have

$$\alpha \approx 0 \Longleftrightarrow \alpha = n|G| \text{ for some } n \in \mathbf{Z}$$

Clearly $\alpha \in \mathrm{End}_{\mathbf{Z}[G]}(\mathbf{Z})$ factorizes through a projective module if and only if it factorizes through some $\mathbf{Z}[G]^m$. Suppose that α factorizes through $\mathbf{Z}[G]^m$, thus

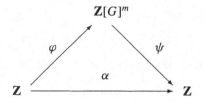

We may write

$$\varphi = (a_1 \Sigma, \ldots, a_m \Sigma) \text{ and } \psi = \begin{pmatrix} b_1 \epsilon \\ \vdots \\ b_m \epsilon \end{pmatrix}$$

so that

$$\alpha = \left(\sum_i a_i b_i \right) \epsilon \Sigma = \left(\sum_i a_i b_i \right) |G|$$

Conversely, if $\alpha = m|G|$, then α factors through $\mathbf{Z}[G]$ since $\alpha = m\epsilon \circ \Sigma$; that is:

$$\alpha \text{ factors through some } \mathbf{Z}[G]^m \iff \alpha = n|G| \text{ for some } n$$

and the result now follows. $\square$

Corollary 33.4: $\mathcal{A}(\mathbf{Z}) \cong (\mathbf{Z}/|G|)^*$

Let $M, D \in \mathcal{F}(\mathbf{Z}[G])$ be such that $\mathrm{Proj}^n(M, D) \neq \emptyset$ and let

$$\mathbf{P} = (0 \to D \to P_{n-1} \to \cdots \to P_0 \to M \to 0)$$

be a projective n-stem over M terminating in D; recall that in Section 26 we showed that $\mathbf{P}$ gives rise to a ring isomorphism $\kappa_{\mathbf{P}} : \mathrm{End}_{\mathcal{D}\mathrm{er}}(D) \xrightarrow{\simeq} \mathrm{End}_{\mathcal{D}\mathrm{er}}(M)$. Moreover, if

$$\mathbf{Q} = (0 \to D \to Q_{n-1} \to \cdots \to Q_0 \to M \to 0)$$

is another element of $\mathbf{Proj}^n(M, D)$, the ring isomorphisms $\kappa_{\mathbf{P}}, \kappa_{\mathbf{Q}}$ are related by

(33.5) $\kappa_{\mathbf{P}} = \lambda_{\mathbf{PQ}} \kappa_{\mathbf{Q}} \lambda_{\mathbf{PQ}}^{-1}$

where $\lambda_{\mathbf{PQ}} = \lambda_{\mathbf{QP}}^{-1}$ is the unique automorphism $\lambda_{\mathbf{PQ}}$ of M in the derived category making the following diagram commute

$$
\begin{array}{ccccccccc}
0 \to D \to & Q_{n-1} & \to & \cdots & \to & Q_0 & \to & M & \to 0 \\
\downarrow \mathrm{Id} & \downarrow \lambda_{n-1} & & & & \downarrow \lambda_0 & & \downarrow \lambda_{\mathbf{PQ}} & \\
0 \to D \to & P_{n-1} & \to & \cdots & \to & P_0 & \to & M & \to 0
\end{array}
$$

In the case where $\mathrm{End}_{\mathcal{D}\mathrm{er}}(M)$ is commutative, conjugation is trivial, and so

(33.6) $\kappa_{\mathbf{P}} = \kappa_{\mathbf{Q}}$

that is, we have a ring isomorphism $\kappa = \kappa_{\mathbf{P}} : \mathrm{End}_{\mathcal{D}\mathrm{er}}(D) \to \mathrm{End}_{\mathcal{D}\mathrm{er}}(M)$ which is independent of the particular choice of projective n-stem $\mathbf{P}$ used to construct it. Though, we have not stressed the point, κ may perhaps still depend on n; that is, the *length* of the projective stem. However, in the case of most interest to us, namely $M \cong \mathbf{Z}$, the independence of κ is absolute. Then the ring $\mathrm{End}_{\mathcal{D}\mathrm{er}}(\mathbf{Z}) \cong \mathbf{Z}/|G|$ is not only commutative, but, as is easily seen, its only ring automorphism is the identity; thus any two ring isomorphisms $\psi, \varphi : \mathrm{End}_{\mathcal{D}\mathrm{er}}(\mathbf{Z}) \to \mathbf{Z}/|G|$, must coincide, and we have:

Proposition 33.7: If $J \in \mathbf{D}_n(\mathbf{Z})$ then there exists a *unique* ring isomorphism $\kappa^J : \mathrm{End}_{\mathcal{D}\mathrm{er}}(J) \to \mathbf{Z}/|G|$.

In particular, (33.7) permits the unambiguous identification $\text{End}_{\mathcal{D}\text{er}}(\mathbf{Z}) \cong \mathbf{Z}/|G|$.

34 k-invariants and the action of $\text{Aut}_{\mathcal{D}\text{er}}(J)$

In Section 26 we showed how to classify projective n-stems $\mathbf{P} \in \mathbf{Proj}^n(M, D)$ by means of $\mathcal{A}(D)$. We now make the classification explicit in the case where $D \in \mathbf{D}_m(\mathbf{Z})$, using the isomorphism $\text{End}_{\mathcal{D}\text{er}}(D) \cong \mathbf{Z}/|G|$.

If M, N are modules in $\mathcal{F}(\mathbf{Z}[G])$ such that $M \in \mathbf{D}_m(\mathbf{Z})$ and $N \in \mathbf{D}_n(M)$, then $\text{Proj}^n(M, N) \neq \emptyset$. Thus, suppose that $\mathbf{P} \in \mathbf{Proj}^n(M, N)$ is a projective n-stem and $\mathbf{E}$ is an arbitrary element of $\mathbf{Ext}^n(M, N)$, the universal property of projective modules leads to a commutative diagram of $\mathbf{Z}[G]$-modules, thus

$$
\begin{array}{c} \mathbf{P} \\ \downarrow \alpha \\ \mathbf{E} \end{array} = \left(\begin{array}{ccccccc} 0 \to & N \to & P_{n-1} \to & \cdots \to & P_0 \to & M \to & 0 \\ & \downarrow \alpha_+ & \downarrow \alpha_{n-1} & & \downarrow \alpha_0 & \downarrow \text{Id} & \\ 0 \to & N \to & E_{n-1} \to & \cdots \to & E_0 \to & M \to & 0 \end{array} \right)
$$

We define the *k-invariant of the transition*, $k(\mathbf{P} \to \mathbf{E}) \in \mathbf{Z}/|G|$, by

$$k(\mathbf{P} \to \mathbf{E}) = \kappa(\alpha_+)$$

It is easy to verify the following properties:

Proposition 34.1: If $\mathbf{E} \in \mathbf{Ext}^n(M, N)$ and $\mathbf{P}, \mathbf{Q} \in \mathbf{Proj}^n(M, N)$ then:

(i) $k(\mathbf{P} \to \mathbf{E}) = k(\mathbf{P} \to \mathbf{Q}) \, k(\mathbf{Q} \to \mathbf{E})$;
(ii) $k(\mathbf{P} \to \mathbf{P}) = 1$;
(iii) $k(\mathbf{P} \to \mathbf{Q}) k(\mathbf{Q} \to \mathbf{P}) = 1$.

Making a specific choice of projective extension $\mathbf{P} \in \mathbf{Proj}^n(M, N)$, if $\mathbf{E} \in \mathbf{Ext}^n(M, N)$, then

$$\mathbf{E} \in \mathbf{Proj}^n(M, N) \Longleftrightarrow k(\mathbf{P} \to \mathbf{E}) \in (\mathbf{Z}/|G|)^*$$

and we obtain:

Proposition 34.2: The correspondence $\mathbf{E} \mapsto k(\mathbf{P} \to \mathbf{E})$ defines a bijective mapping $\text{Proj}^n(M, N) \to (\mathbf{Z}/|G|)^*$.

We observed in Section 26 that $\text{Ext}^n(\mathbf{Z}, J)$ is a left module over the ring $\text{End}_{\mathcal{D}\text{er}}(J)$ when $J \in \mathbf{D}_n(\mathbf{Z})$

$$\text{End}_{\mathcal{D}\text{er}}(J) \times \text{Ext}^n_{\mathcal{C}}(\mathbf{Z}, J) \to \text{Ext}^n_{\mathcal{C}}(\mathbf{Z}, J)$$

$$[\alpha] * [\mathbf{E}] = [\alpha_*(\mathbf{E})]$$

We note the following obvious identity:

Proposition 34.3: Let $\mathbf{P} \in \mathbf{Proj}^n(\mathbf{Z}, J)$ be a projective n-stem and let $\mathbf{E} \in \mathbf{Ext}^n(\mathbf{Z}, J)$; then for all $\alpha \in \mathrm{End}_{\mathcal{D}er}(J)$

$$k(\mathbf{P} \to \alpha_*(\mathbf{E})) = \kappa(\alpha)k(\mathbf{P} \to \mathbf{E})$$

Proof: Choose a lifting of the identity over $\mathbf{Z}$

$$
\begin{matrix}
\mathbf{P} \\
\downarrow \varphi = \\
\mathbf{E}
\end{matrix}
\begin{pmatrix}
0 \to J \xrightarrow{j} P_{n-1} \to \cdots \to P_0 \to \mathbf{Z} \to 0 \\
\quad \downarrow \varphi_+ \; \downarrow \varphi_{n-1} \qquad\qquad \downarrow \varphi_0 \; \downarrow \mathrm{Id} \\
0 \to J \xrightarrow{i} E_{n-1} \to \cdots \to E_0 \to \mathbf{Z} \to 0
\end{pmatrix}
$$

from which we calculate that $k(\mathbf{P} \to \mathbf{E}) = \kappa(\varphi_+)$. Let $\nu_\alpha : \mathbf{E} \to \alpha_*(\mathbf{E})$ denote the canonical mapping; then the composition $\nu_\alpha \circ \varphi$ gives a lifting of the identity on $\mathbf{Z}$ to a morphism : $\mathbf{P} \to \alpha_*(\mathbf{E})$, and takes the form

$$
\begin{matrix}
\mathbf{P} \\
\downarrow \nu_\alpha \circ \varphi = \\
\alpha_*(\mathbf{E})
\end{matrix}
\begin{pmatrix}
0 \to J \xrightarrow{j} P_{n-1} \to \cdots \to P_0 \to \mathbf{Z} \to 0 \\
\quad \downarrow \alpha \circ \varphi_+ \quad \downarrow \qquad\qquad \downarrow \; \downarrow \mathrm{Id} \\
0 \to J \xrightarrow{i} E'_{n-1} \to \cdots \to E'_0 \to \mathbf{Z} \to 0
\end{pmatrix}
$$

from which we calculate that

$$k(\mathbf{P} \to \alpha_*(\mathbf{E})) = \kappa(\alpha \circ \varphi_+) = \kappa(\alpha)\kappa(\varphi_+) = \kappa(\alpha)k(\mathbf{P} \to \mathbf{E}) \qquad \square$$

We previously observed, in Section 26, that the unit group $\mathcal{A}(J) = \mathrm{Aut}_{\mathcal{D}er}(J)$ stabilizes the subcategory $\mathbf{Proj}^n(\mathbf{Z}, J)$. If $\mathbf{P} \in \mathbf{Proj}^n(\mathbf{Z}, J)$ and $\alpha \in \mathcal{A}(J)$, then, since $\alpha_*(\mathbf{P})$ is also a projective n-stem, the k-invariant $k(\alpha_*(\mathbf{P}) \to \mathbf{E})$ is defined for any $\mathbf{E} \in \mathbf{Ext}^n(\mathbf{Z}, J)$. The k-invariant is covariant with respect to the action of $\mathcal{A}(J)$ in the following sense:

Proposition 34.4: Let $\mathbf{P} \in \mathbf{Proj}^n(\mathbf{Z}, J)$ be a projective n-stem and let $\mathbf{E} \in \mathbf{Ext}^n(\mathbf{Z}, J)$; then for all $\alpha \in \mathcal{A}(J)$

$$k(\alpha_*(\mathbf{P}) \to \alpha_*(\mathbf{E})) = k(\mathbf{P} \to \mathbf{E})$$

Proof: We have seen that $k(\mathbf{P} \to \alpha_*(\mathbf{E})) = \kappa(\alpha)k(\mathbf{P} \to \mathbf{E})$. However, by (33.9) we can also calculate $k(\mathbf{P} \to \alpha_*(\mathbf{E}))$ by means of

$$k(\mathbf{P} \to \alpha_*(\mathbf{E})) = k(\mathbf{P} \to \alpha_*(\mathbf{P}))k(\alpha_*(\mathbf{P}) \to \alpha_*(\mathbf{E}))$$

By (34.3), $k(\mathbf{P} \to \alpha_*(\mathbf{P})) = \kappa(\alpha)$ so that we now have

$$\kappa(\alpha)k(\mathbf{P} \to \mathbf{E}) = k(\mathbf{P} \to \alpha_*(\mathbf{E})) = \kappa(\alpha)k(\alpha_*(\mathbf{P}) \to \alpha_*(\mathbf{E}))$$

The desired result follows since $\kappa(\alpha)$ is a unit. $\qquad \square$

35 $\Omega_1(\mathbf{Z})$ and $\Omega_{-1}(\mathbf{Z})$

The simplest stable modules to study are $\Omega_1(\mathbf{Z})$ and its dual $\Omega_{-1}(\mathbf{Z})$. From the exact sequence

$$0 \to \mathbf{I}(G) \xrightarrow{i} \mathbf{Z}[G] \xrightarrow{\epsilon} \mathbf{Z} \to 0$$

defining $\mathbf{I}(G) = \mathrm{Ker}(\epsilon)$ we see that the stable module $\Omega_1(\mathbf{Z})$ coincides with $[\mathbf{I}(G)]$. Moreover, since $\mathrm{rk}_{\mathbf{Z}}(\mathbf{I}(G)) < \mathrm{rk}_{\mathbf{Z}}(\mathbf{Z}[G])$, we see that $\mathbf{I}(G)$ is a minimal representative of $\Omega_1(\mathbf{Z})$.

Proposition 35.1: Let G be a finite group which has the cancellation property for free modules; then $\mathbf{I}(G)$ is the unique minimal representative of $\Omega_1(\mathbf{Z})$, so that $\Omega_1(\mathbf{Z})$ is straight.

Proof: First note that, if $\eta : \mathbf{Z}[G] \to \mathbf{Z}$ is a surjective $\mathbf{Z}[G]$-homomorphism, then $\eta(1) = \pm 1$, so that $\eta = \pm\epsilon$. In particular, $\mathrm{Ker}(\eta) = \mathbf{I}(G)$.

Suppose that there is a module isomorphism $h : J \oplus \mathbf{Z}[G] \cong I \oplus \mathbf{Z}[G]$, and consider the extension

$$0 \to J \oplus \mathbf{Z}[G] \xrightarrow{j} \mathbf{Z}[G] \oplus \mathbf{Z}[G] \xrightarrow{\eta} \mathbf{Z} \to 0$$

where $j = (i \oplus 1) \circ h$ and where $\eta = \epsilon \circ p$, with $p : \mathbf{Z}[G] \oplus \mathbf{Z}[G] \to \mathbf{Z}[G]$ denoting projection on to the first summand. We may factor out the free summand in $\mathrm{Ker}(\eta)$ to get an exact sequence

$$0 \to J \to P \xrightarrow{\eta_*} \mathbf{Z} \to 0$$

where $P = \mathbf{Z}[G] \oplus \mathbf{Z}[G]/j(0 \oplus \mathbf{Z}[G])$. It follows that P is torsion free over $\mathbf{Z}$. However, there is another exact sequence

$$0 \to \mathbf{Z}[G] \to \mathbf{Z}[G] \oplus \mathbf{Z}[G] \to P \to 0$$

which, since P is torsion free over $\mathbf{Z}$, is an exact sequence in $\mathcal{F}(\mathbf{Z}[G])$. From the relative injectivity of free modules it follows that $P \oplus \mathbf{Z}[G] \cong \mathbf{Z}[G] \oplus \mathbf{Z}[G]$, so that P is stably free over $\mathbf{Z}[G]$. By hypothesis on G, P is free, necessarily having $\mathrm{rk}_{\mathbf{Z}[G]}(P) = 1$. Since $\eta_* : P \to \mathbf{Z}$ is surjective, then, by our initial observation, $J \cong I$, and this completes the proof. $\square$

In Chapter 9, we shall see the above conclusion fails in general when the cancellation property for free modules fails to hold. Since duality $J \mapsto J^*$ defines a 1–1 correspondence $\Omega_1(\mathbf{Z}) \leftrightarrow \Omega_{-1}(\mathbf{Z})$ we see from (35.1) that:

Corollary 35.2: If the finite group G has the cancellation property for free modules, then $\mathbf{I}^*(G)$ is the unique minimal representative of $\Omega_{-1}(\mathbf{Z})$, so that $\Omega_{-1}(\mathbf{Z})$ is straight.

Even without the hypothesis of free cancellation, it is possible to say something. Firstly, the dual augmentation sequence

$$0 \to \mathbf{Z} \stackrel{\epsilon^*}{\to} \mathbf{Z}[G] \stackrel{i^*}{\to} \mathbf{I}^*(G) \to 0$$

is still an exact sequence in $\mathcal{F}(\mathbf{Z}[G])$. Moreover, $\epsilon^*(1)$ is the the central element $\Sigma = \sum_{g \in G} \in \mathbf{Z}[G]$. Thus $\mathrm{Im}(\epsilon^*)$ is the two-sided ideal of $\mathbf{Z}[G]$ generated by Σ, and $\mathbf{I}^*(G) \cong \mathbf{Z}[G]/(\Sigma)$. It follows that:

Proposition 35.3: $\mathbf{I}^*(G)$ has a ring structure with respect to which $i^* : \mathbf{Z}[G] \to \mathbf{I}^*(G)$ is a ring homomorphism.

This has a useful consequence:

Proposition 35.4: $\mathbf{I}^*(G)$ is monogenic as a module over $\mathbf{Z}[G]$, being generated by the image of 1 under i^*.

This extends to give a criterion for recognizing $\mathbf{I}(G)^*$ within its stability class:

Proposition 35.5: Let J be stably equivalent to $\mathbf{I}(G)^*$; then $J \cong_{\mathbf{Z}[G]} \mathbf{I}(G)^*$ if and only if there exists a surjective $\mathbf{Z}[G]$-homomorphism $\varphi : \mathbf{Z}[G] \to J$.

Proof: We first show ($\Longleftarrow$). Assume there exists a surjective $\mathbf{Z}[G]$-homomorphism $\varphi : \mathbf{Z}[G] \to J$ and that J is stably equivalent to $\mathbf{I}(G)^*$. Then J is a minimal representative of the stable module $[\mathbf{I}(G)^*]$. Likewise, $\mathbf{I}(G)^*$ is also a minimal representative of $[\mathbf{I}(G)^*]$. It follows by the Swan–Jacobinski Theorem that

$$J \oplus \mathbf{Z}[G] \cong \mathbf{I}(G)^* \oplus \mathbf{Z}[G]$$

By semisimplicity of $\mathbf{Q}[G]$, it follows that $J \otimes \mathbf{Q} \cong_{\mathbf{Q}[G]} \mathbf{I}^*(G) \otimes \mathbf{Q}$. In particular, $J \otimes \mathbf{Q}$ contains no trivial summand, so that the kernel of

$$\varphi \otimes \mathrm{Id} : \mathbf{Q}[G] \to J \otimes \mathbf{Q}$$

is the trivial one-dimensional module over $\mathbf{Q}[G]$. Hence we have a short exact sequence of $\mathbf{Z}[G]$-modules

$$0 \to \mathbf{Z} \to \mathbf{Z}[G] \stackrel{\varphi}{\to} J \to 0$$

Dualizing, we obtain another short exact sequence

$$0 \to J^* \to \mathbf{Z}[G] \to \mathbf{Z} \to 0$$

As we observed in the proof of (35.1), it follows that $J^* \cong \mathbf{I}(G)$, and hence $J \cong \mathbf{I}(G)^*$. This proves the implication ($\Longleftarrow$); the converse follows immediately from (35.3). $\square$

36 Swan modules

Module extensions of the form $0 \to I(G) \to M \to \mathbf{Z} \to 0$ are classified by $\operatorname{Ext}^1(\mathbf{Z}, I(G)) \cong \operatorname{Ext}^1(\mathbf{Z}, \Omega_1(G)) \cong \mathbf{Z}/|G|$. Here we give explicit models, first introduced by Swan in [58] (see also [63]), for the modules corresponding to the various k-invariants.

Since $\mathbf{I} = \mathbf{I}(G)$ is a submodule of $\mathbf{Z}[G]$, the representation to which it corresponds, denoted by $g \mapsto \lambda(g)$, is simply the restriction of the regular representation of G to $\mathbf{I}$. For each $r \in \mathbf{Z}$ we may construct a representation $g \mapsto \lambda_r(g)$ of G on the abelian group $\mathbf{I}(G) \oplus \mathbf{Z}$

$$\lambda_r(g) = \begin{pmatrix} \lambda(g) & rg - r \\ 0 & 1 \end{pmatrix}$$

We denote by $(\mathbf{I}, r)$ the $\mathbf{Z}[G]$-module whose underlying abelian group is $\mathbf{I}(G) \oplus \mathbf{Z}$ and whose associated representation is λ_r. In passing, we note that, in terms of the standard interpretation of cohomology using the 'bar resolution' ([34], Chapter IV), the function $g \mapsto rg - r$ is a 1-cocycle on the $\mathbf{Z}[G]$-module $\mathbf{Z}$ taking values in $\mathbf{I}(G)$. This observation, however instructive, is not necessary in what follows.

The projection from $(\mathbf{I}, r)$ onto $\mathbf{Z}$ is a $\mathbf{Z}[G]$-homomorphism, and defines an extension

$$\mathcal{E}(r) = (0 \to \mathbf{I} \to (\mathbf{I}, r) \to \mathbf{Z} \to 0)$$

λ_1 is equivalent to the regular representation, so that $\mathcal{E}(1)$ is simply the defining extension of the augmentation ideal

$$\mathcal{E}(1) = (0 \to \mathbf{I} \to \mathbf{Z}[G]) \overset{\epsilon}{\to} \mathbf{Z} \to 0)$$

with the consequence that $(\mathbf{I}, 1)$ is isomorphic to the group ring $\mathbf{Z}[G]$. Consequently we may, and do, take $\mathcal{E}(1)$ as the reference extension in calculating k-invariants. With this convention we have:

Proposition 36.1: $\mathcal{E}(r)$ is the extension with k-invariant given by

$$k(\mathcal{E}(1) \to \mathcal{E}(r)) = r$$

Proof: First note that there is a commutative diagram

$$\begin{array}{ccccccccc} 0 \to & \mathbf{I} & \to & \mathbf{Z}[G] & \overset{\epsilon}{\to} & \mathbf{Z} & \to & 0 \\ & \downarrow \times r & & \downarrow \times r & & \downarrow \times r \\ 0 \to & \mathbf{I} & \to & \mathbf{Z}[G] & \overset{\epsilon}{\to} & \mathbf{Z} & \to & 0 \end{array}$$

which gives us $\kappa(\mathbf{I} \xrightarrow{\times r} \mathbf{I}) = r$ where κ is the unique ring isomorphism $\kappa :$ $\mathrm{End}_{\mathcal{D}er}(\mathbf{I}) \to \mathbf{Z}/|G|$. However, there is a commutative diagram, thus

$$
\begin{array}{cc}
\mathcal{E}(1) \\
\downarrow \varphi_r \quad = \\
\mathcal{E}(r)
\end{array}
\left(
\begin{array}{ccccccc}
0 \to & \mathbf{I} \to & (\mathbf{I}, 1) & \to & \mathbf{Z} \to 0 \\
& \downarrow \times r & \downarrow \varphi_r & & \downarrow \mathrm{Id} \\
0 \to & \mathbf{I} \to & (\mathbf{I}, r) & \to & \mathbf{Z} \to 0
\end{array}
\right)
$$

where, relative to the cocycle description of $\mathcal{E}(r)$, φ_r is described by the matrix

$$
\varphi_r = \begin{pmatrix} r & 0 \\ 0 & 1 \end{pmatrix}
$$

By definition, $k(\mathcal{E}(1) \to \mathcal{E}(r)) = \kappa(\mathbf{I} \xrightarrow{\times r} \mathbf{I})$, giving the value $k(\mathcal{E}(1) \to \mathcal{E}(r)) =$ r by the above. This completes the proof. $\square$

It is useful to have an alternative description of the modules $(\mathbf{I}, r)$. Observe that $\epsilon^{-1}(r\mathbf{Z})$ is a two-sided ideal in $\mathbf{Z}[G]$:

Proposition 36.2: $(\mathbf{I}, r) \cong \epsilon^{-1}(r\mathbf{Z})$

Proof: Let $\eta_r : \epsilon^{-1}(r\mathbf{Z}) \to \mathbf{Z}$ be the homomorphism

$$
\eta_r(x) = \frac{\epsilon(x)}{r}
$$

Then η_r is surjective. Let $\mathcal{F}$ denote the exact sequence

$$
\mathcal{F} = (0 \to \mathbf{I} \to \epsilon^{-1}(r\mathbf{Z}) \xrightarrow{\eta_r} \mathbf{Z} \to 0)
$$

We have a commutative diagram

$$
\begin{array}{cc}
\mathcal{E}(1) \\
\downarrow \quad = \\
\mathcal{F}
\end{array}
\left(
\begin{array}{ccccccc}
0 \to & \mathbf{I} \to & \mathbf{Z}[G] & \xrightarrow{\epsilon} & \mathbf{Z} \to 0 \\
& \downarrow \times r & \downarrow \times r & & \downarrow \mathrm{Id} \\
0 \to & \mathbf{I} \to & \epsilon^{-1}(r\mathbf{Z}) & \xrightarrow{\eta_r} & \mathbf{Z} \to 0
\end{array}
\right)
$$

which shows that $k(\mathcal{E}(1) \to \mathcal{F}) = k(\mathcal{E}(1) \to \mathcal{E}(r)) = r$. It follows that $\mathcal{F} \approx \mathcal{E}(r)$ and so $\epsilon^{-1}(r\mathbf{Z}) \cong (\mathbf{I}, r)$. $\square$

Since $\mathrm{Ext}^1_{\mathbf{Z}[G]}(\mathbf{Z}, \mathbf{I}) \cong \mathbf{Z}/|G|$, the congruence class of $\mathcal{E}(r)$ is entirely determined by the residue class of r $(\mathrm{mod}|G|)$; that is:

Proposition 36.3: $\mathcal{E}(r) \approx \mathcal{E}(r') \Longleftrightarrow r \equiv r' \bmod |G|$.

Proof: We describe an explicit congruence

$$
\psi_m : \mathcal{E}(r) \xrightarrow{\approx} \mathcal{E}(r + m|G|)
$$

by means of

$$\psi_m = \begin{pmatrix} \text{Id} & m(\Sigma - |G|) \\ 0 & \text{Id} \end{pmatrix} \qquad \qquad \square$$

37 Swan's isomorphism criterion

In Section 36, we gave a complete description of the congruence classes of Swan modules. We now turn to the more general problem of deciding when two Swan modules are isomorphic.

Since $\mathbf{I} = \mathbf{I}(G) \in \Omega_1(\mathbf{Z})$, it follows quite generally that $\text{End}_{\mathcal{D}\text{er}}(\mathbf{I}) \cong \text{End}_{\mathcal{D}\text{er}}(\mathbf{Z})$ and there is a natural surjective ring homomorphism

$$\nu^{\mathbf{I}} : \text{End}_{\mathbf{Z}[G]}(\mathbf{I}) \to \mathbf{Z}/|G|$$

We show how the monogenicity of $\mathbf{I}^*$ allows for a direct calculation of $\nu^{\mathbf{I}}$. First consider the 'reduced norm' factorization in our context.

Over any ring Λ, a short exact sequence of Λ-modules

$$\mathcal{F} = \left(0 \to K \overset{j}{\to} M \overset{p}{\to} Q \to 0\right)$$

is said to be *fully invariant* when each Λ-endomorphism $\varphi : M \to M$ preserves the exact sequence; that is, when each $\varphi \in \text{End}_\Lambda(M)$ determines a pair

$$(\varphi_K, \varphi_Q) \in \text{End}_\Lambda(K) \times \text{End}_\Lambda(Q)$$

making the following commute

$$0 \to K \overset{j}{\to} M \overset{p}{\to} Q \to 0$$
$$\downarrow \varphi_K \quad \downarrow \varphi \quad \downarrow \varphi_Q$$
$$0 \to K \overset{j}{\to} M \overset{p}{\to} Q \to 0$$

Proposition 37.1: The extensions $\mathcal{E}(1)$ and $\mathcal{E}(1)^*$ are fully invariant.

Proof: If $u \in \mathbf{Z}[G]$, we denote by $\lambda_u : \mathbf{Z}[G] \to \mathbf{Z}[G]$ the $\mathbf{Z}[G]$-endomorphism given by

$$\lambda_u(\mathbf{x}) = u\mathbf{x}$$

Clearly every $\mathbf{Z}[G]$ endomorphism of $\mathbf{Z}[G]$ has this form. However ϵ is a ring homomorphism, so that, if $\mathbf{x} \in \mathbf{I}(G)$, then

$$\epsilon(\lambda_u(\mathbf{x})) = \epsilon(\lambda_u)\epsilon(\mathbf{x}) = 0$$

and so $\lambda_u(\mathbf{I}) \subset \mathbf{I}$. In particular, the following diagram commutes

$$0 \to \mathbf{I}(G) \xrightarrow{\infty} \mathbf{Z}[G] \xrightarrow{\epsilon} \mathbf{Z} \to 0$$
$$\downarrow \lambda_{u|\mathbf{I}} \quad\quad \downarrow \lambda_u \quad\quad \downarrow \epsilon(u)$$
$$0 \to \mathbf{I}(G) \xrightarrow{\infty} \mathbf{Z}[G] \xrightarrow{\epsilon} \mathbf{Z} \to 0$$

that is, $\mathcal{E}(1)$ is fully invariant. Full invariance of $\mathcal{E}(1)^*$ follows by duality. □

Any endomorphism λ of $\mathbf{Z}[G]$ has the form $\lambda = \lambda_u$, and we write $\det(u) = \det_{\mathbf{Z}}(\lambda_u : \mathbf{Z}[G] \to \mathbf{Z}[G])$. Put

$$N(u) = \det(\lambda_{u|\mathbf{I}})$$

By duality, $N(u)$ is also the determinant of the $\mathbf{Z}$-linear mapping $\mathbf{I}^* \to \mathbf{I}^*$ induced by λ_u on $\mathbf{I}^*$. Full invariance of $\mathcal{E}(1)$ or $\mathcal{E}(1)^*$ shows that

(37.2) $$\det(u) = \epsilon(u)N(u)$$

To proceed, it is technically simpler to deal with $\mathcal{E}(1)^*$. By (35.4), the dual augmentation module $\mathbf{I}^*(G)$ is monogenic as a $\mathbf{Z}[G]$-module; it follows that every $\mathbf{Z}[G]$-endomorphism $f : \mathbf{I}^*(G) \to \mathbf{I}^*(G)$ has the form $f = \mu_u$ for some $u \in \mathbf{Z}[G]$; again by duality:

Proposition 37.3: Every $\mathbf{Z}[G]$-endomorphism $\varphi : \mathbf{I}(G) \to \mathbf{I}(G)$ has a representation in the form $\varphi = \lambda_{u|\mathbf{I}}$ for some $u \in \mathbf{Z}[G]$.

This representation need not be unique, however. Consider first the problem of representing the zero homomorphism $0 : \mathbf{I}(G) \to \mathbf{I}(G)$. On dualizing, we have a commutative diagram

$$0 \to \mathbf{Z} \xrightarrow{\epsilon} \mathbf{Z}[G] \xrightarrow{i^*} \mathbf{I}^*(G) \to 0$$
$$\downarrow \epsilon(u) \quad\quad \downarrow \lambda_u \quad\quad \downarrow 0$$
$$0 \to \mathbf{Z} \xrightarrow{\epsilon} \mathbf{Z}[G] \xrightarrow{i^*} \mathbf{I}^*(G) \to 0$$

Since the map at the right-hand end is zero, it follows that the $\mathbf{Z}[G]$-endomorphism $\epsilon(u) : \mathbf{Z} \to \mathbf{Z}$ factors through the free module $\mathbf{Z}[G]$. By (33.1), $\epsilon(u) = n|G|$ for some integer n. It follows that if $u_1, u_2 \in \mathbf{Z}[G]$ represent the same endomorphism of $\mathbf{I}(G)$ then

$$\epsilon(u_1) - \epsilon(u_2) \equiv 0 \bmod(|G|)$$

If $[\ ] : \mathbf{Z} \to \mathbf{Z}/|G|$ denotes the natural epimorphism, a chase of definitions shows that:

Proposition 37.4: If $u \in \mathbf{Z}[G]$ is any element such that $\varphi = \lambda_{u|_I}$, then

$$\nu^{\mathbf{I}}(\varphi) = [\epsilon(u)]$$

If $\varphi = \lambda_{u|\mathbf{I}}$ belongs to the unit group of $\mathrm{End}_{\mathbf{Z}[G]}(\mathbf{I}(G))$, then $\epsilon(u) = \nu^{\mathbf{I}}(\varphi)$ is a unit in $\mathbf{Z}/|G|$, so that:

Corollary 37.5: If $u \in \mathbf{Z}[G]$ satisfies $N(u) = \pm 1$, then $\epsilon(u)$ is coprime to $|G|$.

Proposition 37.6: Let $u \in \mathbf{Z}[G]$ be such that $N(u) = \pm 1$; then λ_u defines an *isomorphism* of extensions

$$\lambda_u : \mathcal{E}(r) \to \mathcal{E}(\epsilon(u)r)$$

Proof: Put $s = \epsilon(u)$. By (37.5), s is coprime to $|G|$. Moreover, one may check that the following diagram commutes

$$
\begin{array}{ccccccccc}
0 & \to & \mathbf{I} & \to & (\mathbf{I}, r) & \overset{\overset{\epsilon}{r}}{\to} & \mathbf{Z} & \to & 0 \\
 & & \downarrow \lambda_{u|_I} & & \downarrow \lambda_u & & \downarrow \mathrm{Id} & & \\
0 & \to & \mathbf{I} & \to & (\mathbf{I}, rs) & \overset{\overset{\epsilon}{rs}}{\to} & \mathbf{Z} & \to & 0
\end{array}
$$

Now $\lambda_u : \mathbf{I} \to \mathbf{I}$ is an isomorphism by the hypothesis $N(u) = \pm 1$, so the result follows from the Five Lemma. $\qquad\square$

We put $\Gamma(G) = \mathrm{Im}(\nu^{\mathbf{I}}) \subset (\mathbf{Z}/|G|)^*$. Observe that $\Gamma(G)$ acts on $(\mathbf{Z}/|G|)^*$ by left translation

$$\Gamma(G) \times (\mathbf{Z}/|G|)^* \to (\mathbf{Z}/|G|)^*$$

$$([r], \qquad [s]) \quad \mapsto \quad [rs]$$

Theorem 37.7: Let $[r], [s] \in \mathbf{Z}/|G|$; the extensions $\mathcal{E}(r)$, $\mathcal{E}(s)$ are isomorphic if and only if there exists $\eta \in \Gamma(G)$ such that $r = \eta s$; that is, if and only if $[r], [s]$ belong to same orbit of $\Gamma(G)$ in $\mathbf{Z}/|G|$.

Proof: ($\Longrightarrow$) Let $\varphi : \mathcal{E}(r) \to \mathcal{E}(s)$ be an isomorphism of extensions

$$
\begin{pmatrix}
0 & \to & \mathbf{I} & \to & (\mathbf{I}, r) & \overset{\overset{\epsilon}{r}}{\to} & \mathbf{Z} & \to & 0 \\
 & & \downarrow \varphi_0 & & \downarrow \varphi & & \downarrow \mathrm{Id} & & \\
0 & \to & \mathbf{I} & \to & (\mathbf{I}, s) & \overset{\overset{\epsilon}{s}}{\to} & \mathbf{Z} & \to & 0
\end{pmatrix}
$$

Choose $u \in \mathbf{Z}[G]$ such that $\lambda_{u \mid I} = \varphi_0^{-1}$ and consider the following diagram

$$
\begin{pmatrix}
0 \to \mathbf{I} \to & (\mathbf{I}, r) \xrightarrow{\frac{\epsilon}{r}} \mathbf{Z} \to 0 \\
\downarrow \varphi_0 \quad \downarrow \varphi \quad \downarrow \mathrm{Id} \\
0 \to \mathbf{I} \to & (\mathbf{I}, s) \xrightarrow{\frac{\epsilon}{s}} \mathbf{Z} \to 0 \\
\downarrow \varphi_0^{-1} \quad \downarrow \lambda_u \quad \downarrow \mathrm{Id} \\
0 \to \mathbf{I} \to & (\mathbf{I}, \eta s) \xrightarrow{\frac{\epsilon}{\eta s}} \mathbf{Z} \to 0
\end{pmatrix}
$$

where $\eta = \epsilon(u)$. Composing the down arrows, we obtain a congruence

$$
\begin{pmatrix}
0 \to \mathbf{I} \to & (\mathbf{I}, r) \xrightarrow{\frac{\epsilon}{r}} \mathbf{Z} \to 0 \\
\downarrow \mathrm{Id} \quad \downarrow \lambda_u \varphi \quad \downarrow \mathrm{Id} \\
0 \to \mathbf{I} \to & (\mathbf{I}, \eta s) \xrightarrow{\frac{\epsilon}{\eta s}} \mathbf{Z} \to 0
\end{pmatrix}
$$

It follows immediately that $\eta s \equiv r \bmod |G|$ and proves ($\Longrightarrow$). However, the converse is just (37.6), so the proof is complete. $\square$

As a consequence, we get the following theorem of Swan which completely describes the isomorphism types of the *modules* $(\mathbf{I}, s)$. In fact, Swan proves the dual result (see (37.11) and (37.12) below).

Theorem 37.8: For $[r], [s] \in \mathbf{Z}/|G|$, $(\mathbf{I}, r) \cong (\mathbf{I}, s)$ if and only if there exists $\eta \in \Gamma(G)$ such that $[r] = \eta[s]$.

Proof: Let $\mathbf{Q}$ denote the trivial $\mathbf{Q}$-module of dimension 1; then $\mathbf{Q}$ occurs in $\mathbf{Q}[G]$ with multiplicity 1. In particular, it does not occur in the rational augmentation ideal $\mathbf{I}_\mathbf{Q}$. Consequently, any $\mathbf{Q}[G]$-homomorphism $\psi : \mathbf{I}_\mathbf{Q} \to \mathbf{Q}$ must be trivial, a fortiori, any $\mathbf{Z}[G]$-homomorphism $\psi : \mathbf{I} \to \mathbf{Z}$ is also trivial.

Suppose that $\varphi : (\mathbf{I}, r) \to (\mathbf{I}, s)$ is an isomorphism; it follows from the above remarks that the restriction of $\frac{\epsilon}{s} \circ \varphi$ to $\mathbf{I}$ is trivial. Hence the following diagram commutes

$$
\begin{pmatrix}
0 \to \mathbf{I} \to (\mathbf{I}, r) \xrightarrow{\frac{\epsilon}{r}} \mathbf{Z} \to 0 \\
\downarrow \varphi \quad \downarrow \varphi \quad \downarrow \pm \mathrm{Id} \\
0 \to \mathbf{I} \to (\mathbf{I}, s) \xrightarrow{\frac{\epsilon}{s}} \mathbf{Z} \to 0
\end{pmatrix}
$$

and φ defines an isomorphism of extensions $\varphi : \mathcal{E}(r) \to \mathcal{E}(s)$. The conclusion follows from (37.7). $\square$

Let r, s be positive integers; we observed in (36.2) above that $(\mathbf{I}, r) = \epsilon^{-1}(r\mathbf{Z})$. Let $\pi_s : \mathbf{Z} \to \mathbf{Z}/s$ denote the canonical surjection, and let $\eta_{r,s} : (\mathbf{I}, r) \to \mathbf{Z}/s$ denote the composite

$$\eta_{r,s} = \pi_s \circ \frac{\epsilon}{r}$$

We see that for any r, s there is an exact sequence

$$0 \to (\mathbf{I}, rs) \to (\mathbf{I}, r) \xrightarrow{\eta_{r,s}} \mathbf{Z}/s \to 0$$

As a consequence, we have:

Theorem 37.9: If r is coprime to $|G|$, then, for any positive integer s, there is an isomorphism:

$$(\mathbf{I}, rs) \oplus \mathbf{Z}[G] \cong (\mathbf{I}, r) \oplus (\mathbf{I}, s)$$

Proof: As a special case of (37.8) we get an exact sequence

$$0 \to (\mathbf{I}, s) \to (\mathbf{I}, 1) \xrightarrow{\eta_{1,s}} \mathbf{Z}/s \to 0$$

whilst the general case of (37.8) gives the exact sequence

$$0 \to (\mathbf{I}, rs) \to (\mathbf{I}, r) \xrightarrow{\eta_{r,s}} \mathbf{Z}/s \to 0$$

$(\mathbf{I}, 1)$ is free, hence projective; since r is coprime to $|G|$, $(\mathbf{I}, r)$ is also projective. Applying Schanuel's Lemma to the above pair of exact sequences gives an isomorphism

$$(\mathbf{I}, rs) \oplus (\mathbf{I}, 1) \cong (\mathbf{I}, r) \oplus (\mathbf{I}, s)$$

The result follows since $(\mathbf{I}, 1) \cong \mathbf{Z}[G]$. □

We consider briefly the corresponding results for $\mathbf{I}^*(G)$ under duality. Since $\mathbf{Z}/|G|$ is commutative, $v^{\mathbf{I}}$ can equally well be regarded as a ring homomorphism

$$v^{\mathbf{I}} : \mathrm{End}_{\mathbf{Z}[G]}(\mathbf{I})^{opp} \to \mathbf{Z}/|G|$$

However, the correspondence $\alpha \mapsto \alpha^*$ induces an isomorphism of rings

$$\mathrm{End}_{\mathbf{Z}[G]}(\mathbf{I})^{opp} \cong \mathrm{End}_{\mathbf{Z}[G]}(\mathbf{I}^*)$$

so that there is a surjective ring homomorphism $v^* : \mathrm{End}_{\mathbf{Z}[G]}(\mathbf{I}^*) \to \mathbf{Z}/|G|$ defined (implicitly) by the formula

$$v^*(\alpha^*) = v^{\mathbf{I}}(\alpha)$$

A straightforward chase of definitions shows that

(37.10) $\tilde{\nu} = \nu^*$

Since $\mathbf{I}^* = \mathbf{I}^*(G)$ has a natural ring structure there is a further simplification; the epimorphism $\mu : \mathbf{Z}[G] \to \mathrm{End}_{\mathbf{Z}[G]}(\mathbf{I}^*); u \mapsto \mu_u$ induces a natural ring isomorphism $\mathbf{I}^* \cong \mathrm{End}_{\mathbf{Z}[G]}(\mathbf{I}^*)$, and hence an isomorphism of groups

$$U(\mathbf{I}^*) \xrightarrow{\cong} \mathrm{Aut}_{\mathbf{Z}[G]}(\mathbf{I}^*)$$

Composition with ν^* gives a homomorphism of groups $\eta : U(\mathbf{I}^*) \to (\mathbf{Z}/|G|)^*$, and a chase of definitions shows that:

Proposition 37.11: Let $a = i^*(v) \in U(\mathbf{I}^*(G))$ where $v \in \mathbf{Z}[G]$; $\eta(a) \in \mathbf{Z}/|G|$ is computed as

$$\eta(a) = [\epsilon(v)]$$

Swan's own expression of the above isomorphism criterion ([58], Lemma (6.3)) is:

(37.12) $(\mathbf{I}, r) \cong_{\mathbf{Z}[G]} (\mathbf{I}, r') \Leftrightarrow r = \eta(a)r'$ for some unit $a \in \mathbf{I}^*(G)$.

In particular:

(37.13) $(\mathbf{I}, r)$ is free $\Longleftrightarrow$ there exists a unit $a \in \mathbf{I}^*(G)$ such that $r = \eta(a)$.

38 Congruence classes and Swan modules

We begin by recalling a basic fact about the pushout construction. This is implicit in the results of Chapter 4, but, for clarity, we spell out the details in a particular case.

Lemma 38.1: Let $\mathbf{E}, \mathbf{F} \in \mathrm{Ext}^n(\mathbf{Z}, J)$, and let $\alpha : J \to J$ be a $\mathbf{Z}[G]$-homomorphism. If there is an elementary congruence $h : \mathbf{E} \to \mathbf{F}$, then there is an elementary congruence $h' : \alpha_*(\mathbf{E}) \to \alpha_*(\mathbf{F})$.

Proof: Decompose $\mathbf{E}$ as a Yoneda product $\mathbf{E} = \mathbf{E}_+ \circ \mathbf{E}_-$ where

$$\mathbf{E}_+ = (0 \to J \xrightarrow{i} E_{n-1} \to K \to 0);$$
$$\mathbf{E}_- = (0 \to K \to E_{n-2} \to \cdots \to E_0 \to \mathbf{Z} \to 0)$$

Likewise, decompose $\mathbf{F} = \mathbf{F}_+ \circ \mathbf{F}_-$ and suppose that $h : \mathbf{E} \to \mathbf{F}$ is an elementary congruence, thus

$$
\begin{array}{c} \mathbf{E} \\ \downarrow h = \\ \mathbf{F} \end{array}
\begin{pmatrix}
0 \to J \to E_{n-1} \to \cdots \to E_0 \to \mathbf{Z} \to 0 \\
\quad\ \downarrow \mathrm{Id}\ \ \downarrow h_{n-1} \qquad\qquad \downarrow h_0\ \ \downarrow \mathrm{Id} \\
0 \to J \to F_{n-1} \to \cdots \to F_0 \to \mathbf{Z} \to 0
\end{pmatrix}
$$

We may split h under Yoneda product

$$
\begin{array}{c} \mathbf{E}_+ \\ \downarrow h_+ = \\ \mathbf{F}_+ \end{array}
\begin{pmatrix}
0 \to \ J \to\ E_{n-1} \to K \to 0 \\
\quad\ \downarrow \mathrm{Id}\ \ \ \downarrow h_{n-1}\ \ \downarrow \eta \\
0 \to \ J \to\ F_{n-1} \to L \to 0
\end{pmatrix}
$$

$$
\begin{array}{c} \mathbf{E}_- \\ \downarrow h_- = \\ \mathbf{F}_- \end{array}
\begin{pmatrix}
0 \to K \to E_{n-2} \to \cdots \to E_0 \to \mathbf{Z} \to 0 \\
\quad\ \downarrow \eta\ \ \downarrow h_{n-2} \qquad\qquad \downarrow h_0\ \ \downarrow \mathrm{Id} \\
0 \to L \to F_{n-1} \to \cdots \to F_0 \to \mathbf{Z} \to 0
\end{pmatrix}
$$

By general properties of pushouts, any morphism $\alpha : J \to J$ induces a morphism on pushouts as below

$$
\begin{array}{c} \alpha_*(\mathbf{E}_+) \\ \downarrow h'_+ = \\ \alpha_*(\mathbf{F}_+) \end{array}
\begin{pmatrix}
0 \to \ J \to\ E'_{n-1} \to K \to 0 \\
\quad\ \downarrow \mathrm{Id}\ \ \ \downarrow h'_{n-1}\ \ \downarrow \eta \\
0 \to \ J \to\ F'_{n-1} \to L \to 0
\end{pmatrix}
$$

where $E'_{n-1} = (J \oplus E_{n-1})/(\alpha \times -i)$, $F'_{n-1} = (J \oplus F_{n-1})/(\alpha \times -j)$ and h'_{n-1} is the map induced by the matrix

$$
\begin{pmatrix} \mathrm{Id} & 0 \\ 0 & h_{n-1} \end{pmatrix}
$$

By definition, $\alpha_*(\mathbf{E}) = \alpha_*(\mathbf{E}_+) \circ \mathbf{E}_-$, and $\alpha_*(\mathbf{F}) = \alpha_*(\mathbf{F}_+) \circ \mathbf{F}_-$. Since the right-hand end map of h'_+ is still η, and the left-hand end map is still the identity on J, h'_+ and h_- glue together to give the required elementary congruence

$$
\begin{array}{c} \alpha_*(\mathbf{E}) \\ \downarrow h' = \\ \alpha_*(\mathbf{F}) \end{array}
\begin{pmatrix}
0 \to J \to E'_{n-1} \to \cdots \to E_0 \to \mathbf{Z} \to 0 \\
\quad\ \downarrow \mathrm{Id}\ \ \downarrow h'_{n-1} \qquad\qquad \downarrow h_0\ \ \downarrow \mathrm{Id} \\
0 \to J \to F'_{n-1} \to \cdots \to F_0 \to \mathbf{Z} \to 0
\end{pmatrix} \qquad \square
$$

Suppose that $J \in D_n(\mathbf{Z})$; by (26.1), the set $\mathbf{Proj}^n(\mathbf{Z}, J)$ of projective n-stems

$$
(0 \to J \to P_{n-1} \to \cdots \to P_0 \to \mathbf{Z} \to 0)
$$

is nonempty. We wish to find a convenient parametrization of the set of congruence classes in $\mathbf{Proj}^n(\mathbf{Z}, J)$. First observe that $J \in \mathbf{D}_{n-1}(\mathbf{I}(G))$, so that, by (26.1), we may choose a projective $(n - 1)$-stem $\mathcal{Q} \in \mathbf{Proj}^{n-1}(\mathbf{I}(G), J)$ which, by re-indexing, we may write for convenience, thus

$$\mathcal{Q} = (0 \to J \to Q_{n-1} \to \cdots \to Q_1 \overset{\mu}{\to} \mathbf{I}(G) \to 0$$

The Yoneda product $\mathcal{Q} \circ \mathcal{E}(t)$ takes the form

$$\mathcal{Q} \circ \mathcal{E}(t) = \big(0 \to J \to Q_{n-1} \to \cdots \to Q_1 \overset{i \circ \mu}{\to} (I, t) \overset{\eta_t}{\to} \mathbf{Z} \to 0\big)$$

From a straightforward chase of definitions we get

(38.2) $\qquad k(\mathcal{Q} \circ \mathcal{E}(1) \to \mathcal{Q} \circ \mathcal{E}(t)) = [t] \in (\mathbf{Z}/|G|)^*$

Proposition 38.3: Let $J \in \mathbf{D}_n(\mathbf{Z})$ where $n \geq 2$, and let $\mathcal{Q}$ be *any* projective $(n - 1)$-stem in $\mathbf{Proj}^{n-1}(\mathbf{I}(G), J)$; then:

(i) for $\mathcal{P} \in \mathbf{Proj}^n(\mathbf{Z}, J)$ there exists a unique $t \in (\mathbf{Z}/|G|)^*$ such that $\mathcal{P} \approx \mathcal{Q} \circ \mathcal{E}(t)$; furthermore

(ii) if $\alpha : J \to J$ is any $\mathbf{Z}[G]$-homomorphism, then $\alpha_*(\mathcal{P}) \approx \mathcal{Q} \circ \mathcal{E}(t\kappa(\alpha))$.

Proof: To prove (i) choose $\mathcal{Q} \in \mathbf{Proj}^{n-1}(\mathbf{I}(G), J)$ and put $[1] = \mathcal{Q} \circ \mathcal{E}(1) \in \mathbf{Proj}^n(\mathbf{Z}, J)$. Suppose $k([1] \to \mathcal{P}) = t \in (\mathbf{Z}/|G|)^*$, then by (38.2), $t = k([1] \to \mathcal{Q} \circ \mathcal{E}(t))$, so that $\mathcal{P} \approx \mathcal{Q} \circ \mathcal{E}(t)$. Uniqueness is clear.

To prove (ii), let $\nu_\alpha : \mathcal{P} \to \alpha_*(\mathcal{P})$ be the canonical morphism associated with the pushout construction; ν_α takes the form

$$\begin{matrix} \mathcal{P} \\ \downarrow \nu_\alpha = \\ \alpha_*(\mathcal{P}) \end{matrix} \begin{pmatrix} 0 \to J \to P_{n-1} \to \cdots \to P_0 \to \mathbf{Z} \to 0 \\ \quad \downarrow \alpha \quad \downarrow \nu_{n-1} \qquad \downarrow \nu_0 \downarrow \mathrm{Id} \\ 0 \to J \to P'_{n-1} \to \cdots \to P_0 \to \mathbf{Z} \to 0 \end{pmatrix}$$

and tautologously $k(\mathcal{P} \to \alpha_*(\mathcal{P})) = \kappa(\alpha)$. We calculate the k-invariant of the transition $k([1] \to \alpha_*(\mathcal{P}))$ thus

$$k([1] \to \alpha_*(\mathcal{P})) = k([1] \to \mathcal{P})k(\mathcal{P} \to \alpha_*(\mathcal{P})) = [t]\kappa(\alpha)$$

However, $[t]\kappa(\alpha) = k([1] \to \mathcal{Q} \circ \mathcal{E}(t\kappa(\alpha))$, so that $\alpha_*(\mathcal{P}) \approx \mathcal{Q} \circ \mathcal{E}(t\kappa(\alpha))$, completing the proof. $\qquad \square$

39 The Tate ring

Let G be a finite group and let

$$\mathcal{P} = \left(\cdots \to P_{n+1} \overset{\partial_{n+1}}{\to} P_n \overset{\partial_n}{\to} P_{n-1} \overset{\partial_{n-1}}{\to} \cdots \overset{\partial_1}{\to} P_0 \overset{\epsilon}{\to} \mathbf{Z} \to 0\right)$$

be a resolution of $\mathbf{Z}$ by finite generated projectives P_i. Then the dual sequence

$$\mathcal{P}^* = \left(0 \to \mathbf{Z} \xrightarrow{\epsilon^*} P_0^* \xrightarrow{\partial_1^*} P_1 \xrightarrow{\partial_2^*} \cdots \xrightarrow{\partial_n^*} P_n^* \xrightarrow{\partial_{n+1}^*} \to \cdots\right)$$

is also exact. For convenience, we write

$$\mathcal{P}^* = \left(0 \to \mathbf{Z} \xrightarrow{\epsilon^*} P_{-1} \xrightarrow{\partial_{-1}} P_{-2} \xrightarrow{\partial_{-2}} \cdots \xrightarrow{\partial_n^*} P_{-n} \xrightarrow{\partial_{-n}} \to \cdots\right)$$

where $P_{-n} = P_{n-1}^*$ and $\partial_{-n} = \partial_n^*$. Splicing $\mathcal{P}$ and $\mathcal{P}^*$ together gives a so-called 'complete resolution'

$$\widehat{\mathcal{P}} = \left(\cdots \to P_n \xrightarrow{\partial_n} P_{n-1} \xrightarrow{\partial_{n-1}} \cdots \to P_1 \xrightarrow{\partial_1} P_0 \xrightarrow{\epsilon^*\epsilon} P_{-1} \to \cdots\right)$$

If A is a $\mathbf{Z}[G]$-module, the Tate cohomology groups $\hat{H}^k(G, A)$ are defined by

$$\hat{H}^n(G, A) = \mathrm{Ker}\left(\partial_{n+1}^A\right)/\mathrm{Im}(\partial_n^A)$$

where, as usual, $\partial_n^A : \mathrm{Hom}_\Lambda(P_{n-1}, A) \to \mathrm{Hom}_\Lambda(P_n, A)$ is the induced map $\partial_n^A(\alpha) = \alpha \circ \partial_n$. When A is a $\mathbf{Z}[G]$-lattice, we have:

Theorem 39.1: For all $n \in \mathbf{Z}$

$$\hat{H}^n(G, A) \cong \mathrm{Hom}_{\mathcal{D}er}(\mathbf{D}_n(\mathbf{Z}), A)$$

Proof: By (25.7), it suffices to show that

$$\hat{H}^{-n}(G, A) \cong \mathrm{Hom}_{\mathcal{D}er}(\mathbf{D}_{-n}(\mathbf{Z}), A)$$

when $n > 0$. Take a truncated resolution

$$\mathcal{P} = \left(\cdots \to J_N \to P_{N-1} \xrightarrow{\partial_{N-1}} P_{N-2} \xrightarrow{\partial_{N-2}} \cdots \xrightarrow{\partial_1} P_0 \xrightarrow{\epsilon} \mathbf{Z} \to 0\right)$$

and its (re-indexed) dual

$$\mathcal{P}^* = \left(0 \to \mathbf{Z} \xrightarrow{\epsilon^*} P_{-1} \xrightarrow{\partial_{-1}} P_{-2} \xrightarrow{\partial_{-2}} \cdots \xrightarrow{\partial_{-(N-1)}^*} P_{-N} \to J_N^* \to 0\right)$$

Here $J_N \in \mathcal{F}(\mathbf{Z}[G])$ represents $\mathbf{D}_N(\mathbf{Z})$, and so $\mathbf{D}_{-N}(\mathbf{Z})$ is represented by J_N^*. Assume that $n < N$; then $\hat{H}^{-n}(G, A) \cong \mathrm{Ext}^{N-n}(J_N^*, A) \cong \mathrm{Hom}_{\mathcal{D}er}(\mathbf{D}_{N-n}(J_N^*), A)$, and hence $\hat{H}^{-n}(G, A) \cong \mathrm{Hom}_{\mathcal{D}er}(\mathbf{D}_{N-n}\mathbf{D}_{-N}(\mathbf{Z}), A) \cong \mathrm{Hom}_{\mathcal{D}er}(\mathbf{D}_{-n}(\mathbf{Z}), A)$ as claimed. $\square$

We now specialize to the case $A = \mathbf{Z}$. Then, for all $m, n \in \mathbf{Z}$, there are dimension shifting isomorphisms (depending on the choice of the complete resolution $\widehat{\mathcal{P}}$)

$$\mathrm{Hom}_{\mathcal{D}er}(\mathbf{D}_n(\mathbf{Z}), \mathbf{Z}) \cong \mathrm{Hom}_{\mathcal{D}er}(\mathbf{D}_{n+m}(\mathbf{Z}), \mathbf{D}_m(\mathbf{Z}))$$

Moreover, composition $(f, g) \mapsto f \circ g$ gives bilinear maps for any $a, b, c \in \mathbf{Z}$.

$$\circ : \mathrm{Hom}_{\mathcal{D}\mathrm{er}}(\mathbf{D}_b(\mathbf{Z}), \mathbf{D}_c(\mathbf{Z})) \times \mathrm{Hom}_{\mathcal{D}\mathrm{er}}(\mathbf{D}_a(\mathbf{Z}), \mathbf{D}_b(\mathbf{Z})) \to \mathrm{Hom}_{\mathcal{D}\mathrm{er}}(\mathbf{D}_a(\mathbf{Z}), \mathbf{D}_c(\mathbf{Z}))$$

Re-interpreting the Tate groups $\hat{\mathrm{H}}^k(G, \mathbf{Z})$ via the dimension shifting isomorphisms above, we get

$$\hat{\mathrm{H}}^n(G, \mathbf{Z}) \cong \mathrm{Hom}_{\mathcal{D}\mathrm{er}}(\mathbf{D}_{n+m}(\mathbf{Z}), \mathbf{D}_m(\mathbf{Z}))$$

and hence bilinear pairings

$$\circ : \hat{\mathrm{H}}^k(G, \mathbf{Z}) \times \hat{\mathrm{H}}^l(G, \mathbf{Z}) \to \hat{\mathrm{H}}^{k+l}(G, \mathbf{Z})$$

It is known that the product is anti-commutative; that is

$$x_2 x_1 = (-1)^{k_1 k_2} x_1 x_2$$

for $x_i \in \hat{\mathrm{H}}^{k_i}(G, \mathbf{Z})$ It follows from standard universal coefficient theorems that

$$\hat{\mathrm{H}}^{-k}(G, \mathbf{Z}) = \mathrm{Hom}_{\mathrm{Ab}}(\hat{\mathrm{H}}^k(G, \mathbf{Z}); \mathbf{Z}/|G|)$$

Moreover the pairing

$$\hat{\mathrm{H}}^{-k}(G, \mathbf{Z}) \times \hat{\mathrm{H}}^k(G, \mathbf{Z}) \to \hat{\mathrm{H}}^0(G, \mathbf{Z}) = \mathbf{Z}/|G|$$

is equivalent to the evaluation pairing

$$\mathrm{Hom}_{\mathrm{Ab}}(\hat{\mathrm{H}}^k(G, \mathbf{Z}), \mathbf{Z}/|G|) \times \hat{\mathrm{H}}^k(G, \mathbf{Z}) \to \mathbf{Z}/|G|$$

Finally, for any $\mathbf{Z}[G]$-lattice A, the Tate groups $\hat{\mathrm{H}}^*(G, A)$ possess the structure of a graded module over the Tate ring. In fact, writing $\mathbf{D}_n = \mathbf{D}_n(\mathbf{Z})$, composition

$$\mathrm{Hom}_{\mathcal{D}\mathrm{er}}(\mathbf{D}_p, A) \times \mathrm{Hom}_{\mathcal{D}\mathrm{er}}(\mathbf{D}_{p+q}, \mathbf{D}_p) \to \mathrm{Hom}_{\mathcal{D}\mathrm{er}}(\mathbf{D}_{p+q}, A)$$

becomes the pairing

$$\hat{\mathrm{H}}^p(G, A) \times \hat{\mathrm{H}}^q(G, \mathbf{Z}) \to \hat{\mathrm{H}}^{p+q}(G, A)$$

defining the action of $\hat{\mathrm{H}}^*(G, \mathbf{Z})$ on $\hat{\mathrm{H}}^*(G, A)$.

Chapter 7
Groups of periodic cohomology

For a finite group G there are *a priori* two possibilities; within the derived category of $\mathbf{Z}[G]$: *either*

  (i) the derived modules $(D_n(\mathbf{Z}))_{n \in \mathbf{Z}}$ are isomorphically distinct *or*
 (ii) $D_n(\mathbf{Z}) \cong D_m(\mathbf{Z})$ for some $m, n \in \mathbf{Z}$ with $m \neq n$.

In this chapter, we show that the categorization of finite groups within the two types depends only upon the Sylow subgroup structure. In particular, 'most' finite groups are in case (i), and G is of type (ii) when for each odd prime p, the Sylow p-subgroup is cyclic and the Sylow 2-subgroup is either cyclic or generalized quaternion.

Zassenhaus [84] has classified all finite soluble groups satisfying these restrictions, and his classification has been extended by Suzuki [56] to include nonsoluble groups. In a final section we give a brief outline of their results.

40 Periodicity conditions

We say that $n > 0$ is a *cohomological period* of G if $D_{n+k}(\mathbf{Z}) \cong D_k(\mathbf{Z})$ for all $k \in \mathbf{Z}$. There are a number of ways of characterizing this possibility. Consider the following conditions which can be placed on a finite group G:

$\mathcal{P}_1(n)$: $D_{n+k}(\mathbf{Z}) \cong D_k(\mathbf{Z})$ for all integers k;

$\mathcal{P}_2(n)$: $D_{n+k}(\mathbf{Z}) \cong D_k(\mathbf{Z})$ for at least one nonzero integer k;

$\mathcal{P}_3(n)$: There exists an exact sequence in $\mathcal{F}(\mathbf{Z}[G])$ of the form

$$0 \to \mathbf{Z} \to P_{n-1} \to \cdots \to P_0 \to \mathbf{Z} \to 0$$

where each P_i is finitely generated projective.

$\mathcal{P}_4(n) : \mathrm{H}^n(G; \mathbf{Z}) \cong \mathbf{Z}/|G|$.

Theorem 40.1: If G is a finite group and n is a positive integer, then the conditions $\mathcal{P}_1(n)$, $\mathcal{P}_2(n)$, $\mathcal{P}_3(n)$, $\mathcal{P}_4(n)$ are equivalent.

Proof: The implication $\mathcal{P}_1(n) \implies \mathcal{P}_2(n)$ is clear. To prove that $\mathcal{P}_2(n) \implies \mathcal{P}_3(n)$, suppose that $D_{n+k}(\mathbf{Z}) \cong D_k(\mathbf{Z})$ for some nonzero integer k. Since D_{-k} is a self-equivalence of the derived category, we see that there are isomorphisms in the derived category

$$D_n(\mathbf{Z}) \cong D_{-k}D_{n+k}(\mathbf{Z}) \cong D_{-k}D_k(\mathbf{Z}) \cong \mathbf{Z}$$

Using the description of isomorphism types in the derived category given in Section 20 we can express this in terms of exact sequences as follows. There is an exact sequence

$$0 \to M \to P_{n-1} \to \cdots \to P_0 \to \mathbf{Z} \to 0$$

where each P_i is finitely generated projective, and where $M \oplus Q_1 \cong \mathbf{Z} \oplus Q_2$ for some finitely generated projectives Q_1, Q_2. We may modify the above exact sequence in stages: firstly, to an exact sequence of the form

$$0 \to M \oplus Q_1 \to P_{n-1} \oplus Q_1 \to P_{n-2} \to \cdots \to P_0 \to \mathbf{Z} \to 0$$

since $M \oplus Q_1 \cong \mathbf{Z} \oplus Q_2$, we then obtain an exact sequence of the form

$$0 \to \mathbf{Z} \oplus Q_2 \to P_{n-1} \oplus Q_1 \to P_{n-2} \to \cdots \to P_0 \to \mathbf{Z} \to 0$$

Finally, from the relative injectivity of projectives, we see easily that the obvious quotient $P' = (P_{n-1} \oplus Q_1)/Q_2$ is projective. Hence we get an exact sequence

$$0 \to \mathbf{Z} \to P' \to P_{n-2} \to \cdots \to P_0 \to \mathbf{Z} \to 0$$

in which $P_0 \cdots P_{n-2}$, P' are all finitely generated projective, and $\mathcal{P}_3(n)$ is satisfied.

To prove $\mathcal{P}_3(n) \implies \mathcal{P}_1(n)$, suppose that $\mathcal{P}_3(n)$ is true, then there is an isomorphism (in the derived category) $D_n(\mathbf{Z}) \cong \mathbf{Z}$. Since each D_k is a self-equivalence of the derived category, we see that $D_{n+k}(\mathbf{Z}) \cong D_k(\mathbf{Z})$ for all integers k, and this is condition $\mathcal{P}_1(n)$.

We have now shown that conditions $\mathcal{P}_1(n)$, $\mathcal{P}_2(n)$, $\mathcal{P}_3(n)$ are equivalent. However, if $\mathbf{D}_n(\mathbf{Z})$ is isomorphic in the derived category to $\mathbf{Z}$, we see that

$$H^n(G, \mathbf{Z}) \cong \mathrm{Hom}_{\mathcal{D}er}(\mathbf{D}_n(\mathbf{Z}), \mathbf{Z}) \cong \mathrm{Hom}_{\mathcal{D}er}(\mathbf{Z}, \mathbf{Z}) \cong \mathbf{Z}/|G|$$

and this proves that $\mathcal{P}_1(n) \implies \mathcal{P}_4(n)$.

To complete the proof, it suffices to show that $\mathcal{P}_4(n) \implies \mathcal{P}_1(n)$. The pairing

$$\circ : \hat{H}^{-n}(G, \mathbf{Z}) \times \hat{H}^n(G; \mathbf{Z}) \to \hat{H}^0(G, \mathbf{Z}) \cong \mathbf{Z}/|G|$$

is equivalent to the evaluation pairing

$$\text{Hom}_{\text{Ab}}(\hat{H}^n(G, \mathbf{Z}), \mathbf{Z}/|G|) \times \hat{H}^n(G, \mathbf{Z}) \to \mathbf{Z}/|G|$$

Since $n > 0$, there is no distinction between $H^n(G; \mathbf{Z})$ and $\hat{H}^n(G; \mathbf{Z})$, so that, if $u \in \hat{H}^n(G, \mathbf{Z}) \cong \mathbf{Z}/|G|$ is a generator, then there exists $u^{-1} \in \hat{H}^n(G, \mathbf{Z})$ such that $u^{-1}u = 1 \in \hat{H}^0(G, \mathbf{Z}) \cong \mathbf{Z}/|G|$. For any $\mathbf{Z}[G]$-lattice A, $\hat{H}^*(G, A)$ is a module over the Tate ring $\hat{H}^*(G; \mathbf{Z})$, thus

$$\circ : \hat{H}^k(G, A) \times \hat{H}^n(G; \mathbf{Z}) \to \hat{H}^{k+n}(G, A)$$

If $\xi \in \hat{H}^n(G, \mathbf{Z})$, we denote by ρ_ξ the $\mathbf{Z}$-homomorphism

$$\rho_\xi : \hat{H}^k(G, A) \to \hat{H}^{k+n}(\mathbf{Z}, A); \quad \rho_\xi(v) = v \circ \xi$$

If $u \in \hat{H}^n(G, \mathbf{Z}) \cong \mathbf{Z}/|G|$ is a generator, then $\rho_u : H^k(\mathbf{Z}, A) \to H^{k+n}(\mathbf{Z}, A)$ is an isomorphism with inverse ρ_u^{-1} for all $\mathbf{Z}[G]$-lattices A. Make a choice

$$\varphi \in \text{Hom}_{\mathcal{D}er}(\mathbf{D}_k(\mathbf{Z}), \mathbf{D}_{k+n}(\mathbf{Z})) = H^k(\mathbf{Z}, \mathbf{D}_{k+n}(\mathbf{Z}))$$

so that $\rho_u(\varphi) = \text{Id} \in \text{Hom}_{\mathcal{D}er}(\mathbf{D}_{k+n}(\mathbf{Z}), \mathbf{D}_{k+n}(\mathbf{Z})) = H^{k+n}(\mathbf{Z}, \mathbf{D}_{k+n}(\mathbf{Z}))$; then $\varphi : \mathbf{D}_k(\mathbf{Z}) \to \mathbf{D}_{k+n}(\mathbf{Z})$ is an isomorphism in the derived category. (We note that this is simply Yoneda's Lemma [45], *in its original context* [83]). Thus $\mathcal{P}_4(n) \implies \mathcal{P}_1(n)$, and this completes the proof. $\quad\square$

We saw in Section 22 that elements of $\mathbf{D}_n(\mathbf{Z})$ can be represented by modules of the form $N \oplus P$ where $N \in \Omega_n(\mathbf{Z})$ and P is finitely generated projective. Passing to the corresponding rational theory, we may identify $\Omega_n(\mathbf{Z}) \otimes \mathbf{Q}$ with $\Omega_n(\mathbf{Q})$. Also, by (16.1), $P_{\mathbf{Q}}$ is free over $\mathbf{Q}[G]$. Thus elements of $\mathbf{D}_n(\mathbf{Q})$ are represented by modules of the form $N_{\mathbf{Q}} \oplus P_{\mathbf{Q}}$ where $N_{\mathbf{Q}} \in \Omega_n(\mathbf{Q})$ and $P_{\mathbf{Q}}$ is free over $\mathbf{Q}[G]$. Previously we saw, (29.1), that $\Omega_n(\mathbf{Q})$ is represented by $\mathbf{Q}$ if n is even, and by $\mathbf{I}_{\mathbf{Q}}(G)$ if n is odd. It follows easily that:

Proposition 40.2: A cohomological period of a finite group is necessarily even.

We say that $n > 0$ is a *free cohomological period* (or just *free period*) of G if $\Omega_{n+k}(\mathbf{Z}) \cong \Omega_k(\mathbf{Z})$ for all $k \in \mathbf{Z}$. Likewise, there are a number of ways of characterizing this possibility. Consider the following conditions on G:

$\mathcal{F}_1(n)$: $\Omega_{n+k}(\mathbf{Z}) \cong \Omega_k(\mathbf{Z})$ for all integers k;
$\mathcal{F}_2(n)$: $\Omega_{n+k}(\mathbf{Z}) \cong \Omega_k(\mathbf{Z})$ for at least one nonzero integer k;
$\mathcal{F}_3(n)$: There exists an exact sequence in $\mathcal{F}(\mathbf{Z}[G])$ of the form

$$0 \to \mathbf{Z} \to F_{n-1} \to \cdots \to F_0 \to \mathbf{Z} \to 0$$

where each F_i is finitely generated free over $\mathbf{Z}[G]$.

Using the stable module 'constructions' $M \mapsto \Omega_k(M)$ rather than the functors $M \mapsto D_k(M)$, a similar argument to (40.1), with only slightly different justifications, gives:

Proposition 40.3: If G is a finite group and n is a positive integer, then the conditions $\mathcal{F}_1(n)$, $\mathcal{F}_2(n)$, $\mathcal{F}_3(n)$ are equivalent.

The relation between the notions of cohomological period and free cohomological period will be considered again in Chapter 8; in particular, we will show that, if k is a cohomological period of the finite group G, then, for some positive integer N, Nk is a free cohomological period of G (c.f. Section 46 *infra*). For the moment, we are content to consider some examples.

41 Examples

By (40.2), the smallest possible non-trivial cohomological period is $k = 2$. This is realized in the case of cyclic groups.

(i) Cyclic groups: In fact, taking the cyclic group C_n in its balanced presentation

$$C_n = \langle x \mid x^n \rangle$$

there is a free resolution of period 2

$$0 \to \mathbf{Z} \xrightarrow{\epsilon^*} \mathbf{Z}[C_n] \xrightarrow{x-1} \mathbf{Z}[C_n] \xrightarrow{\epsilon} \mathbf{Z} \to 0$$

Swan has shown that the converse statement is also true, namely, if $k = 2$ is a cohomological period of G then G must be cyclic ([60] Lemma (5.2), p. 205).

(ii) Dihedral groups of order $\equiv 2 \bmod(4)$: The dihedral group D_{4n+2} of order $4n + 2$ can be defined by means of the balanced presentation

$$D_{4n+2} = \langle x, y \mid x^{2n+1} = y^2, \, yx^n = x^{n+1}y \rangle$$

and D_{4n+2} admits a free resolution of period 4 (this is false for the dihedral groups of order $4n$)

$$0 \to \mathbf{Z} \xrightarrow{\epsilon^*} \mathbf{Z}[D_{4n+2}] \xrightarrow{\delta} \mathbf{Z}[D_{4n+2}]^2 \xrightarrow{\partial_2} \mathbf{Z}[D_{4n+2}]^2 \xrightarrow{\partial_1} \mathbf{Z}[D_{4n+2}] \xrightarrow{\epsilon} \mathbf{Z} \to 0$$

Here

$$\delta = \begin{pmatrix} 1 + x - x^{n+1} - y \\ -x + x^n y \end{pmatrix}; \quad \partial_2 \sim \begin{pmatrix} \Sigma_x & \theta_1 - \theta_2 y \\ -\Sigma_y & x^n - 1 \end{pmatrix}; \quad \partial_1 \sim (x - 1, y - 1)$$

where

$$\Sigma_x = 1 + x + \cdots + x^{2n}; \quad \Sigma_y = 1 + y; \quad \theta_1 = 1 + x + \cdots + x^{n-1}; \text{ and}$$
$$\theta_2 = 1 + x + \cdots + x^n = \theta_1 + x^n$$

whilst ϵ is the augmentation map, and ϵ^* is its dual.

(iii) The quaternion groups $Q(4n)$ ($n \geq 2$): For each $n \geq 2$, the quaternion group $Q(4n)$ is defined by the balanced presentation

$$Q(4n) = \langle x, y \mid x^n = y^2, xyx = y \rangle$$

It is well known that there is a faithful imbedding $\iota : D_{2n} \to SO(3)$ of the dihedral group D_{2n} in $SO(3)$, the (proper) rotation group of Euclidean 3-space. For example, if $\varphi : D_{2n} \to O(2)$ is the faithful representation of D_{2n} as the symmetry group of a two-sided regular n-gon, and $\sigma : D_{2n} \to O(1) = \{\pm 1\}$ is the sign representation, $\sigma(x) = +1; \sigma(y) = -1$, then $\iota = \varphi \times \sigma : D_{2n} \to O(2) \times O(1) \subset O(3)$ takes values in $SO(3)$. If $\nu : \text{Spin}(3) \to SO(3)$ is the spin double covering, then $Q(4n) \cong \nu^{-1}(\iota(D_{2n}))$. That is, $Q(4n)$ is the spin double cover of the dihedral group D_{2n}.

The algebraic Cayley complex of the presentation given above extends one step to the left to give a resolution of free period 4: (see [12], Chapter XII)

$$0 \to \mathbf{Z} \xrightarrow{\epsilon^*} \mathbf{Z}[Q(4n)] \xrightarrow{\delta} \mathbf{Z}[Q(4n)]^2 \xrightarrow{\partial_2} \mathbf{Z}[Q(4n)]^2 \xrightarrow{\partial_1} \mathbf{Z}[Q(4n)] \xrightarrow{\epsilon} \mathbf{Z} \to 0$$

where

$$\delta = \begin{pmatrix} x - 1 \\ 1 - xy \end{pmatrix}; \quad \partial_2 = \begin{pmatrix} \Sigma_x & yx + 1 \\ -(y+1) & x - 1 \end{pmatrix}; \quad \partial_1 = (x - 1, y - 1),$$

where $\Sigma_x = 1 + x + \cdots + x^{n-1}$, whilst ϵ is the augmentation map, and ϵ^* is its dual.

Proposition 41.1: $Q(4n)$ has free period 4 for $n \geq 2$.

In particular:

Proposition 41.2: $Q(2^{m+2})$ has free period 4 for $m \geq 1$.

(iv) Groups of order pq where p, q are distinct primes: Let p, q be distinct primes, and suppose without loss of generality that $q < p$. If q does not divide $p - 1$, then there is a unique group of order pq, namely the cyclic group C_{pq}. When q *does divide* $p - 1$, there is, in addition, a unique non-abelian group $\mathcal{G}(pq)$ of order pq, which is described as a semidirect product

$$\mathcal{G}(pq) \cong C_p \rtimes C_q$$

where the generator of C_q acts on C_p as an automorphism of order q. It can be shown that:

Proposition 41.3: Let p,q be distinct primes with $q < p$, and such that q divides $p - 1$; then $\mathcal{G}(pq)$ has cohomological period $2q$.

The case of the dihedral group $2p$ is the first case of this. Beyond that, the next case is the non-abelian group of order 21 which has cohomological period 6.

(v) $C_p \times C_p$ **where p is prime:** By contrast, the product $C_p \times C_p$ is not periodic. To see this, let $\pi : C_p \times C_p \to C_p$ be the projection $\pi(x_1, x_2) = x_1$, and let $j : C_p \to C_p \times C_p$ be the inclusion homomorphism $j(x) = (x, 1)$; then $\pi \circ j = \mathrm{Id} : C_p, \to C_p$, and so, taking induced maps in cohomology

$$j^* \circ \pi^* = \mathrm{Id}^* : H^q(C_p; \mathbf{Z}) \to H^q(C_p; \mathbf{Z})$$

In particular, for each q, $H^q(C_p; \mathbf{Z})$ is a direct summand of $H^q(C_p \times C_p; \mathbf{Z})$. If $C_p \times C_p$ were periodic, then for some $n > 0$

$$H^{2n}(C_p \times C_p; \mathbf{Z}) \cong \mathbf{Z}/p^2$$

This is a contradiction since $H^{2n}(C_p; \mathbf{Z}) \cong \mathbf{Z}/p$ is not a direct summand of $\mathbf{Z}/p^2$.

42 Subgroup structure at odd primes

From the Shapiro Lemma it follows that:

Proposition 42.1: Let G be a finite group; if G has cohomological period $2n$, then so does every subgroup.

Since $C_p \times C_p$ is not periodic, it now follows that:

Proposition 42.2: Let G be a finite abelian group; then G has periodic cohomology if and only if G is cyclic.

Likewise:

Proposition 42.3: Let G be a finite group of periodic cohomology, and let H be a subgroup of order p^2 where p is a prime; then H is cyclic.

Let p be a prime, and let G be a group of order p^n where $(n \geq 1)$; if $n = 1$, then G is cyclic and so has periodic cohomology. Thus suppose that $n \geq 2$; as is well known, every group of order p^2 is abelian, so that, by (42.3), every subgroup of order p^2 must be cyclic. In fact, this necessary condition is also sufficient as we shall see. We first establish some preliminary results.

Let G be a group and let x, $y \in$; we denote by $[x, y] \in G$ the commutator

$$[x, y] = xyx^{-1}y^{-1}$$

and we denote by $[G, G]$ the commutator subgroup of G, that is, the subgroup generated by all commutators. We denote by $\mathcal{Z}(G)$ the *centre* of G.

We say that G satisfies condition $\mathcal{Z}$ when $[G, G] \subset \mathcal{Z}(G)$.

Proposition 42.4: Let G satisfy $\mathcal{Z}$; then for each x, $y \in G$

$$[x, y]^n = [x^n, y]$$

Proof: By induction on n; for $n = 1$ there is nothing to prove. Suppose proved for n; then

$$\begin{aligned}
[x^{n+1}, y] &= x[x^n, y]yx^n x^{-(n+1)}y^{-1} \\
&= x[x^n, y]yx^{-1}y^{-1} \\
&= [x^n, y]xyx^{-1}y^{-1} \\
&= [x^n, y][x, y] \\
&= [x, y]^n[x, y] \\
&= [x, y]^{n+1}
\end{aligned}$$

$\square$

Put

$$t(n) = \frac{n(n-1)}{2}$$

Proposition 42.5: Let G satisfy $\mathcal{Z}$; then for each x, $y \in G$

$$(xy)^n = [y, x]^{t(n)}x^n y^n.$$

Proof: By induction on n. For $n = 1$ there is nothing to prove. Suppose proved for n; then

$$(xy)^{n+1} = xy[y, x]^{t(n)}x^n y^n = [x, y]yx[y, x]^{t(n)}x^n y^n$$

and, since $[y, x] \in \mathcal{Z}(G)$ and $[x, y] = [y, x]^{-1}$, we get

$$(xy)^{n+1} = [y, x]^{t(n)-1}yx^{n+1}y^n = [y, x]^{t(n)-1}[y, x^{n+1}]x^{n+1}y^{n+1}$$

However, by (42.4), $[y, x^{n+1}] = [y, x]^{n+1}$, so that

$$(xy)^{n+1} = [y, x]^{t(n)+n}x^{n+1}y^{n+1}$$

the result follows since $t(n) + n = t(n + 1)$. $\square$

For the rest of the discussion, p will denote an odd prime.

Theorem 42.6: Let G be a non-abelian group of order p^n where p is an odd prime; then G has a subgroup isomorphic to $C_p \times C_p$.

Proof: Put $K = \mathcal{Z}(G)$, and let $\pi : G \to G/K$ be the projection; K is an abelian group of order p^a for some $a \geq 1$. If K is not cyclic then

$$K \cong C_{p^{e_1}} \times \cdots \times C_{p^{e_m}}$$

for some $m \geq 2$ and some $e_i \geq 1$, so that K, and hence G, has a subgroup isomorphic to $C_p \times C_p$. Thus we may assume that K is cyclic, $K \cong C_{p^m}$ say, so that G/K is a group of order p^N for some $N \geq 1$, and $\mathcal{Z}(G/K)$ is a non-trivial abelian group of order p^b where $1 \leq b \leq N$. Moreover, it cannot then be true that $\mathcal{Z}(G/K)$ is also cyclic, so that $\mathcal{Z}(G/K)$ contains a subgroup $Q \cong C_p \times C_p$. Putting $H = \pi^{-1}(Q)$, we have a central extension

$$1 \to K \to H \xrightarrow{\pi} Q \to 1$$

where $K \cong C_{p^m}$ and $Q \cong C_p \times C_p$.

It follows easily now that H satisfies condition $\mathcal{Z}$, and, by (42.4), (42.5), the map $H \to H$ defined by $\varphi(x) = x^p$ gives a homomorphism $\varphi : H \to K$.

For $x \in Q$ we denote by $\langle x \rangle$ the subgroup of Q generated by x, and put $G(x) = \pi^{-1}(\langle x \rangle)$. By centrality of the extension, it is clear that each $G(x)$ is abelian. From the classification of abelian groups we see that for a given $x \in Q$, *either*

(i) $G(x) \cong C_{p^{m+1}}$ *or*
(ii) $G(x) \cong C_{p^m} \times C_p$.

If each $G(x) \cong C_{p^{m+1}}$, then it is straightforward to see that $\mathrm{Ker}(\varphi) \subset K$ and, in fact, $\mathrm{Ker}(\varphi) \cong C_p$, from which it would follow, from the Noether Isomorphism Theorem, that $|\mathrm{Im}(\varphi)| = p^{m+1}$. However, $\mathrm{Im}(\varphi)$ is a subgroup of K so that $|\mathrm{Im}(\varphi)| \leq p^m$, which is a contradiction. It follows that $G(x) \cong C_{p^m} \times C_p$ for at least one $x \in Q$. In particular, some $G(x)$, and hence G, has a subgroup isomorphic to $C_p \times C_p$, and this completes the proof.   □

Proposition 42.7: Let G be a group of order p^n where p is an odd prime; if each subgroup of G of order p^2 is cyclic, then G is cyclic.

Proof: Observe that G must be abelian, since by (42.6) G would otherwise have a noncyclic subgroup of order p^2. The conclusion now follows immediately from the classification of finite abelian groups.   □

43 Subgroup structure at the prime 2

For notational convenience, we denote the cyclic group of order 2^n by $C(2^n)$ rather than C_{2^n}.

If G has periodic cohomology then every abelian subgroup is cyclic. In the case where $|G| = 2^n$ then $\mathcal{Z}(G) \cong C(2^m)$ for some $m \geq 1$; in particular, G has a unique central subgroup $Z \cong C(2)$. In fact, this is the only subgroup of G isomorphic to $C(2)$, for suppose that G has another subgroup $W \cong C(2)$, then $W \cap Z = \{1\}$ and Z centralizes W so that WZ is a subgroup of G isomorphic to $C(2) \times C(2)$. This is a contradiction since WZ is then a subgroup of a group with periodic cohomology, but does not itself have periodic cohomology. Thus we have shown:

Proposition 43.1: Let G be a group of order 2^n for some $n \geq 1$; if G has periodic cohomology, then G has a unique subgroup isomorphic to $C(2)$.

In this section, we show the converse is true by giving Zassenhaus' classification of groups of order 2^n having periodic cohomology. Thus suppose that $|G| = 2^n$ and that G has a unique subgroup Z of order 2, then it is clear that Z is central in G.

Let A be a maximal abelian normal subgroup of G. By classification of abelian groups of order 2^N, we have

$$A \cong A_1 \times \cdots \times A_m$$

where each A_i is cyclic, of order 2^{a_i}, say. However Z is the unique subgroup of A of order 2, so that $m = 1$ and $A \cong A_i$ is cyclic. Consider the 'operator homomorphism' $c : G/A \to \mathrm{Aut}(A)$ given by

$$c([g])(x) = gxg^{-1}$$

Since A is a maximal normal abelian subgroup, it follows easily that c is injective. Since A is cyclic, $\mathrm{Aut}(A)$ is abelian, and we get:

Proposition 43.2: Let G be a group of order 2^n, and suppose that G contains a unique subgroup of order 2; then G occurs as an extension

$$1 \to A \to G \to Q \to 1$$

where A is cyclic and Q is a subgroup of $\mathrm{Aut}(A)$; in particular, Q is abelian.

We may therefore put $A = C(2^n) = \{1, x, \ldots, x^{2^n-1}\}$; there are a number of special cases:

(i) $Q = \{1\}$; in this case $G = A$ is cyclic.
(ii) $n = 1$; then $\mathrm{Aut}(A) = \{1\}$ so that $Q = \{1\}$ and again $G = A$ is cyclic.

(iii) $n = 2$ and $Q \neq \{1\}$; in this case, $\text{Aut}(A) \cong C(2)$ and the non-trivial element of $\text{Aut}(A)$ is the involution $\tau(x) = x^{-1}$. Consider the extension

$$1 \to A \to G \xrightarrow{\pi} C(2) \to 1$$

and let $y \in G$ be such that $\pi(y) = \tau$, so that $yxy^{-1} = x^{-1}$. Since $\pi(y^2) = 1$, then y^2 is one of $1, x, x^2, x^3$. If y^2 is either x or x^3, then $\text{ord}(y) = 8$ and $G \cong C(8)$, which contradicts the fact that $yxy^{-1} = x^{-1}$. Thus y^2 is either 1 or x^2. However, if $y^2 = 1$, then the extension splits, contradicting the assumption that G has a unique subgroup of order 2. Thus the only possibility we are left with is $y^2 = x^2$; that is, G is generated by two elements $\{x, y\}$, satisfying the conditions

$$x^4 = 1; \quad x^2 = y^2; \quad yxy^{-1} = x^{-1}$$

and in this case $G = Q(8)$, the quaternion group of order 8.

We can now consider the general case where $n \geq 3$ and $Q \neq \{1\}$. Then $\text{Aut}(A) \cong C(2^{n-2}) \times C(2)$; where the factor $C(2)$ is generated by τ where $\tau(x) = x^{-1}$, and the factor $C(2^{n-2})$ is generated by α where $\alpha(x) = x^5$. It follows that G contains a subgroup H of the form

$$1 \to A \to H \xrightarrow{\pi} \Psi \to 1$$

where $\Psi \subset \text{Aut}(A)$ is a subgroup of order 2. Let $\psi \in \Psi$ denote a generator; there are precisely three candidates for ψ, namely:

(i) $\psi = \tau$;
(ii) $\psi = \alpha^{2^{n-2}}$;
(iii) $\psi = \tau\alpha^{2^{n-2}}$.

In fact neither (ii) nor (iii) can occur because the corresponding extension splits, contradicting the assumption that G has a unique subgroup of order 2. Since cases (ii) and (iii) can only occur when there is a non-trivial projection of Q on to the factor $C(2^{n-2})$ generated by α, the only possibility left for G is as an extension

$$1 \to A \to G \xrightarrow{\pi} \langle \tau \rangle \to 1$$

Let $y \in G$ be such that $\pi(y) = \tau$, so that $yxy^{-1} = x^{-1}$, let B denote the subgroup of A generated by y^2. Observe that $B \neq A$, since otherwise y^2 would be a generator of A, contradicting the relation $yxy^{-1} = x^{-1}$. It follows that y^2 is already a square in A, say $y^2 = x^{2k}$. The relation $yxy^{-1} = x^{-1}$ can be written $xyx = y$, from which we see that

$$x^{2k}yx^{2k} = y$$

Since x^{2k} commutes with y, we now have $x^{4k} = 1$. Hence $4k = 2^n$ or, more conveniently, $2k = 2^{n-1}$, so that G is generated by elements x, y subject to the relation that

$$x^{2^{n-1}} = y^2; \quad xyx = y$$

and so $G \cong Q(2^{n+1})$. Thus to summarize:

Theorem 43.3: Let G be a group of order 2^m for some $m \geq 1$; then the following conditions on G are equivalent:

(i) G has periodic cohomology;
(ii) G has a unique subgroup isomorphic to $C(2)$;
(iii) Every subgroup of order 4 is cyclic;
(iv) G is isomorphic either to $Q(2^m)$ or to $C(2^m)$.

Evidently, the case $G \cong Q(2^m)$ can only occur when $m \geq 3$.

44 The Artin–Tate Theorem

Let G be a finite group, let p be a prime dividing $|G|$, and write $|G| = kp^n$ where k is coprime to p; $G(p)$ will denote a Sylow p-subgroup of G, and $H^m(G; A)_{(p)}$ will denote the p-primary component of $H^m(G; A)$.

When $m > 0$, the image of the transfer map

$$t : H^m(G(p); \mathbf{Z}) \to H^m(G; \mathbf{Z})$$

lies in the p-primary subgroup $H^m(G; \mathbf{Z})_{(p)}$ (or we could use Tate cohomology and allow any value of m). With this notation, we have:

Proposition 44.1: For each $m > 0$, there is an isomorphism

$$H^m(G(p); \mathbf{Z}) \cong H^m(G; \mathbf{Z})_{(p)} \oplus \mathrm{Ker}(t)$$

Proof: Let $i : G(p) \subset G$ be the inclusion of the Sylow p-subgroup; then the composite

$$t \circ i^* : H^m(G; \mathbf{Z}) \to H^m(G; \mathbf{Z})$$

has the effect of multiplying by $k = [G; G(p)]$, and, since k is coprime to p, $t \circ i^*$ gives an automorphism

$$t \circ i^* : H^m(G; \mathbf{Z})_{(p)} \to H^m(G; \mathbf{Z})_{(p)}$$

of the p-primary part. Since $H^m(G; \mathbf{Z})_{(p)}$ is finite, some power of $t \circ i^*$ restricted to $H^m(G; \mathbf{Z})_{(p)}$ is the identity, say

$$(t \circ i^*)^{e+1} = \text{Id}$$

when $e \geq 0$. Now put $j = i^* \circ (t \circ i^*)^e$. Then $t \circ j$ is the identity on $H^m(G; \mathbf{Z})_{(p)}$, and $j : H^m(G; \mathbf{Z})_{(p)} \to H^m(G(p); \mathbf{Z})$ is a right splitting for the surjection

$$t : H^m(G(p); \mathbf{Z}) \to H^m(G; \mathbf{Z})_{(p)}$$

The result follows.   □

Corollary 44.2: Suppose that $H^m(G(p); \mathbf{Z}) \cong \mathbf{Z}/p^a$ for some $a > 0$; if the transfer map $t : H^m(G(p); \mathbf{Z}) \to H^m(G; \mathbf{Z})_{(p)}$ is nonzero then $H^m(G; \mathbf{Z})_{(p)} \cong \mathbf{Z}/p^k$, and $t : H^m(G(p); \mathbf{Z}) \to H^m(G; \mathbf{Z})_{(p)}$ is an isomorphism.

Proof: We have seen that t is surjective, and that $H^m(G(p); \mathbf{Z}) \cong H^m(G; \mathbf{Z})_{(p)} \oplus \text{Ker}(t)$. Since, by hypothesis, $H^m(G(p); \mathbf{Z}) \cong \mathbf{Z}/p^a$, then either $H^m(G; \mathbf{Z})_{(p)} = 0$ or $\text{Ker}(t) = 0$. If $H^m(G; \mathbf{Z})_{(p)} \neq 0$ then $\text{Ker}(t) = 0$, and so

$$t : H^m(G(p); \mathbf{Z}) \to H^m(G; \mathbf{Z})_{(p)}$$

is both injective and surjective.   □

Proposition 44.3: Suppose that $H^m(G(p); \mathbf{Z}) \cong \mathbf{Z}/p^n$; then there exists $e \geq 1$ such that for any multiple d of e

$$t : H^{dm}(G(p); \mathbf{Z}) \to H^m(G; \mathbf{Z})_{(p)}$$

is an isomorphism.

Proof: By (44.2), it suffices to show that there exists $e \geq 1$ such that for any multiple d of e

$$t : H^{dm}(G(p); \mathbf{Z}) \to H^m(G; \mathbf{Z})_{(p)}$$

is nonzero. Let e be the exponent of $\text{Aut}(\mathbf{Z}/p^n) \cong (\mathbf{Z}/p^n)^*$; then if d is a multiple of e

$$\xi^d = 1$$

for all $\xi \in (\mathbf{Z}/p^n)^*$. G acts as a group of automorphisms of $H^m(G(p); \mathbf{Z}) \cong \mathbf{Z}/p^n$ via the Eckmann–Shapiro identity

$$H^m(G(p); \mathbf{Z}) \cong H^m(G; \mathbf{Z} \otimes_{\mathbf{Z}[G(p)]} \mathbf{Z}[G]).$$

Let $z \in H^m(G(p); \mathbf{Z})$ be a generator. If c_*^g is the conjugation automorphism of $H^m(G(p); \mathbf{Z}) \cong H^m(G(p); \mathcal{E}\mathbf{Z})$ induced by $g \in G$, then we may write

$$c_*^g(z) = \xi_g z$$

for some $\xi_g \in (\mathbf{Z}/p^n)^*$. When d is a multiple of e, we have, by naturality of the G-action

$$c_*^g(z^d) = c_*^g(z)^d = \xi_g^d z^d = z^d \in \mathrm{H}^{dm}(G(p); \mathbf{Z})$$

Thus z^d is invariant under the action of G, so that $i^* t(z^d) = k z^d$ where $k = [G : G(p)]$. Since k is coprime to p, then $t(z^d) \neq 0$, and the result follows. $\square$

Corollary 44.4: Let G be a finite group, let $p_1, \ldots, p_N$ be the distinct primes dividing $|G|$; then G has periodic cohomology if and only if $G(p_i)$ has periodic cohomology for each i.

Proof: We may write the prime factorization of $|G|$ in the form

$$|G| = p_1^{n_1} \cdots p_N^{n_N}$$

If G has periodic cohomology, then so does every subgroup. In particular, so also does $G(p_i)$.

Suppose, conversely, that $G(p_i)$ has periodic cohomology; then, for some m_i, $\mathrm{H}^{m_i}(G(p_i); \mathbf{Z}) \cong \mathbf{Z}/p_i^{n_i}$. Then by (44.3) there exists $e_i \geq 1$ such that, for every *nonzero, positive* multiple d_i of e_i the transfer map

$$t : \mathrm{H}^{d_i m_i}(G(p_i); \mathbf{Z}) \to \mathrm{H}^{d_i m_i}(G; \mathbf{Z})_{(p_i)}$$

is an isomorphism, each side being isomorphic to $\mathbf{Z}/p_i^{n_i}$ Thus when m is a multiple of each $e_i m_i$

$$t : \mathrm{H}^m(G(p_i); \mathbf{Z}) \to \mathrm{H}^m(G; \mathbf{Z})_{(p_i)}$$

is an isomorphism for each i. It follows easily that

$$\mathrm{H}^m(G; \mathbf{Z}) \cong \mathbf{Z}/p_1^{n_1} \times \cdots \times \mathbf{Z}/p_N^{n_N} \cong \mathbf{Z}/|G|$$

and so G has periodic cohomology, by (40.1). $\square$

From our previous discussion, in Sections 42–43 of the periodicity conditions for groups of prime power order, we get the Artin–Tate–Zassenhaus characterization of groups of periodic cohomology:

Theorem 44.5: Let G be a finite group, and let $p_1, \ldots, p_N$ be the distinct primes dividing $|G|$; then G has a finite cohomological period if and only if $G(2)$ is either cyclic or generalized quaternionic and $G(p_i)$ is cyclic whenever p_i is odd.

Alternatively expressed, G has periodic cohomology if and only if each subgroup of order p^2 is cyclic.

45 The Zassenhaus–Suzuki classification

The Zassenhaus–Suzuki Theorem classifies all finite groups whose odd Sylow subgroups are cyclic and whose Sylow 2-subgroup contains a cyclic subgroup of index 2. This is slightly larger than the class of groups of periodic cohomology, since it allows the possibility of $C_{2^m} \times C_2$ as a Sylow 2-subgroup. However, the subclass of groups of periodic cohomology is easily retrieved from this classification.

We treat the soluble case first. Wolf's reworking of Zassenhaus gives four such classes ([82] tabulated on p. 179). We adapt his treatment to suit our purpose:

$\mathcal{A}(m, n; r)$: Let $C_n = \langle Y : Y^n = 1 \rangle$ be the cyclic group of order $n \geq 2$, and let $r \geq 1$ be an integer coprime to n; then there is an automorphism $\varphi_r : C_n \to C_n$ defined by

$$\varphi_r(Y^s) = Y^{rs}$$

When $\mathrm{ord}(\varphi_r)$ divides m, that is, when $r^m \equiv 1 \bmod n$), there is a split extension

$$1 \to C_n \to \mathcal{A}(m, n; r) \to C_m \to 1$$

of C_n by C_m, so that $\mathcal{A}(m, n; r)$ is a semidirect product

$$\mathcal{A}(m, n; r) \cong C_n \underset{(r)}{\times\!\!\!\!|} C_m$$

where the generator X of C_m acts by conjugation on C_n as

$$XYX^{-1} = \varphi_r(Y) = Y^r$$

We define this formally by the presentation

$$\mathcal{A}(m, n; r) = \langle X, Y : X^m = Y^n = 1; \quad XYX^{-1} = Y^r \rangle$$

where the parameters m, n, r are subject to the conditions:

 (i) m, n are coprime, with m odd; and
(ii) $r^m \equiv 1 \pmod{n}$.

In order for $\mathcal{A}(m, n; r)$ to satisfy the p^2 conditions it is necessary and sufficient that m and n be coprime. Moreover, since $\mathrm{Aut}(C_{2^u})$ is a 2-group, any extension

$$1 \to C_{2^u} \to G \to C_v \to 1$$

in which v is odd is necessarily a direct product $G \cong C_{2^u} \times C_v$. Thus in the semidirect product

$$\mathcal{A}(m, n; r) \cong C_n \underset{(r)}{\rtimes} C_m$$

we may suppose, without loss of generality, that the 2-torsion of G lies in the factor C_m; that is, we may suppose that n is odd.

$\mathcal{B}(2^{u+2}, v, n; r, s, t)$: Put $m = 2^{u+1}v$ and form the nonsplit extension

$$1 \to \mathcal{A}(m, n; r) \to \mathcal{B}(2^{u+2}, v, n; r, s, t) \to C_2 \to 1$$

where a generator R which projects on to the generator of C_2 acts by conjugation on the factors of the metacyclic subgroup $\mathcal{A}(m, n; r)$ by

$$RXR^{-1} = X^s; \quad RYR^{-1} = Y^t,$$

and satisfies $R^2 = X^{\frac{m}{2}}$. To achieve this, in addition to the conditions required for $\mathcal{A}(m, n; r)$, we also require:

(iii) $u \geq 1$ and v is odd;
(iv) $s \equiv -1 \bmod 2^u$ and $s^2 \equiv 1 \bmod m$;
(v) $t^2 \equiv r^{s-1} \equiv 1 \bmod n$.

In this case, the Sylow 2-subgroup is $Q(2^{u+2})$ and the cohomological period is $\geq \text{LCM}(4, \text{ord}(\varphi_r))$.

$\mathcal{C}(8; m, n; r)$: Take the quaternion group of order 8 in its standard presentation

$$Q(8) = \langle P, Q : P^2 = Q^2; PQP = Q \rangle.$$

Let $\alpha : Q(8) \to Q(8)$ be the automorphism of order 3 given by

$$\alpha(P) = PQ; \quad \alpha(Q) = P^{-1}$$

and take the split extension

$$\mathcal{C}(8; m, n; r) \cong (Q(8) \times C_n) \underset{(\alpha, r)}{\rtimes} C_m$$

in which a generator X of C_m acts on $Q(8) \times C_n$ by means of

$$XPX^{-1} = PQ; \quad XQX^{-1} = P^{-1}; \quad XYX^{-1} = Y^r$$

The conditions that the parameters must satisfy here are:

(i) $m \equiv 0 \pmod 3$;
(ii) $r^m \equiv 1 \pmod n$;
(iii) m, n are coprime;
(iv) n is odd.

Conditions (i) and (ii) are required to construct the extension; conditions (iii) and (iv) are then necessary and sufficient to guarantee the p^2 conditions. The Sylow subgroup is $Q(8)$, and the cohomological period is LCM(4, ord(φ_r)).

$\mathcal{D}(16; m, n; r, s, t)$: We construct the nonsplit extension

$$1 \to \mathcal{C}(8; m, n; r) \to \mathcal{D}(16; m, n; r, s, t) \to C_2 \to 1$$

where a generator R which projects on to the generator of C_2 acts by conjugation on $\mathcal{C}(8; m, n; r)$ by

$$RPR^{-1} = PQ; \quad RQR^{-1} = P^{-1}; \quad RXR^{-1} = X^s; \quad RYR^{-1} = Y^t$$

and satisfies $R^2 = P^2$. The conditions required to guarantee this are:

  (i) $m \equiv 0 \pmod 3$;
  (ii) (ii) $r^m \equiv 1 \pmod n$;
 (iii) m, n are coprime;
 (iv) n is odd;
  (v) $s^2 \equiv 1 \pmod m$;
 (vi) $s \equiv -1 \pmod 3$;
(vii) $r^{t-1} \equiv s^2 \equiv 1 \pmod n$.
The Sylow 2-subgroup is $Q(16)$ and the cohomological period is $\geq$ LCM(4, ord(φ_r)).

Zassenhaus' classification of soluble groups satisfying the p^2 conditions can be stated in two parts; the first, when the Sylow 2-subgroup is cyclic, is actually due to Burnside:

Theorem 45.1: Let G be a soluble finite group in which all Sylow subgroups are cyclic ; then $G \cong \mathcal{A}(m, n; r)$ for some suitable m, n, r.

Theorem 45.2: Let G be a soluble finite group in which the Sylow 2-subgroup is generalized quaternion and all odd Sylow subgroups are cyclic; then G is isomorphic to one of the groups $\mathcal{B}(2^{u+2}, v, n; r, s, t)$, $\mathcal{C}(8; m, n; r)$, $\mathcal{D}(16; m, n; r, s, t)$.

We now restrict attention to groups of period four. The natural analysis is in terms of the structure of the Sylow 2-subgroup, beginning with the case where it is trivial. At a number of points, we refer to the notation of Milnor's paper [42], where the question of which finite groups G can act freely on S^n is considered. Such a group must have $n + 1$ as a cohomological period, and Milnor showed in addition that either G is of odd order or else contains a unique element of order 2, which must necessarily be central. In connection with the question for $n = 3$, Milnor listed all such groups of period four.

Odd order groups of period 4

An odd order group G with periodic cohomology must be of the form $G \cong \mathcal{A}(m, n; r)$ where both m and n are odd; moreover, the order of φ_r must also be odd. Since the minimal period of $\mathcal{A}(m, n; r)$ is $2 \operatorname{ord}(\varphi_r)$, it must then be true that $\varphi_r = \operatorname{Id}$. Thus we may take $r = 1$ and write $G \cong \mathcal{A}(m, n; 1) \cong C_m \times C_n \cong C_{mn}$. We obtain:

Proposition 45.3: A finite group of odd order and period ≤ 4 is necessarily cyclic.

Groups of period four with cyclic Sylow 2-subgroup

In general, a finite group G with periodic cohomology and cyclic Sylow 2-subgroup has the form $G \cong \mathcal{A}(2^a; v, n; r)$ where v, n are both odd; such a group has a subgroup of the form $H = \mathcal{A}(v, n; s)$. If G has period ≤ 4 so also does H, and so H is cyclic, by (45.3). Without loss of generality, we may describe G in the form $G \cong \mathcal{A}(2^a, 2b + 1; t)$ where the order of t mod $2b + 1$ is of the form 2^c where $0 \leq c \leq a$. In fact, it is easy to calculate that the minimal period of $G \cong \mathcal{A}(2^a, 2b + 1; t)$ is 2^{c+1}. Thus the condition that G has period ≤ 4 forces c to be either 0 or 1. In the case $c = 0$, G is necessarily cyclic. In the case $c = 1$, we see that $G \cong \mathcal{A}(2^a, 2b + 1; t)$ where $t^2 = 1$ mod $2b + 1$.

Let $\tau_N : C_N \to C_N$ denote the canonical involution $\tau_N(x) = x^{-1}$. An easy argument using prime decomposition shows that, if $\theta : C_{2b+1} \to C_{2b+1}$ is a non-trivial automorphism of order 2 then *either* $\theta = \tau_{2b+1}$, or else θ is equivalent to

$$\tau_{2r+1} \times \operatorname{Id} : C_{2r+1} \times C_{2s+1} \to C_{2r+1} \times C_{2s+1}$$

for some factorization $(2r + 1)(2s + 1) = (2b + 1)$. We now see easily that:

Proposition 45.4: Let G be a finite group of period 4; if the Sylow 2-subgroup of G is cyclic, then *either*

(i) G is cyclic $\cong C_{2^k}$ where $k \geq 1$, *or*
(ii) $G \cong \mathcal{A}(2^a, 2r + 1; -1) \times C_{2s+1}$ for some $k \geq 1$, $r \geq 1$ and $s \geq 0$.

The special cases $k = 1$ and $k = 2$ merit further attention. In the case $k = 1$, recall from Section 41, that the dihedral group

$$D_{4n+2} = \langle \xi, \eta : \xi^{2n+1} = \eta^2 = 1; \ \eta\xi\eta^{-1} = \xi^{-1} \rangle$$

has period 4. D_{4n+2} occurs as $\mathcal{A}(2, 2n + 1; -1)$; the 'quasi-dihedral' groups $D_{4n+2} \times C_{2m+1}$ also occur as $\mathcal{A}(2(2m + 1), 2n + 1; -1)$. Observe that D_{4n}

does not have periodic cohomology, since it has a subgroup $\{1, \xi^n, \eta, \xi^n y\}$ isomorphic to $C_2 \times C_2$.

In the case $k = 2$, we have seen in Section 41 that the quaternion group

$$Q(4m) = \langle \xi, \eta : \xi^m = \eta^4 = 1, \eta\xi\eta^{-1} = \xi^{-1} \rangle$$

has period 4. When m is odd, the Sylow 2-subgroup is C_4, and it occurs in the above list as $\mathcal{A}(4, m; -1)$.

For $k \geq 3$, there does not seem to be any generally agreed name for the groups $\mathcal{A}(2^k, 2m + 1; -1)$. Milnor ([42]) calls them '$D_{2^k(2m+1)}$' but this clashes with the now standard nomenclature for dihedral groups.

Groups with quaternionic Sylow 2-subgroup of order ≥ 32

The groups $Q(4m)$ when m is odd are more accurately described as *binary dihedral groups*. Here we consider the true quaternion group

$$Q(2^{u+2}) = \langle P, Q : P^{2^u} = Q^2; QPQ^{-1} = P^{-1} \rangle$$

of order 2^{u+2}. This occurs as $\mathcal{B}(2^{u+2}, 1, 1; 1, 1, 1)$ under a notation change in which the 'P' of $Q(2^{u+2})$ is actually the 'R' of $\mathcal{B}(2^{u+2}, 1, 1; 1, 1, 1)$.

The groups $Q(2^{u+2}; m, n)$: when m, n are odd of coprime order, we describe $C_m \times C_n \cong C_{mn}$ in the form

$$C_m \times C_n = \langle X, Y : X^m = Y^n = 1; \quad XY = YX \rangle$$

and let $Q(2^{u+2}; m, n)$ denote the split extension

$$1 \to C_m \times C_n \to Q(2^{u+2}; m, n) \to Q(2^{u+2}) \to 1$$

where the action of $Q(2^{u+2})$ on $C_m \times C_n$ is given by

$$PXP^{-1} = X^{-1}; \quad PYP^{-1} = Y; \quad QXQ^{-1} = X; \quad QYQ^{-1} = Y^{-1}$$

It is easy to see that this group occurs as $\mathcal{B}(2^{u+2}, m, n; 1, -1, -1)$.

$Q(2^{u+2}; m, n)$ has period 4. Observe that the abelianization $Q(2^{u+2})^{\mathrm{ab}}$ is isomorphic to $C_{2^u} \times C_2$ where P projects on to the first factor and Q on to the second. In principal, the generator P can act with any order 2^k for $0 \leq k \leq u$. However, if $2 \leq k$ the resulting extension has period > 4. More generally, if k is odd and coprime to both m and n, then the direct product $Q(2^{u+2}; m, n) \times C_k$ has period 4, and occurs in the above list as $\mathcal{B}(2^{u+2}, m, kn; 1, -1, t)$ where $t^2 = 1 \pmod{kn}$, $t = 1 \pmod{k}$, and $t = -1 \pmod{n}$; moreover, this is the most general group of period 4 with quaternionic Sylow subgroup of order ≥ 32.

Groups with Sylow 2-subgroups of type Q(8)

Within the general class of periodic groups $Q(8)$ can arise as a Sylow 2-subgroup in two ways, as *either*

(I) $\mathcal{B}(2^3, v, n; r, k, l)$, *or*
(II) $\mathcal{C}(8; m, n, r)$.

In period 4, groups of type (I) take the form $Q(8; m, n) \times C_k$ for some suitable odd k, m, n. In type (II) one case is of special interest, namely the split extension

$$1 \to Q(8) \to \mathcal{C}(8; 3^k, 1; 1) \to C_{3^k} \to 1$$

where a generator Z of C_{3^k} acts on $Q(8)$ by means of

$$ZPZ^{-1} = PQ; \quad ZQZ^{-1} = P^{-1}$$

In Milnor's notation, $\mathcal{C}(8; 3^k, 1; 1) = P'_{8 \cdot 3^k}$. More generally, when n is odd and coprime to 3, the product $P'_{8 \cdot 3^k} \times C_n = \mathcal{C}(8; 3^k, n; 1) = \mathcal{C}(8; 3^k n, 1; 1)$ has period 4; this exhausts the groups of period 4 which have $Q(8)$ as Sylow 2-subgroup.

Groups with Sylow 2-subgroups of type Q(16)

Within the general class of periodic groups $Q(16)$ can arise as a Sylow 2-subgroup in two ways, as *either*

(III) $\mathcal{B}(2^4, v, n; r, k, l)$, *or*
(IV) $\mathcal{D}(16; m, n; r, s, t)$.

In period 4, groups of type $\mathcal{B}(2^4, v, n; r, k, l)$ take the form $Q(16; m, n) \times C_k$ for some suitable odd k, m, n. Again following Milnor, if n is odd and coprime to 3, we write $P''_{16 \cdot 3^k \cdot n} = \mathcal{D}(16; 3^k, n; 1, 1, -1)$. Then the most general group period 4 group of type $\mathcal{D}$ is $P''_{48n} \times C_m$, where m is odd and coprime to both 3 and n; we note that $P''_{48n} \times C_m$ can be described in the form $\mathcal{D}(16; 3^k, mn; 1, 1, t)$, where $t \equiv 1 \bmod (m)$, $t \equiv -1 \bmod (n)$, and $t^2 \equiv 1 \bmod (mn)$.

Nonsoluble groups of period 4

Suzuki has classified of nonsoluble groups in which the p^2-conditions are satisfied [56]. In this case, the condition of nonsolubility implies that the Sylow 2-subgroup cannot be cyclic, and so is forced to be generalized quaternion; then we have:

Theorem 45.5: Let G be a nonsoluble finite group in which the Sylow 2-subgroup is generalized quaternion and all odd Sylow subgroups are cyclic; then G has a subgroup G_0 of index ≤ 2 such that

$$G_0 \cong \mathcal{A}(m, n; r) \times \mathrm{SL}_2(\mathbf{F}_p)$$

for some prime p and suitable m, n, r; here $\mathbf{F}_p$ is the field with p elements.

From this we get one completely new example of a group of period 4, namely $\mathrm{SL}_2(\mathbf{F}_5)$ which has order 120; it arises geometrically as the binary icosahedral group $\mathbf{I}^*$, that is, the spin double covering of the simple group A_5, of order 60. Note that the groups $\mathrm{SL}_2(\mathbf{F}_2)$, $\mathrm{SL}_2(\mathbf{F}_3)$ are soluble and already accounted for, whilst the groups $\mathrm{SL}_2(\mathbf{F}_p)$ for $p \geq 7$ have period > 4. In addition, when m is odd and coprime to 3,5 the groups $C_m \times \mathbf{I}^*$ also have period 4.

Milnor's list can be retrieved by eliminating the dihedral and quasi-dihedral groups from the above discussion.

Chapter 8

Algebraic homotopy theory

In this chapter, we encounter homotopy theory in both its geometric and algebraic aspects. We begin by reviewing the homotopy theory of projective chain complexes, as outlined, for example, by Wall in [69]. To ensure consistency of notation, we start from elementary considerations. One of our primary aims is to introduce the finiteness obstruction χ of Swan [58] and Wall [68] which, viewed algebraically, detects when a projective chain complex is homotopy equivalent to a free chain complex.

If X is a CW complex we say that X is *reduced* when it has a single 0-cell. Clearly a reduced complex is automatically connected. If X is a connected CW complex and $T \subset X^{(1)}$ is a maximal tree then the quotient X/T is a reduced CW complex which is homotopy equivalent to X. Without further mention we assume that in a reduced complex the basepoint is the unique 0-cell, and allow ourselves to write $\pi_1(X)$ rather than $\pi_1(X, *)$.

We shall fix a finitely presented group G, described in some specific way, and consider all CW complexes with the fundamental group isomorphic to G. In working with distinct spaces whose fundamental groups though isomorphic are nevertheless distinct, it is necessary to keep track of the different ways elements in the various fundamental groups can be identified. We do this by assuming at the outset that each CW complex X under consideration is given a specific isomorphism (sometimes called a polarization), $p_X : \pi_1(X, *) \to G$. If $f : X \to Y$ is a cellular map of based complexes, there is a natural induced map $f_* : \pi_1(X, *) \to \pi_1(Y, *)$. We say that f_* is a *mapping over* G when $p_Y \circ f_* = \pi_X$. In cases where the nature of the identifications p_X and p_Y is clear, we shall abuse notation and say that f induces the identity on π_1.

46 Chain complexes, homotopy and homology

Let Λ be a ring; by a Λ-*chain complex* we mean a sequence $C_* = (C_r, \partial_r^C)_{r \in \mathbf{Z}}$ where $(C_r)_{r \in \mathbf{Z}}$ is a graded Λ-module, and for each r, $\partial_r^C : C_r \to C_{r-1}$ is a Λ-homomorphism, the rth *boundary map* such that

$$\partial_r^C \partial_{r+1}^C = 0$$

If $C_* = (C_r, \partial_r^C)_r$, $D_* = (D_r, \partial_r^D)_r$ are Λ-chain complexes, by a chain homomorphism over Λ, $f = (f_r) : C_* \to D_*$, we mean a graded Λ-homomorphism $f_* : C_* \to D_*$ such that the following diagram commutes for each r

$$
\begin{array}{ccc}
C_{n+1} & \to & C_n \\
\downarrow f_{n+1} & & \downarrow f_n \\
D_{n+1} & \to & D_n
\end{array}
$$

We say that a chain complex C_* is *bounded below* when, for some $N \in \mathbf{Z}$, $C_r = 0$ for all $r < N$. Likewise, we say that a chain complex C_* is *bounded above* when, for some $N \in \mathbf{Z}$, $C_r = 0$ for all $N < r$. Without further mention, all chain complexes will henceforth be bounded below.

We denote by $\mathcal{C}\text{hain}(\Lambda)$ the category whose objects are Λ-chain complexes C_* which are bounded below and whose morphisms are chain homomorphisms over Λ. If C_* is a Λ-chain complex we denote by $\mathcal{Z}_n(C) = \text{Ker}(\partial_n^C)$ the *module of n-cycles*, and by $\mathcal{B}_n(C) = \text{Im}(\partial_{n+1}^C)$ the *module of n-boundaries*. Since $\partial_r^C \partial_{r+1}^C = 0$ it follows that

$$\mathcal{B}_n(C) \subset \mathcal{Z}_n(C)$$

A Λ-chain complex C_* gives rise to graded Λ-module $H_*(C) = (H_n(C))_{n \in \mathbf{Z}}$, the *homology* of C_*, by means of

$$H_n(C) = \mathcal{Z}_n(C)/\mathcal{B}_n(C)$$

The correspondence $C_* \mapsto H_*(C)$ is functorial; given a chain mapping $f : C_* \to D_*$, for each $n \in \mathbf{Z}$ there is an induced Λ-homomorphism $H_n(f) : H_n(C) \to H_n(D)$ given by

$$H_n(f)(z + \mathcal{B}_n(C)) = f_n(z) + \mathcal{B}_n(D)$$

where $z \in Z_n(C)$. Fix a short exact sequence of chain complexes

$$0 \to A_* \xrightarrow{i} B_* \xrightarrow{i} C_* \to 0$$

For each n, there is a Λ-linear mapping $\delta : \mathcal{Z}_{n+1}(C) \to H_n(A)$ defined by the formula

$$\delta(c) = \left[i^{-1} \partial^B_{n+1} p^{-1}(c) \right]$$

where $[a]$ denotes the class in $H^n(A)$ of $a \in \mathcal{Z}(A)$; δ vanishes on $\mathcal{B}_{n+1}(C)$, and so induces a homomorphism, $\delta : H^{n+1}(C) \to H^n(A)$.

Proposition 46.1: With this notation, the sequence

$$H_{n+1}(B) \overset{H_{n+1}(p)}{\to} H_{n+1}(C) \overset{\delta}{\to} H_n(A) \overset{H_n(i)}{\to} H_n(B) \overset{H_n(p)}{\to} H_n(C)$$

is exact for all n.

If $f, g : C_* \to D_*$ are chain maps, we say that f *is homotopic to* g (written $f \simeq g$) when there are Λ-homomorphisms $\eta_n : C_n \to D_{n+1}$ such that, for each n

$$f_n - g_n = \eta_{n-1}\partial_n + \partial_{n+1}\eta_n$$

Proposition 46.2: Let $f, g : C_* \to D_*$ be chain maps; if $f \simeq g$, then $H_n(f) = H_n(g)$ for all n.

Proof: It suffices to show that for all $z \in \mathcal{Z}_n(C)$

$$f_n(z) + \mathcal{B}_n(D) = g_n(z) + \mathcal{B}_n(D)$$

We know that

$$f_n(z) - g_n(z) = \partial_{n+1}\eta_n(z) + \eta_{n-1}\partial_n(z)$$

for all $z \in C_n$. If $z \in \mathcal{Z}_n(C)$, that is, if $\partial_n(z) = 0$, then

$$f_n(z) - g_n(z) = \partial_{n+1}\eta_n(z)$$

so that $f_n(z) - g_n(z) \in \mathcal{B}_n(D)$, and the desired conclusion follows. □

Example: The cellular chains on a CW complex

For any, not necessarily finite, group G, we denote by $\mathbf{CW}_G$ the category whose objects are pairs (X, p_X), where X is a finite reduced CW complex with $\pi_1(X) \cong G$, and where $p_X : \pi_1(X) \cong G$ is a specific isomorphism, and whose morphisms are cellular maps *which induce the identity on* π_1. For any $n \geq 1$, we denote by $\mathbf{CW}^n_G$ the full subcategory of $\mathbf{CW}_G$ consisting of complexes of dimension $\leq n$. If K is a connected CW complex with $\pi_1(K) \cong G$, we denote by $C_*(K)$ the complex of free $\mathbf{Z}[G]$-modules constructed from the universal covering $\tilde{K}$.

Formally, we construct $C_*(K)$ from standard singular homology in the manner of Milnor [44] : take $C_n(K)$ to be the relative singular homology group

$$C_n(K) = H_n\big(\tilde{K}^{(n)}, \tilde{K}^{(n-1)}; \mathbf{Z}\big)$$

The boundary operator of the triple $(\tilde{K}^{(n)}, \tilde{K}^{(n-1)}, \tilde{K}^{(n-2)})$ gives a natural boundary map $\partial_n : C_n(K) \to C_{n-1}(K)$. On taking the homology of this complex in the standard way one simply retrieves the singular homology of $\tilde{K}$

$$H_n(C_*(K)) = H_n(\tilde{K}; \mathbf{Z})$$

However, the covering action of G on $\tilde{K}$ extends to an action of $\mathbf{Z}[G]$ on each $C_r(K)$, under which $C_r(K)$ becomes a free module over $\mathbf{Z}[G]$ with basis elements of $C_r(K)$ corresponding to liftings to $\tilde{K}$ of r-cells in K. Moreover, the boundary maps of $C_*(\tilde{K})$ are equivariant with respect to the covering action of G on $\tilde{K}$, so that $C_*(\tilde{K})$ can be regarded as a chain complex over $\mathbf{Z}[G]$. Since $\tilde{K}$ is simply connected, then $H_1(\tilde{K}; \mathbf{Z}) = 0$, and the sequence

$$C_2(K) \xrightarrow{\partial_2} C_1(K) \xrightarrow{\partial_1} C_0(K)$$

is exact at $C_1(K)$.

Since $\tilde{K}$ is its own universal covering, we may equally write $C_*(\tilde{K}) = C_*(K)$, though there is a pedantic difference in that $C_r(\tilde{K})$ is now a free module over $\mathbf{Z}[G]$, not because G is the fundamental group of $\tilde{K}$, but rather because G is a group which acts freely on $\tilde{K}$.

We say that a chain complex C_* is *contractible* when $\mathrm{Id}_{C_*} \simeq 0$. If C_* is contractible, then from the identity $\mathrm{Id}_{P_n} = \partial_{n+1}\eta_n + \eta_{n-1}\partial_n$ it follows easily that $\mathcal{Z}_n(C) = \mathcal{B}_n(C)$; hence $H_n(C) = 0$ for each n, and C_* is exact. When C_* is projective, the converse is true; thus suppose $C_* = (\cdots \to C_n \xrightarrow{\partial_n} C_{n-1} \xrightarrow{\partial_{n-1}} \cdots \to C_1 \xrightarrow{\partial_1} C_0 \to 0)$ is an exact sequence of Λ-modules in which each C_i is projective. We construct a contraction for C_* as follows: since ∂_1 is surjective, then by the universal property of projective modules there exists a homomorphism $\eta_0 : C_0 \to C_1$ making the following diagram commute

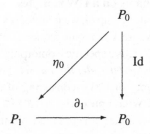

that is, $\partial_1 \eta_0 = \mathrm{Id}_{C_0}$. Taking, conventionally, $C_{-1} = 0$ and $\eta_{-1} : C_{-1} \to C_0$ $\partial_0 : C_0 \to C_{-1}$ to be the zero maps, we see that

$$\mathrm{Id}_{C_0} = \partial_1 \eta_0 + \eta_{-1}\partial_0$$

Suppose inductively that there are maps $\eta_r : C_r \to C_{r+1}$ such that

$$\mathrm{Id}_{C_r} = \partial_{r+1}\eta_r + \eta_{r-1}\partial_r$$

for $r < n$. In particular

$$\mathrm{Id}_{C_{n-1}} = \partial_n \eta_{n-1} + \eta_{n-2}\partial_{n-1}$$

so that

$$\partial_n = \partial_n \eta_{n-1}\partial_n + \eta_{n-2}\partial_{n-1}\partial_n$$

However, $\partial_{n-1}\partial_n = 0$, so that $\partial_n = \partial_n \eta_{n-1}\partial_n$, whence $\partial_n(\mathrm{Id}_{C_n} - \eta_{n-1}\partial_n) = \partial_n - \partial_n \eta_{n-1}\partial_n$. By exactness, it follows that $\mathrm{Im}(\mathrm{Id}_{C_n} - \eta_{n-1}\partial_n) \subset \mathrm{Im}(\partial_{n+1})$. Thus by the universal property for the projective module C_n, there exists a homomorphism $\eta_n : C_n \to C_{n+1}$ making the following diagram commute

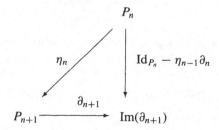

that is, $\mathrm{Id}_{C_n} - \eta_{n-1}\partial_n = \partial_{n+1}\eta_n$ or, equivalently

$$\mathrm{Id}_{C_n} = \partial_{n+1}\eta_n + \eta_{n-1}\partial_n$$

This completes the induction, and the construction of the contraction $\eta = (\eta_r)_{r \in \mathbb{N}}$, and we have shown:

Proposition 46.3: If C_* is a projective chain complex, then

$$C_* \text{ is contractible} \Longleftrightarrow H_*(C) = 0$$

If $f : C_* \to D_*$ is a chain mapping, we define a chain complex $M(f)_*$, the so-called *mapping cone of f*

$$M(f)_n = C_{n-1} \oplus D_n; \quad \partial_n^M = \begin{pmatrix} -\partial_{n-1}^C & 0 \\ f_{n-1} & \partial_n^D \end{pmatrix}$$

where we represent the elements of $C_{n-1} \oplus D_n$ as column vectors

$$\begin{pmatrix} c_{n-1} \\ d_n \end{pmatrix}$$

where $c_{n-1} \in C_{n-1}$ and $d_n \in D_n$.

Proposition 46.4: Let $f_* : C_* \to D_*$ be a chain mapping; if $M(f)$ is chain contractible, then there exists a chain mapping $g_* : D_* \to C_*$ such that $g \circ f \simeq \mathrm{Id}_C$ and $f \circ g \simeq \mathrm{Id}_D$.

Proof: Any Λ-linear mapping $\Psi_n : M(f)_n \to M(f)_{n+1}$ is represented by a matrix

$$\Psi_n = \begin{pmatrix} \xi_{n-1} & g_n \\ \gamma_{n-1} & \eta_n \end{pmatrix}$$

of Λ-linear mappings

$$\begin{aligned} \xi_{n-1} &: C_{n-1} \to C_n; \\ g_n &: D_n \to C_n; \\ \gamma_{n-1} &: C_{n-1} \to D_{n+1}; \\ \eta_n &: D_n \to D_{n+1}. \end{aligned}$$

A straightforward calculation gives the following matrix for $\Psi_{n-1}\partial_n^M + \partial_{n+1}^M \Psi_n$

$$\begin{pmatrix} -\left(\xi_{n-2}\partial_{n-1}^C + \partial_n^C \xi_{n-1}\right) + g_{n-1}f_{n-1} & g_{n-1}\partial_n^D - \partial_n^C g_n \\ f_n\xi_{n-1} + \eta_{n-1}f_{n-1}) + \partial_{n+1}^D \gamma_{n-1} - \gamma_{n-2}\partial_{n-1}^C & \eta_{n-1}\partial_n^D + \partial_{n+1}^D \eta_n + f_n g_n \end{pmatrix}$$

If Ψ is a chain contraction, then for each n

$$\Psi_{n-1}\partial_n^M + \partial_{n+1}^M \Psi_n = \begin{pmatrix} \mathrm{Id} & 0 \\ 0 & \mathrm{Id} \end{pmatrix}$$

so that, equating entries in the $(1, 2)$ position, we get

$$g_{n-1}\partial_n^D - \partial_n^C g_n = 0$$

that is, $g_* : D_* \to C_*$ is chain mapping. Equating entries in the $(1, 1)$ position gives

$$\xi_{n-2}\partial_{n-1}^C + \partial_n^C \xi_{n-1} = g_{n-1}f_{n-1} - \mathrm{Id}$$

that is, $\xi = (\xi_n)_n$ gives a chain homotopy $g \circ f \overset{\sim}{\to} \mathrm{Id}_{C_*}$. Finally, equating entries in the $(2, 2)$ position gives

$$\eta_{n-1}\partial_n^D + \partial_{n+1}^D \eta_n = \mathrm{Id} - f_n g_n$$

that is, $\eta = (\eta_n)_n$ gives a chain homotopy $g \circ f \xrightarrow{\sim} \mathrm{Id}_{C_*}$. This completes the proof. $\qquad\qquad\qquad\qquad\qquad\qquad\qquad\qquad\qquad\qquad\qquad\qquad\qquad\qquad\square$

For any chain complex C_*, we denote by $C_*(-1)$ the chain complex given by

$$C_n(-1) = C_{n-1}; \quad \partial_n^{C(-1)} = \partial_{n-1}^C$$

A chain mapping $f : C_* \to D_*$ gives rise to an obvious 'mapping cone' exact sequence

$$0 \to D_* \xrightarrow{i} M(f)_* \xrightarrow{p} C_*(-1) \to 0$$

Proposition 46.5: If $f : C_* \to D_*$ is a chain mapping, the long exact sequence in homology of the mapping cone sequence takes the following form

$$\cdots \to H_{n+1}(M(f)) \xrightarrow{p_*} H_n(C) \xrightarrow{f_*} H_n(D) \xrightarrow{i_*} H_n(M(f))$$
$$\xrightarrow{p_*} H_{n-1}(C) \xrightarrow{f_*} H_{n-1}(D) \cdots$$

Proof: In view of the identity $H_n(C(-1)) = H_{n-1}(C)$, the long exact sequence obtained from the mapping cone sequence has the form

$$\cdots \xrightarrow{i_*} H_{n+1}(M(f)) \xrightarrow{p_*} H_n(C) \xrightarrow{\delta} H_n(D) \xrightarrow{i_*} H_n(M(f))$$
$$\xrightarrow{p_*} H_{n-1}(C) \xrightarrow{\delta} H_{n-1}(D) \cdots$$

However, an easy calculation shows the connecting homomorphism $\delta : H_{n+1}(C(-1)) \equiv H_n(C) \to H_n(D)$ coincides with the homomomorphism $f_* : H_n(C) \to H_n(D)$ induced by $f : C_* \to D_*$. $\qquad\qquad\qquad\qquad\qquad\square$

Theorem 46.6: Let $f : C_* \to D_*$ be a chain map between (nonnegative) projective chain complexes, then the following statements are equivalent

(i) $H_*(f)$ is an isomorphism for each n;
(ii) $M(f)$ is exact;
(iii) $M(f)$ is chain contractible;
(iv) there exists a chain mapping $g : D_* \to C_*$ such that $f \circ g \simeq \mathrm{Id}$ and $g \circ f \simeq \mathrm{Id}$.

Proof: It suffices to show that (i) $\Rightarrow$ (ii) $\Rightarrow$ (iii) $\Rightarrow$ (iv) $\Rightarrow$ (i). Since C_* and D_* are assumed to be projective, so is $M(f)$, and the implications (ii) $\Rightarrow$ (iii) and (iii) $\Rightarrow$ (iv) are simply restatements of (46.3) and (46.4).

To show that (i) $\Rightarrow$ (ii), note that, if $H_n(f)$ is an isomorphism for each n, then from the exact sequence of (46.5) it follows that $H_n(M(f)) = 0$ for each

n, and so $M(f)$ is exact. Finally, the implication (iv) $\Rightarrow$ (i) is an immediate consequence of functoriality.   □

A chain complex C_* is said to be of *finite type* when C_* is bounded both above and below, and, in addition, each C_r is finitely generated. To any projective chain complex P_* of finite type we associate its *generalized Euler characteristic* (or *Wall obstruction*) defined by

$$\chi(P_*) = \sum_r (-1)^r [P_r] \in \tilde{K}_0(\mathbf{Z}[G])$$

The following is usually known as 'Whitehead's Trick' [69].

Proposition 46.7: Let P_* be a projective chain complex of finite type, and suppose that P_* is chain contractible; then $\bigoplus_r P_{2r+1} \cong \bigoplus_r P_{2r}$.

Proof: If Q_0 is projective then any exact sequence of the form

$$0 \to K \to Q_1 \to Q_0 \to 0$$

splits and $Q_1 \cong Q_0 \oplus K$. If Q_1 is projective, then so also is K. A straightforward induction, using cutting and splicing, shows that if

$$0 \to K \to Q_{n-1} \to \cdots \to Q_1 \to Q_0 \to 0$$

is an exact sequence where each Q_r is projective, then so also is K.

Now suppose that P_* is a contractible projective chain complex of finite type. By dimension shifting, if necessary, we may write P_* in the form

$$P_* = \left(0 \to P_n \xrightarrow{\partial_n} \cdots \xrightarrow{\partial_2} P_1 \xrightarrow{\partial_1} P_0 \to 0 \right)$$

Since P_* is chain contractible this sequence is exact. It suffices to show that

$$\bigoplus_{0 \le k} P_{n-2k} \cong \bigoplus_{0 \le k} P_{n-2k-1}$$

The proof goes by induction on n. When $n = 1$, $\partial_1 : P_1 \to P_0$ is the desired isomorphism. In general, suppose true for $n - 1$, and split the above exact sequence thus

$$0 \to P_n \xrightarrow{\partial_n} P_{n-1} \xrightarrow{\partial_1} \mathrm{Im}(\partial_{n-1}) \to 0$$

and

$$P_*' = \left(0 \to \mathrm{Im}(\partial_{n-1}) \subset P_{n-2} \xrightarrow{\partial_{n-2}} \cdots \xrightarrow{\partial_2} P_1 \xrightarrow{\partial_1} P_0 \to 0 \right)$$

By the observation above, $\mathrm{Im}(\partial_{n-1})$ is projective, and by induction

$$\bigoplus_{1\leq k} P_{n-2k} \cong \mathrm{Im}(\partial_{n-1}) \oplus \left(\bigoplus_{1\leq k} P_{n-2k-1}\right)$$

However, again since $\mathrm{Im}(\partial_{n-1})$ is projective, the sequence

$$0 \to P_n \xrightarrow{\partial_n} P_{n-1} \xrightarrow{\partial_1} \mathrm{Im}(\partial_{n-1}) \to 0$$

splits and

$$P_{n-1} \cong P_n \oplus \mathrm{Im}(\partial_{n-1})$$

It follows that

$$P_n \oplus \left(\bigoplus_{1\leq k} P_{n-2k}\right) \cong P_{n-1} \oplus \left(\bigoplus_{1\leq k} P_{n-2k-1}\right)$$

which is the required result. $\qquad\square$

Theorem 46.8: Let C_*, D_* be projective chain complexes of finite type, and let $f : C_* \to D_*$ be a chain map which induces an isomorphism on homology in each dimension; then

$$\chi(C_*) = \chi(D_*)$$

Proof: Let M_* denote the mapping cone of f. Clearly M_* is projective of finite type, and M_* is chain contractible, by (46.6). It follows that

$$\bigoplus_{0\leq k} M_{2k+1} \cong \bigoplus_{0\leq k} M_{2k}$$

However, $M_n \cong C_{n-1} \oplus D_n$, so that

$$\left(\bigoplus_{0\leq k} C_{2k}\right) \oplus \left(\bigoplus_{0\leq k} D_{2k+1}\right) \cong \left(\bigoplus_{0\leq k} C_{2k+1}\right) \oplus \left(\bigoplus_{0\leq k} D_{2k}\right)$$

and hence, taking formal differences in $\tilde{K}_0(\mathbf{Z}[G])$, we obtain

$$\sum_{0\leq k}([C_{2k}] - [C_{2k+1}]) + \sum_{0\leq k}([D_{2k+1}] - [D_{2k}]) = 0$$

so that $\chi(C_*) = \chi(D_*)$ as required. $\qquad\square$

Example: Cohomological period and free cohomological period
Recall that if G is a finite group then the reduced projective class group $\widetilde{K}_0(\mathbf{Z}[G])$ is finite. This has the following consequence.

Proposition 46.9: If k is a cohomological period of the finite group G, then Nk is a free period of G for some positive integer N.

Proof: If k is a cohomological period of G, then there exists an exact sequence

$$P_* = (0 \to \mathbf{Z} \to P_{k-1} \to \cdots \to P_0 \to \mathbf{Z} \to 0)$$

where each P_i is finitely generated projective. Consider the Wall obstruction $\chi(P_*) \in \widetilde{K}_0(\mathbf{Z}[G])$. Since $\widetilde{K}_0(\mathbf{Z}[G])$ is finite there exists a positive integer N such that $N\chi(P_*) = 0$. However $N\chi(P_*)$ is just the Wall obstruction of

$$\underbrace{P_* \circ \cdots \circ P_*}_{N}$$

the N-fold Yoneda product of P_* with itself. It follows from (30.2) that

$$\underbrace{P_* \cdots P_*}_{N}$$

can be replaced up to congruence by an exact sequence of the form

$$0 \to \mathbf{Z} \to F_{Nk-1} \to \cdots \to F_0 \to \mathbf{Z} \to 0$$

where each F_i is finitely generated free over $\mathbf{Z}[G]$; that is, Nk is a free period of G.

47 Two-dimensional complexes

We now restrict attention to complexes of dimension ≤ 2; observe that for $K \in \mathbf{CW}_G^2$, the kernel of $\partial_2 : C_2(K) \to C_1(K)$ is precisely $H_2(\tilde{K}; \mathbf{Z})$ and so the chain complex

$$0 \to C_2(K) \xrightarrow{\partial_2} C_1(K) \xrightarrow{\partial_1} C_0(K) \to 0$$

extends to an exact sequence of $\mathbf{Z}[G]$-modules

$$0 \to H_2(\tilde{K} : \mathbf{Z}) \to C_2(K) \xrightarrow{\partial_2} C_1(K) \xrightarrow{\partial_1} C_0(K) \to H_0(\tilde{K} : \mathbf{Z}) \to 0$$

Since $\tilde{K}$ is connected then $H_0(\tilde{K}; \mathbf{Z}) \cong \mathbf{Z}$. Moreover there is a canonical identification coming from the augmentation map $\epsilon : C_0(K) \to \mathbf{Z}$ obtained by projecting to the base point in K. If K is a finite reduced 2-complex with $\pi_1(K) = G$ we choose, once and for all, a lifting of the basepoint to a 0-cell $*_K \in \tilde{K}$. The map to the basepoint $\tilde{K} \to *$ induces an augmentation $\epsilon : C_0(\tilde{K}) = \mathbf{Z}[G] \to \mathbf{Z}$. If L is another such complex, by an *augmented* *chain map* $\varphi : C_*(\tilde{K}) \to C_*(\tilde{L})$ we mean a commutative diagram of $\mathbf{Z}[G]$-homomorphisms

$$\begin{array}{c} C_*(\tilde{K}) \\ \downarrow \varphi \\ C_*(\tilde{L}) \end{array} = \begin{pmatrix} C_2(\tilde{K}) \to C_1(\tilde{K}) \to \mathbf{Z}[G] \overset{\epsilon}{\to} \mathbf{Z} \to 0 \\ \downarrow \varphi_2 \qquad \downarrow \varphi_1 \qquad \downarrow \varphi_0 \quad \downarrow Id \\ C_2(\tilde{L}) \to C_1(\tilde{L}) \to \mathbf{Z}[G] \overset{\epsilon}{\to} \mathbf{Z} \to 0 \end{pmatrix}$$

Such a commutative diagram automatically induces a $\mathbf{Z}[G]$-homomorphism $\varphi_* : \pi_2(K) \to \pi_2(L)$ where we make the identifications $\pi_2(K) = \mathrm{Ker}(C_2(K) \to C_1(K))$, $\pi_2(L) = \mathrm{Ker}(C_2(L) \to C_1(L))$.

Furthermore, since $\tilde{K}$ is simply connected, $H_2(\tilde{K}; \mathbf{Z})$ may, by the Hurewicz Theorem, be identified with the second homotopy group $\pi_2(\tilde{K}) = \pi_2(K)$. The inclusion $\pi_2(K) \to C_2(K)$ is the *co-augmentation*, and it is useful to describe $C_*(K)$ as an exact sequence of $\mathbf{Z}[G]$-modules displaying both augmentation and co-augmentation, thus

$$C_*(K) = \left(0 \to \pi_2(K) \to C_2(K) \overset{\partial_2}{\to} C_1(K) \overset{\partial_1}{\to} C_0(K) \overset{\epsilon}{\to} \mathbf{Z} \to 0 \right)$$

We wish to give a purely algebraic approximation to two-dimensional geometric homotopy. We use the above as a model. To start with a two-dimensional chain complex

$$\mathbf{E} = \left(E_2 \overset{\partial_2}{\to} E_1 \overset{\partial_1}{\to} E_0 \right)$$

over $\mathbf{Z}[G]$ would be too general. The first restriction we impose is that $H_0(\mathbf{E}) \cong \mathbf{Z}$, corresponding to connectivity; the second is exactness at E_1, corresponding to simple connectivity of $\tilde{K}$.

In the case of a cellular chain complex $C_*(K)$, the $C_n(K)$ are free over $\mathbf{Z}[G]$. As we shall see, this is slightly too restrictive, and we allow E_i to be *stably free*. Thus, by an *algebraic 2-complex* over G we mean a sequence of $\mathbf{Z}[G]$ modules and homomorphisms

$$\mathbf{E} = \left(E_2 \overset{\partial_2}{\to} E_1 \overset{\partial_1}{\to} E_0 \right)$$

for which:

(i) $H_0(\mathbf{E}) \cong \mathbf{Z}$ and
(ii) $\mathrm{Ker}(\partial_1) = \mathrm{Im}(\partial_2)$.
(iii) E_i is finitely generated stably free.

We denote by $\mathbf{Alg}_G$ the full subcategory of $\mathcal{C}\text{hain}(\mathbf{Z}[G])$ whose objects are algebraic 2-complexes; that is, morphisms take the form

$$
\begin{matrix}
\mathbf{E} \\
\downarrow h \\
\mathbf{F}
\end{matrix}
=
\begin{pmatrix}
0 \to E_2 \to E_1 \to E_0 \to 0 \\
\quad \downarrow h_2 \quad \downarrow h_1 \quad \downarrow h_0 \\
0 \to F_2 \to F_1 \to F_0 \to 0
\end{pmatrix}
$$

Clearly the correspondence $K \mapsto C_*(K)$ defines a functor $C_* : \mathbf{CW}_G^2 \to \mathbf{Alg}_G$. By analogy with the geometrical case, it is useful to think of $H_2(\mathbf{E})$ as an 'algebraic π_2', and we put

$$\pi_2(\mathbf{E}) = \mathrm{Ker}(\partial_2 : E_2 \to E_1)$$

enabling us to write $\mathbf{E}$ in augmented, co-augmented form as an exact sequence

$$\mathbf{E} = \left(0 \to \pi_2(\mathbf{E}) \to E_2 \xrightarrow{\partial_2} E_1 \xrightarrow{\partial_1} E_0 \to \mathbf{Z} \to 0 \right)$$

When so written, the appropriate notion of homotopy equivalence is *weak homotopy equivalence*; if $\mathbf{E}$, $\mathbf{F}$ are algebraic 2-complexes, a chain mapping $h : \mathbf{E} \to \mathbf{F}$ is said to be a weak homotopy equivalence when the induced maps $h_* : H_0(\mathbf{E}) \to H_0(\mathbf{F})$ and $h_* : \pi_2(\mathbf{E}) \to \pi_2(\mathbf{F})$ are isomorphisms. By hypothesis, $H_1(\mathbf{E}) = H_1(\mathbf{F}) = 0$ so the isomorphism $h_* : H_1(\mathbf{E}) \to H_1(\mathbf{F})$ is automatic. From (46.6) it follows that:

Proposition 47.1: Let $h : \mathbf{E} \to \mathbf{F}$ be a morphism of algebraic 2-complexes over G; then

h is a weak homotopy equivalence $\iff$ h is a chain homotopy equivalence.

Thus the question of whether or not to include augmentation and co-augmentation in the description is essentially a matter of taste. It also follows directly from (46.6) that:

Proposition 47.2: Weak homotopy equivalence is an equivalence relation on $\mathbf{Alg}_G$.

This has the practical consequence that when $h : \mathbf{P}_1 \to \mathbf{P}_2$ is a morphism of algebraic 2-complexes inducing isomorphisms $h_r : H_r(\mathbf{P}_1) \to H_r(\mathbf{P}_2)$, then there exists a morphism $g : \mathbf{P}_2 \to \mathbf{P}_1$ inducing $g_r = h_r^{-1} : H_r(\mathbf{P}_2) \to H_r(\mathbf{P}_1)$.

There is the generalization which it is occasionally necessary to consider; denote by $\mathbf{Proj}_G$ the full subcategory of $\mathcal{C}\text{hain}(\mathbf{Z}[G])$ whose objects are *projective 2-complexes*

$$\mathbf{P} = \left(P_2 \xrightarrow{\partial_2} P_1 \xrightarrow{\partial_1} P_0 \right)$$

in which, apart from the previous conditions $\operatorname{Coker}(\partial_1) \cong \mathbf{Z}$ and $\operatorname{Ker}(\partial_1) = \operatorname{Im}(\partial_2)$, we require also that each P_i be finitely generated projective. Likewise there is a specialization, $\mathbf{Free}_G$, in which objects

$$\mathbf{F} = \left(F_2 \xrightarrow{\partial_2} F_1 \xrightarrow{\partial_1} F_0 \right)$$

must have F_i finitely generated. One sees easily that any algebraic 2-complex is congruent, hence homotopy equivalent, to a free 2-complex, whereas a projective 2-complex

$$\mathbf{P} = \left(P_2 \xrightarrow{\partial_2} P_1 \xrightarrow{\partial_1} P_0 \right)$$

is homotopy equivalent to an algebraic 2-complex if and only if $\chi(\mathbf{P}) = 0$.

Let $\mathbf{E} =$ be an algebraic 2-complex over G. We say that $\mathbf{E}$ is *geometrically realizable* when there is an object $X \in \mathbf{CW}_G^2$ and a weak homotopy equivalence $\varphi : C_*(X) \to \mathbf{E}$. By symmetry, this is equivalent to the existence of a weak homotopy equivalence $\psi : \mathbf{E} \to C_*(X)$.

Writing an algebraic 2-complex in augmented and co-augmented form

$$\mathbf{E} = \left(0 \to \pi_2(\mathbf{E}) \to E_2 \xrightarrow{\partial_2} E_1 \xrightarrow{\partial_1} E_0 \to \mathbf{Z} \to 0 \right)$$

gives (at least in those cases which come within the scope of our previous definitions, that is, when G is finite), an element of $\mathbf{Stab}^3(\mathbf{Z}, \pi_2(\mathbf{E}))$. In the case of a geometrical cellular chain complex

$$C_*(K) = \left(0 \to \pi_2(K) \to C_2(K) \xrightarrow{\partial_2} C_1(K) \xrightarrow{\partial_1} C_0(K) \xrightarrow{\epsilon} \mathbf{Z} \to 0 \right)$$

we get an element of $\mathbf{Free}^3(\mathbf{Z}, \pi_2(K))$.

48 Cayley complexes

The above discussion can be carried out in the context of group presentations; to any finite presentation $\mathcal{G} = \langle x_1, \dots, x_g \mid W_1, \dots, W_r \rangle$ of a group G, we associate a canonical two-dimensional complex $K_{\mathcal{G}}$ with $\pi_1(K_{\mathcal{G}}) \cong G$ by regarding the generators x_i as 1-cells and the relators W_j as 2-cells. For any such presentation $\mathcal{G}$, we write $\pi_2(\mathcal{G})$ for the second homotopy group $\pi_2(K_{\mathcal{G}})$. Put $K = K_{\mathcal{G}}$, and let $\tilde{K}$ denote the universal cover; then $\pi_2(\mathcal{G}) = \pi_2(K) = \pi_2(\tilde{K})$ is a module over $\mathbf{Z}[G]$ via the action of G on $\tilde{K}$. We have a free algebraic 2-complex

$$0 \to \pi_2(\mathcal{G}) \to C_2(K_{\mathcal{G}}) \xrightarrow{\partial_2} C_1(K_{\mathcal{G}}) \xrightarrow{\partial_1} C_0(K_{\mathcal{G}}) \xrightarrow{\epsilon} \mathbf{Z} \to 0$$

the algebraic Cayley complex of the presentation. When G is finite, $\Omega_3(\mathbf{Z})$ is defined, and we obtain:

Proposition 48.1: If $\mathcal{G}$ is a finite presentation of G, then $\pi_2(\mathcal{G}) \in \Omega_3(\mathbf{Z})$.

Again, when G is finite, $\tilde{K}$ is a finite 2-complex and, by simple connectivity, we have

$$\tilde{K} \simeq \underbrace{S^2 \vee \cdots \vee S^2}_{N}$$

A simple argument using the multiplicativity of the Euler characteristic on finite coverings allows us to calculate the $\mathbf{Z}$-rank of $\pi_2(\mathcal{G})$:

Proposition 48.2: Let $\mathcal{G} = \langle x_1, \ldots, x_g; W_1, \ldots, W_r \rangle$ be a finite presentation of the finite group G and $K = K_{\mathcal{G}}$ the canonical 2-complex associated with $\mathcal{G}$; then $\pi_2(\mathcal{G})$ is a free abelian group of rank $(r - g + 1)|G| - 1$.

This yields a well-known fact:

Proposition 48.3: Let $\mathcal{G} = \langle x_1, \ldots, x_g; W_1, \ldots, W_r \rangle$ be a finite presentation of the finite group G; then $g \leq r$.

From the Hurewicz Theorem we see that $\pi_2(K_{\mathcal{G}}) \cong H_2(\tilde{K}_{\mathcal{G}}; \mathbf{Z})$. Thus calculation of $\pi_2(K_{\mathcal{G}})$, and hence of $\Omega_3(\mathbf{Z})$, comes down to being able to give an effective description of $\text{Ker}(\partial_2)$. This is possible in proportion to the extent that we can describe G effectively. Fox gave a formal method, free differential calculus [19], which we illustrate by an example.

The quaternion group of order 16

We take $Q(16)$ in its standard presentation

$$Q(16) = \langle x, y : x^4 = y^2; xyx = y \rangle$$

The two generators x, y give rise to two 1-cells ϵ_1, ϵ_2, which, when lifted to the universal covering, we may portray as follows

$$
\begin{array}{ccc}
[1] & \epsilon_1 & [x] \\
\bullet\!\!\longrightarrow\!\!\bullet & &
\end{array}
\qquad
\begin{array}{ccc}
[1] & \epsilon_2 & [y] \\
\bullet\!\!\longrightarrow\!\!\bullet & &
\end{array}
$$

If ϵ_1 is orientated by starting from $[1]$ and ending at $[x]$, then its boundary is given by

$$\partial(\epsilon_1) = x - 1$$

Likewise

$$\partial(\epsilon_2) = y - 1$$

Observe that $\partial(\epsilon_1)$ and $\partial(\epsilon_2)$ belong to, and indeed together generate, the augmentation ideal $\mathbf{I}(Q(16))$.

The first relation $x^4 = y^2$ gives a 2-cell $\mathbf{E_1}$, which, when lifted to the universal covering, becomes a 6-sided polygon bounded by translates of the basic 1-cells as follows

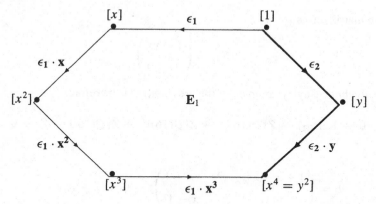

Expressing the boundary of the 2-cell $\mathbf{E_1}$ in terms of the translates of the basic 1-cells we obtain

$$\partial(\mathbf{E_1}) = \epsilon_1(1 + x + x^2 + x^3) - \epsilon_2(1 + y)$$

or, in matrix terms

$$\partial(\mathbf{E_1}) = \begin{pmatrix} 1 + x + x^2 + x^3 \\ -(1 + y) \end{pmatrix}$$

Similarly, the second relation $xyx = y$ gives a 2-cell $\mathbf{E_2}$, which, when lifted to the universal covering, becomes the following 4-sided polygon

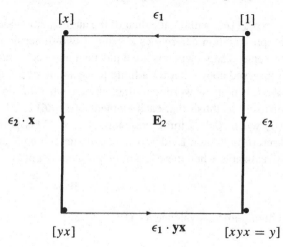

The boundary $\mathbf{E}_2$ is then expressed as

$$\partial(\mathbf{E}_2) = \epsilon_1(1 + yx) + \epsilon_2(x - 1)$$

or, in matrix terms

$$\partial(\mathbf{E}_2) = \begin{pmatrix} 1 + yx \\ -1 + x \end{pmatrix}$$

The algebraic Cayley complex of the presentation is therefore

$$0 \to \mathrm{Ker}(\partial_2) \to \mathbf{Z}[Q(16)]^2 \overset{\partial_2}{\to} \mathbf{Z}[Q(16)]^2 \overset{\partial_1}{\to} \mathbf{Z}[Q(16)] \overset{\epsilon}{\to} \mathbf{Z} \to 0$$

where

$$\partial_2 = \begin{pmatrix} 1 + x + x^2 + x^3 & yx + 1 \\ -(y + 1) & x - 1 \end{pmatrix}; \quad \partial_1 = (x - 1, y - 1)$$

and where ϵ is the augmentation map. Normally speaking, it is still a step to go from a description of ∂_2 to a description of $\pi_2(\mathcal{G}) = \mathrm{Ker}(\partial_2)$. In this case we are aided by the fact that we can extend the Cayley complex to a complete resolution of period 4 (compare the complete resolution given in Section 41 and also [12] Chapter XII) as follows

$$0 \to \mathbf{Z} \overset{\epsilon^*}{\to} \mathbf{Z}[Q(16)] \overset{\delta}{\to} \mathbf{Z}[Q(16)]^2 \overset{\partial_2}{\to} \mathbf{Z}[Q(16)]^2 \overset{\partial_1}{\to} \mathbf{Z}[Q(16)] \overset{\epsilon}{\to} \mathbf{Z} \to 0$$

where

$$\delta = \begin{pmatrix} x - 1 \\ 1 - yx \end{pmatrix}$$

This identifies $\pi_2(\mathcal{Q}(16))$ with I^*, the dual of the augmentation ideal.

Every finite presentation determines a reduced two-dimensional cell complex, and vice versa. The viewpoints are equivalent and occasionally complementary. We observed above that for a finite presentation of a finite group G, $\pi_2(\mathcal{G}) \in \Omega_3(\mathbf{Z})$. It is perhaps worth pointing out here what this does *not* mean. It does not of itself establish that each element of $\Omega_3(\mathbf{Z})$ can be realized in the form $\pi_2(\mathcal{G})$ where $\mathcal{G}$ is a finite presentation of G. This point is crucial, and we shall return to it later. Evidently, an algebraic 2-complex $\mathcal{E} \in \mathbf{Alg}_G$ is geometrically realizable when there is a finite presentation of G

$$\mathcal{G} = \langle x_1, \ldots, x_g \mid W_1, \ldots, W_r \rangle$$

and a (weak) homotopy equivalence $h : C_*(\mathcal{G}) \overset{\sim}{\to} \mathcal{E}$.

When G is finite, the general fact that $\pi_2(\mathcal{G}) \in \Omega_3(\mathbf{Z})$ means that, although the isomorphism class of $\pi_2(\mathcal{G})$ in $\mathcal{F}(\mathbf{Z}[G])$ depends upon the particular presentation $\mathcal{G}$, its *stable class depends only upon* G. In fact, this is true for all finitely presented groups, as was first demonstrated by Tietze [66]. Tietze's Theorem shows that any two finite presentations of G are related by a finite chain of transformations of the following type (or their inverses):

(I) Add a new generator x and a new relator R of the form xw^{-1}, where w is a word in the existing generators.

(II) Add a relator R which is a word in the existing relators.

If $\mathcal{G} = \langle x_1, \ldots, x_g \mid W_1, \ldots, W_r \rangle$ is a finite presentation of G, then under a Tietze transformation of type (I), $\pi_2(K_{\mathcal{G}})$ remains the same, whilst under a Tietze transformation of type (II), $\pi_2(K_{\mathcal{G}})$ adds on a direct summand $\mathbf{Z}[G]$. It follows that:

Theorem 48.4: Let $\mathcal{G}_1 = \langle x_1, \ldots, x_g \mid R_1, \ldots, R_r \rangle$ and $\mathcal{G}_2 = \langle y_1, \ldots, y_h \mid S_1, \ldots, S_s \rangle$ be finite presentations for G; then $\pi_2(K_{\mathcal{G}_1})$ and $\pi_2(K_{\mathcal{G}_2})$ are stably equivalent.

We point out that $\Omega_2(\mathbf{Z})$ also plays a role in free differential calculus; $\mathrm{Ker}(\partial_1)$ (sometimes called the *second Fox ideal*, despite the fact that it is not normally an ideal) is a representative of $\Omega_2(\mathbf{Z})$.

49 Algebraicity of two-dimensional homotopy

The main result of this section applies to all finitely presented groups, not merely finite groups. Throughout this section, until further notice, G will denote a finitely presented group.

The functor $C_* : \mathbf{CW}_G^2 \to \mathbf{Alg}_G$ transforms geometric homotopy into chain homotopy; that is, if $K, L \in \mathbf{CW}_G^2$ and $f : K \to L$ is a homotopy equivalence over the identity in G, then $C(f)_* : C_*(K) \to C_*(L)$ is a chain homotopy equivalence. If $K, L \in \mathbf{CW}_G^2$, and $f : K \to L$ is a cellular homotopy equivalence which induces the identity on $G = \pi_1(K) = \pi_1(L)$, then the induced maps $f_* : H_0(\tilde{K}) \to H_0(\tilde{L})$ and $f_* : \pi_2(K) \to \pi_2(L)$ are isomorphisms, so that, by (46.6), $C_*(K)$ and $C_*(L)$ are weakly homotopy equivalent; that is:

Proposition 49.1: If $K, L \in \mathbf{CW}_G^2$, then

$$K \simeq_G L \Longrightarrow C_*(K) \simeq C_*(L)$$

The converse to this is also true, as we now proceed to show.

We can make some simplifying transformations. Let

$$\mathcal{G} = \langle x_1, \ldots, x_g \mid W_1, \ldots, W_r \rangle$$

be a finite presentation for G, and let $y = y(x_1, \ldots, x_g)$ be a word in the generators. Consider the presentation

$$\mathcal{G}(y) = \langle x_1, \ldots, x_{g+1} \mid W_1, \ldots, W_{r+1} \rangle$$

where $W_{r+1} = x_{g+1}^{-1} y$. Then $\mathcal{G}(y)$ is also a presentation of G, the passage from $\mathcal{G}$ to $\mathcal{G}(y)$ being effected a Tietze transformation of type (I). If $K = \mathcal{K}_\mathcal{G}$ is the Cayley complex of $\mathcal{G}$ and $K(y)$ is that of $\mathcal{G}(y)$ then $K(y)$ is homotopy equivalent to K by a simple homotopy equivalence over the identity on $\pi_1 = G$.

Proposition 49.2: Let K, L be finite reduced 2-complexes with $\pi_1(K) = \pi_1(L) = G$; then there exist finite reduced 2-complexes K_1, L_1 such that

(i) $K_1 \simeq_G K$ and $L_1 \simeq_G L$;
(ii) $K_1^{(1)} \equiv L_1^{(1)}$.

Proof: Let $\{x_1, \ldots, x_m\}$ (resp. $\{y_1, \ldots, y_n\}$) be the generating sets of G corresponding to the 1-cells of K (resp. L), and regard y_i as a word in $\{x_1, \ldots, x_m\}$. We can regard K as the Cayley complex of some presentation $\mathcal{G}$. Start with the generating set $\{x_1, \ldots, x_m\}$; by adding the elements $y_1, \ldots, y_n$ successively, we obtain a sequence of presentations

$$\mathcal{G}, \ \mathcal{G}(y_1), \ \mathcal{G}(y_1, y_2), \ldots, \mathcal{G}(y_1, y_2, \ldots, y_n)$$

and a sequence of (simple) homotopy equivalences over the identity on $\pi_1 = G$

$$K \subset K(y_1) \subset K(y_1, y_2) \subset \cdots \subset K(y_1, y_2, \ldots, y_n) = K_1$$

Likewise, regarding L as the Cayley complex $\mathcal{L} = \mathcal{L}_\mathcal{H}$ of some presentation $\mathcal{H}$, and adding the elements $x_1, \ldots, x_m$ successively, starting with the generating set $\{y_1, \ldots, y_n\}$, we obtain a sequence of presentations

$$\mathcal{H}, \ \mathcal{H}(x_1), \ \mathcal{H}(x_1, x_2), \ldots, \mathcal{H}(x_1, x_2, \ldots, x_m)$$

and a sequence of (simple) homotopy equivalences again over the identity on G

$$L \subset L(x_1) \subset L(x_1, x_2) \subset \cdots \subset L(x_1, x_2, \ldots, x_m) = L_1$$

Then in some ordering, the 1-cells of K_1 and those of L_1 correspond to the generating elements $(x_1, x_2, \ldots, x_m, y_1, y_2, \ldots, y_n)$; that is, we may identify $K_1^{(1)}$ with $L_1^{(1)}$.   □

Proposition 49.3: Let K, L be finite reduced 2-complexes with $\pi_1(K) = \pi_1(L) = G$, and suppose that $K^{(1)} = L^{(1)}$. If $v : C_*(K) \to C_*(L)$ is an augmented $\mathbf{Z}[G]$-chain map; then v is chain homotopic to a chain map μ which restricts to the identity on $C_1(\tilde{K}) = C_1(\tilde{L})$.

Proof: Begin with the chain map

$$
\begin{array}{cc}
C_*(K) \\
\downarrow v \;\; = \\
C_*(L)
\end{array}
\left(
\begin{array}{ccccccccc}
C_2(K) & \overset{\partial_2^K}{\to} & C_1(K) & \overset{\partial_1}{\to} & \mathbf{Z}[G] & \overset{\epsilon}{\to} & \mathbf{Z} & \to & 0 \\
\downarrow v_2 & & \downarrow v_1 & & \downarrow v_0 & & \downarrow Id \\
C_2(L) & \overset{\partial_2^L}{\to} & C_1(L) & \overset{\partial_1}{\to} & \mathbf{Z}[G] & \overset{\epsilon}{\to} & \mathbf{Z} & \to & 0
\end{array}
\right)
$$

where, by hypothesis, $C_1(L) = C_1(K)$ and $\partial_1^K = \partial_1^L = \partial_1$. Then $\mathrm{Im}(Id - v_0) \subset \mathrm{Ker}(\epsilon) = \mathrm{Im}(\partial_1)$. Choose a $\mathbf{Z}[G]$-homomorphism $h_0 : \mathbf{Z}[G] \to C_1(L)$ such that

$$\partial_1 h_0 = Id - v_0$$

and put $\lambda_1 = h_0 \partial_1 + v_1$, $\lambda_2 = v_2$. We obtain a chain map

$$
\begin{array}{cc}
C_*(K) \\
\downarrow \lambda \;\; = \\
C_*(L)
\end{array}
\left(
\begin{array}{ccccccccc}
C_2(K) & \overset{\partial_2}{\to} & C_1(K) & \overset{\partial_1}{\to} & \mathbf{Z}[G] & \overset{\epsilon}{\to} & \mathbf{Z} & \to & 0 \\
\downarrow \lambda_2 & & \downarrow \lambda_1 & & \downarrow Id & & \downarrow Id \\
C_2(L) & \overset{\partial_2^L}{\to} & C_1(L) & \overset{\partial_1}{\to} & \mathbf{Z}[G] & \overset{\epsilon}{\to} & \mathbf{Z} & \to & 0
\end{array}
\right)
$$

Clearly $\mathrm{Im}(Id_1 - \lambda_1) \subset \mathrm{Ker}(\partial_1) = \mathrm{Im}(\partial_2^L)$. We can now choose a $\mathbf{Z}[G]$-homomorphism $h_1 : C_1(K) \to C_2(L)$ such that $\partial_2^L h_1 = Id_1 - \lambda_1$. Put $\mu_2 = h_1 \partial_2 + \lambda_2$. Then

$$
\begin{aligned}
\partial_2^L \mu_2 &= \partial_2^L (h_1 \partial_2 + \lambda_2) \\
&= \partial_2^L h_1 \partial_2^K + \partial_2^L \lambda_2 \\
&= \partial_2 - \lambda_1 \partial_2 + \partial_2^L \lambda_2 \\
&= \partial_2
\end{aligned}
$$

The commutative diagram

$$
\begin{array}{ccccccccc}
C_2(K) & \overset{\partial_2^K}{\to} & C_1(K) & \overset{\partial_1}{\to} & \mathbf{Z}[G] & \overset{\epsilon}{\to} & \mathbf{Z} & \to & 0 \\
\downarrow \mu_2 & & \downarrow Id & & \downarrow Id & & \downarrow Id \\
C_2(L) & \overset{\partial_2^L}{\to} & C_1(L) & \overset{\partial_1}{\to} & \mathbf{Z}[G] & \overset{\epsilon}{\to} & \mathbf{Z} & \to & 0
\end{array}
$$

defines the desired chain map μ, and $h = (h_1, h_0)$ is a chain homotopy from μ to v. $\qquad\square$

If $\varphi : C_*(K) \rightarrow C_*(L)$ is an algebraic chain map, by a *realization* of φ we mean a cellular map $f : K \rightarrow L$ such that $C_*(f) = \varphi$. The following Lemma, which is a special case of a slightly more general result of Gutierrez and Latiolais ([21] Prop. (2.3)), gives conditions under which some chain maps are realizable.

Lemma 49.4: Let G be a finitely presented group and let $K, L \in \mathbf{CW}_G^2$; suppose that

(i) $K^{(1)} \equiv L^{(1)}$ and
(ii) there exists $\varphi : C_*(K) \rightarrow C_*(L)$ such that $\varphi_{|C_i(K)} = \mathrm{Id}$ for $i = 0, 1$;

then there exists a cellular map $f : K \rightarrow L$ such that

(iii) f extends the identity of $K^{(1)} \equiv L^{(1)}$ and induces the identity on π_1, and
(iv) $C_*(f) = \varphi$.

Proof: The hypothesis on φ ensures that the following diagram commutes

$$
\begin{array}{ccc}
C_2(K) & \xrightarrow{\partial_2^K} & C_1(K) \\
\downarrow \varphi_2 & & \| \\
C_2(L) & \xrightarrow{\partial_2^L} & C_1(L)
\end{array}
$$

First choose some cellular map $\lambda : K \rightarrow L$ with the property that λ extends the identity of $K^{(1)} \equiv L^{(1)}$ and induces the identity on π_1. This can be done, for example, by first attaching cells of dimension ≥ 3 to L to form a space B_G of homotopy type $K(G, 1)$. By construction, $B_G^{(2)} = L$. Then the identity map on $K^{(1)}$ extends to a map $\lambda : K \rightarrow B_G$ which classifies the universal G-bundle $\tilde{K} \rightarrow K$ and induces the identity on π_1. Without loss of generality, the cellular approximation theorem allows us to choose λ to be cellular without altering its definition on $K^{(1)}$ where it already is cellular. In particular $\lambda : K \rightarrow \lambda(K) \subset B_G^{(2)} = L$.

Now the chain map $C_*(\lambda)$ induced by λ makes the following diagram commute

$$
\begin{array}{ccc}
C_2(K) & \xrightarrow{\partial_2^K} & C_1(K) \\
\downarrow \lambda_2 & & \| \\
C_2(L) & \xrightarrow{\partial_2^L} & C_1(L)
\end{array}
$$

In particular

$$
\partial_2^L(\varphi_2 - \lambda_2) = 0
$$

By the Hurewicz Theorem, $\mathrm{Ker}(\partial_2^L) = H_2(\tilde{L}; \mathbf{Z})$ can be identified with $\pi_2(L)$, so that $\varphi_2 - \lambda_2$ gives a $\mathbf{Z}[G]$-homomorphism

$$\varphi_2 - \lambda_2 : C_2(K) \to \pi_2(L)$$

Let $\{E_1, \ldots, E_r\}$ be an enumeration of the 2-cells of K. After lifting to $\tilde{K}$, we get a $\mathbf{Z}[G]$-basis $\{\hat{E}_1, \ldots, \hat{E}_r\}$ of $C_2(K)$. For each i, let $h_i : S^2 \to L$ be a map whose homotopy class $[h_i]$ satisfies

$$[h_i] = (\varphi_2 - \lambda_2)(\hat{E}_i)$$

Let E denote the standard 2-cell

$$E = \{\mathbf{x} \in \mathbf{R}^2 : ||\mathbf{x}|| \leq 1\}$$

and put

$$\sigma = \left\{ \mathbf{x} \in \mathbf{R}^2 : ||\mathbf{x}|| = \frac{1}{2} \right\}.$$

The quotient E/σ is homeomorphic to the one point union $E \vee S^2$, so that we have a 'pinch map' $c : E \to E \vee S^2$, obtained by collapsing σ to a point. Doing this for each of the 2-cells of K gives a collection of pinch maps $c_i : E_i \to E_i \vee S_i^2$. Now we may write

$$K = \left(\coprod_i E_i \right) / \sim$$

where '$\sim$' makes identifications in the boundaries of the E_i. Put

$$X = \left(\coprod_i E_i \vee S_i^2 \right) / \sim$$

where '$\sim$' makes the same identifications as before; observe that E_i and $E_i \vee S_i^2$ have the same boundary. One may think of X as formed from K by attaching a copy of S^2 at a single interior point within each 2-cell of K.

Put $\lambda_i = \lambda_{|E_i}$ and define

$$c = \left(\coprod_i c_i \right) / \sim \; : K \to X; \quad \Psi = \left(\coprod_i \lambda_i \vee h_i \right) / \sim \; : X \to L$$

and put $f = \Psi \circ c : K \to L$. f is a basepoint preserving map; it is straightforward to see that f extends the identity on $K^{(1)}$ and induces the identity on π_1. Furthermore, one calculates easily that, if $f_n : C_n(K) \to C_n(L)$ denotes the induced map on n-chains, then

$$f_2(\hat{E}_i) = \lambda_2(\hat{E}_i) + [h_i]$$

where we identify $[h_i] \in \pi_2(L)$ with its image in $C_2(L)$. However, by choice

$$(\varphi_2 - \lambda_2)(\hat{E}_i) = [h_i]$$

Thus

$$f_2(\hat{E}_i) = \lambda_2(\hat{E}_i) + (\varphi_2 - \lambda_2)(\hat{E}_i)$$
$$= \varphi_2(\hat{E}_i)$$

and so $f_2 = \varphi_2$ as required.   □

Theorem 49.5: Let G be a finitely presented group, and let $K, L \in \mathbf{CW}_G^2$; suppose that there exists an algebraic chain map $\varphi : C_*(K) \to C_*(L)$ which is a chain homotopy equivalence, then there exists a cellular homotopy equivalence $f : K \to L$ whose induced map on the fundamental group is the identity.

Proof: Let $\varphi : C_*(K) \to C_*(L)$ be the postulated chain homotopy equivalence. Since φ induces an isomorphism on $H_0 \cong \mathbf{Z}$, and since $\mathrm{Aut}(\mathbf{Z}) = \pm\mathrm{Id}$, then by replacing φ by $-\varphi$, if necessary, we may assume without loss of generality that φ induces the identity on $H_0 \cong \mathbf{Z}$.

By (49.2), we may choose complexes $K_1, L_1 \in \mathbf{CW}_G^2$ such that $K^{(1)} = L^{(1)}$, and cellular homotopy equivalences $\alpha : K \xrightarrow{\sim} K_1$, $\beta : L_1 \xrightarrow{\sim} K$, inducing the identity on π_1. Let $\delta : K_1 \xrightarrow{\sim} K$, $\gamma : L \xrightarrow{\sim} L_1$ be homotopy inverses for α, β, and put

$$\mu = C_*(\gamma) \circ \varphi \circ C_*(\delta) : C_*(K_1) \to C_*(L_1)$$

Then μ is an algebraic chain homotopy equivalence inducing the identity on H_0. By (49.3), we may further assume that μ restricts to the identity on $C_r(K_1) = C_r(L_1)$ for $r = 0, 1$. Applying (49.4), we obtain a cellular map $g : K_1 \to L_1$ inducing the identity on the fundamental group and such that $g_* = \mu : H_2(\tilde{K}_1; \mathbf{Z}) \to H_2(\tilde{L}_1; \mathbf{Z})$. In particular, $g_* : H_2(\tilde{K}_1; \mathbf{Z}) \to H_2(\tilde{L}_1; \mathbf{Z})$ is an isomorphism, so that g is a homotopy equivalence, by Whitehead's Theorem. Then $f = \beta \circ g \circ \alpha : K \xrightarrow{\sim} L$ is the desired cellular homotopy equivalence.   □

Corollary 49.6: If $K, L \in \mathbf{CW}_G^2$, then

$$K \simeq_G L \Longleftrightarrow C_*(\tilde{K}) \simeq C_*(\tilde{L})$$

When $\mathcal{H}$ is a category which possesses a suitable notion of homotopy, we will denote by $\widehat{\mathcal{H}}$ its class of homotopy types. Then (49.5) can be paraphrased as saying:

(49.7) The mapping induced by the cellular chain functor

$$C_* : \widehat{\mathbf{CW}_G^2} \to \widehat{\mathbf{Alg}_G}$$

is injective.

Chapter 9
Stability theorems

In this chapter we prove three of the six theorems stated in the Introduction. We have seen that stabilization induces a tree structure on the stable module $\Omega_3(\mathbf{Z})$. Two-dimensional homotopy types possess an analogous tree structure; homotopy types at the bottom level are said to be minimal. We say that G has the *realization property* when all algebraic 2-complexes over $\mathbf{Z}[G]$ are geometrically realizable. The first of our results is then:

Theorem II: The finite group G has the realization property if and only if all minimal algebraic 2-complexes are realizable.

Theorem II is a general condition on homotopy types. It is more useful to have a criterion for realizability in terms of homotopy *groups*. The second result of this Chapter gives a sufficient condition on the minimal modules in $\Omega_3(\mathbf{Z})$ for G to possess the realization property. We say that a module $J \in \Omega_3(\mathbf{Z})$ is *realizable* when for some finite 2-complex $K \in \mathbf{CW}_G^2$ there is an isomorphism of $\mathbf{Z}[G]$-modules $J \cong \pi_2(K)$; J is said to be *full* when the natural map $\mathrm{Aut}_{\mathbf{Z}[G]}(J) \to \mathrm{Ker}(S^J) \subset \mathrm{Aut}_{\mathcal{D}\mathrm{er}}(J)$ is surjective; we also prove:

Theorem III: If each minimal module $J \in \Omega_3(\mathbf{Z})$ is both realizable and full, then G has the realization property.

We proceed to both results via stabilization; at the geometric level this is simply the operation

$$X \mapsto X \vee S^2$$

At the algebraic level, the straightforward analogue of geometric stabilization simply adds a copy of $\mathbf{Z}[G]$ to the 'algebraic π_2', replacing

$$0 \to J \xrightarrow{i} E_2 \xrightarrow{\partial_2} E_1 \xrightarrow{\partial_1} E_0 \xrightarrow{\epsilon} \mathbf{Z} \to 0$$

by

$$0 \to J \oplus \mathbf{Z}[G] \xrightarrow{i \oplus \mathrm{Id}} E_2 \oplus \mathbf{Z}[G] \xrightarrow{(\partial_2, 0)} E_1 \xrightarrow{\partial_1} E_0 \xrightarrow{\epsilon} \mathbf{Z} \to 0$$

Algebraically, the situation is slightly more complicated, in that it is also necessary to consider 'internal stabilizations' which do not alter congruence classes. This is simply the 'linear algebra' version of altering a group presentation by introducing a new generator and a cancelling relation.

Finally, we give the classification of minimal two-dimensional homotopy types over a finite group G of period 4. It assumes its simplest form when G has free period four. Then there is a natural 1–1 correspondence between algebraic homotopy types and stably free modules; that is:

Theorem IV: Let G be a finite group of free period 4; there is an isomorphism of directed graphs

$$\widehat{\mathbf{Alg}_G} \longleftrightarrow SF(\mathbf{Z}[G])$$

At the minimal level, this gives a parametrization of minimal homotopy types by stably free modules of rank 1, $\mathbf{Alg}_G^{\min} \longleftrightarrow SF_1(\mathbf{Z}[G])$.

To describe the general case, we must first re-interpret the finiteness obstruction as a map, the 'Swan map'

$$S : (\mathbf{Z}/|G|)^* \to \tilde{K}_0(\mathbf{Z}[G])$$

Given a projective resolution of period 4

$$\mathbf{P} = (0 \to \mathbf{Z} \to P_3 \to P_2 \to P_1 \to P_0 \to \mathbf{Z} \to 0)$$

the class of $\chi(\mathbf{P}) \in \tilde{K}_0(\mathbf{Z}[G])$ is not in itself an invariant of G. Rather, if

$$\mathbf{P}' = (0 \to \mathbf{Z} \to P_3' \to P_2' \to P_1' \to P_0' \to \mathbf{Z} \to 0)$$

is another projective resolution of period 4, the difference $\chi(\mathbf{P}) - \chi(\mathbf{P}')$ lies in $\mathrm{Im}(S)$, and allows us to define an absolute invariant $\chi(G) \in \tilde{K}_0(\mathbf{Z}[G])/\mathrm{Im}(S)$. In the general case, minimal homotopy types are parametrized by rank 1 projectives within any stable class representing $\chi(G)$.

Without exception, throughout this chapter, G will denote a finite group.

50 Algebraic stabilization

Fix a finite group G, put $\Lambda = \mathbf{Z}[G]$, and let $M, N \in \mathcal{F}(\Lambda)$. If K is also a module in $\mathcal{F}(\Lambda)$ there are functors

$$\Sigma_-^K : \mathbf{Ext}^1(M, N) \to \mathbf{Ext}^1(M \oplus K, N)$$
$$\Sigma_+^K : \mathbf{Ext}^1(M, N) \to \mathbf{Ext}^1(M, N \oplus K)$$

defined on objects thus

$$\Sigma_-^K \left(0 \to N \xrightarrow{i} A \xrightarrow{p} M \to 0\right) = \left(0 \to N \xrightarrow{\lambda(i)} A \oplus K \xrightarrow{\sigma(p)} M \oplus K \to 0\right)$$

and

$$\Sigma_+^K \left(0 \to N \xrightarrow{i} A \xrightarrow{p} M \to 0\right) = \left(0 \to N \oplus K \xrightarrow{\sigma(i)} A \oplus K \xrightarrow{\rho(p)} M \to 0\right)$$

where

$$\lambda(a) = \begin{pmatrix} a \\ 0 \end{pmatrix}; \quad \sigma(b) = \begin{pmatrix} b & 0 \\ 0 & \mathrm{Id} \end{pmatrix}; \quad \rho(c) = (c \quad 0)$$

and where Σ_-^K, Σ_+^K act on morphisms in the obvious way. More generally, an extension

$$\mathcal{E} = \left(0 \to N \xrightarrow{i} A_n \xrightarrow{\partial_n} \cdots \xrightarrow{\partial_1} A_0 \xrightarrow{p} M \to 0\right) \in \mathbf{Ext}^{n+1}(M, N)$$

can be decomposed as a Yoneda product of short exact sequences

$$\mathcal{E} = \mathcal{E}_n \circ \mathcal{E}_{n-1} \circ \cdots \circ \mathcal{E}_1 \circ \mathcal{E}_0$$

where, for $0 < r < n$

$$\mathcal{E}_r = \left(0 \to \mathrm{Ker}(\partial_r) \to A_r \xrightarrow{\partial_r} \mathrm{Ker}(\partial_{r-1}) \to 0\right)$$

with

$$\mathcal{E}_n = \left(0 \to N \xrightarrow{i} A_n \xrightarrow{\partial_n} \mathrm{Ker}(\partial_{n-1}) \to 0\right)$$

and

$$\mathcal{E}_0 = \left(0 \to \mathrm{Ker}(p) \to A_0 \xrightarrow{p} M \to 0\right)$$

For $1 \le r \le n$, we define

$$\Sigma_r^K(\mathcal{E}) = \mathcal{E}_n \circ \cdots \circ \mathcal{E}_{r+1} \circ \Sigma_-^K(\mathcal{E}_r) \circ \Sigma_+^K(\mathcal{E}_{r-1}) \circ \mathcal{E}_{r-2} \circ \cdots \circ \mathcal{E}_0$$

with the special cases at the ends given by

$$\Sigma_-^K(\mathcal{E}) = \mathcal{E}_n \circ \cdots \circ \mathcal{E}_1 \circ \Sigma_-^K(\mathcal{E}_0)$$

and

$$\Sigma_+^K(\mathcal{E}) = \Sigma_+^K(\mathcal{E}_n) \circ \mathcal{E}_{n-1} \circ \cdots \circ \mathcal{E}_1 \circ \mathcal{E}_0$$

With obvious action on morphisms, there are 'external addition' functors

$$\Sigma_-^K : \mathbf{Ext}^{n+1}(M, N) \to \mathbf{Ext}^{n+1}(M \oplus K, N);$$
$$\Sigma_+^K : \mathbf{Ext}^{n+1}(M, N) \to \mathbf{Ext}^{n+1}(M, N \oplus K)$$

determined respectively by the correspondences $\mathcal{E} \mapsto \Sigma_+^K(\mathcal{E})$ and $\mathcal{E} \mapsto \Sigma_-^K(\mathcal{E})$. By contrast the correspondence $\mathcal{E} \mapsto \Sigma_r^K(\mathcal{E})$ for $0 \leq r \leq n$ gives an 'internal addition' functor

$$\Sigma_r^K : \mathbf{Ext}^{n+1}(M, N) \to \mathbf{Ext}^{n+1}(M, N)$$

Internal addition does not alter the congruence class, as we now see:

Proposition 50.1: If $\mathcal{E} \in \mathbf{Ext}^{n+1}(M, N)$, then

$$\Sigma_r^K(\mathcal{E}) \approx \mathcal{E}$$

for all r in the range $1 \leq r \leq n$.

Proof: If $\mathcal{E} = (0 \to N \xrightarrow{i} A_1 \oplus K \xrightarrow{\partial_1} A_0 \oplus K \xrightarrow{p} M \to 0)$, then $\Sigma_1^K(\mathcal{E})$ is the extension $(0 \to N \xrightarrow{\lambda(i)} A_1 \oplus K \xrightarrow{\sigma(\partial_1)} A_0 \oplus K \xrightarrow{\rho(p)} M \to 0)$. For $r = 1, 2$, the inclusion maps

$$j_r : A_r \to A_r \oplus K, \quad a \mapsto (a, 0)$$

induce an elementary congruence $j : \mathcal{E} \to \Sigma_1^K(\mathcal{E})$, the formal inverse $\pi : \Sigma_1^K(\mathcal{E}) \to \mathcal{E}$ being induced by the projection maps

$$\pi_r : A_r \oplus K \to A_r, \quad (a, \lambda) \mapsto a$$

The general case follows by dimension shifting. $\qquad\qquad\qquad\qquad\square$

In the special case where $K \cong \Lambda^k$, the internal addition functor $\Sigma_r^{\Lambda^k}$ is said to be an *internal stabilization functor*. Likewise $\Sigma_-^{\Lambda^k}$, $\Sigma_+^{\Lambda^k}$ are called *external stabilization functors*, and to simplify notation we write

$$\Sigma_r^k = \Sigma_r^{\Lambda^k}; \quad \Sigma_-^k = \Sigma_-^{\Lambda^k}; \quad \Sigma_+^k = \Sigma_+^{\Lambda^k}$$

If $\mathcal{E} \in \mathbf{Ext}^{n+1}(M, N)$ then by a *internal stabilization* of $\mathcal{E}$ we mean an object of the form

$$\Sigma_{r_m}^{k_m} \circ \cdots \circ \Sigma_{r_1}^{k_m}(\mathcal{E})$$

where $1 \leq r \leq n$.

Recall that $\mathbf{Stab}^{n+1}(M, N)$ is the full subcategory of $\mathbf{Ext}^{n+1}(M, N)$ consisting of extensions of the form

$$\mathcal{E} = \left(0 \to N \xrightarrow{i} E_n \xrightarrow{\partial_n} \cdots \xrightarrow{\partial_1} E_0 \xrightarrow{p} M \to 0 \right)$$

where each E_r is finitely generated stably free. The internal stabilization functors Σ_r^k preserve $\mathbf{Stab}^{n+1}(M, N)$ and, by (50.1), leave congruence classes unchanged. We allow the trivial case where each $v_i = 0$, so that $\mathcal{E}$ can be regarded as a trivial stabilization of itself.

More generally, if S is stably free, the functors Σ_r^S, Σ_-^S, Σ_+^S will be referred to as *quasi-stabilizations*. In this case, internal quasi-stabilization maps $\mathbf{Stab}^{n+1}(M, N)$ to itself and preserves congruence classes.

51 Stable isomorphism of stably free extensions

Recall the notion of isomorphism of extensions. Let $M, M', N, N' \in \mathcal{F}(\Lambda)$, let $\mathcal{E} \in \mathbf{Ext}^{n+1}(M, N)$ and $\mathcal{E}' \in \mathbf{Ext}^{n+1}(M', N')$, and let $\varphi : M \to M'$ be an isomorphism. We say that a morphism of the following type

$$
\begin{array}{c}
\mathcal{E} \\
\downarrow \tilde{\varphi} = \\
\mathcal{E}'
\end{array}
\begin{pmatrix}
0 \to N \to & A_n \to & \cdots \to & A_0 \to & M \to & 0 \\
& \downarrow \varphi_+ \; \downarrow \varphi_n & & \downarrow \varphi_0 \; \downarrow \varphi_- = \varphi \\
0 \to N' \to & A_n' \to & \cdots \to & A_0' \to & M' \to & 0
\end{pmatrix}
$$

is a *lifting* of φ. Moreover, when φ_+ and all φ_r are isomorphisms, $\tilde{\varphi}$ is said to be an isomorphism over φ. We write $\mathcal{E} \cong_\varphi \mathcal{E}'$ when there exists an isomorphism $\tilde{\varphi} : \mathcal{E} \to \mathcal{E}'$ lifting φ. We shall be interested principally in the notion of isomorphism over the identity and its relation to congruence and homotopy equivalence.

Proposition 51.1: Suppose we have a commutative diagram in $\mathcal{F}(\Lambda)$ as below

$$
\begin{array}{c}
\mathcal{E} \\
\downarrow \tilde{\varphi} = \\
\mathcal{E}'
\end{array}
\begin{pmatrix}
0 \to N \xrightarrow{i} E \xrightarrow{p} M \to 0 \\
\quad \downarrow \varphi_+ \; \downarrow \varphi_0 \; \cong \downarrow \varphi \\
0 \to N' \xrightarrow{j} Q \xrightarrow{\pi} M' \to 0
\end{pmatrix}
$$

in which (i) both rows are exact; (ii) $\varphi : M \to M'$ is an isomorphism; (iii) $\varphi_+ : N \to N'$ is surjective; and (iv) Q is projective. Then $\varphi_0 : E \to Q$ is surjective and there is an isomorphism $\psi : \mathcal{E} \to \Sigma_+^K(\mathcal{E}')$ where $K = \mathrm{Ker}(\varphi_0)$.

Proof: The surjectivity of φ_0 follows from an easy diagram chase. Since Q is projective, the exact sequence

$$0 \to K \xrightarrow{k} E \xrightarrow{\varphi_0} Q \to 0$$

defining $K = \mathrm{Ker}(\varphi_0)$ splits. In particular there exists a homomorphism $\rho : E \to K$ which splits the sequence on the left; that is, $\rho \circ k = \mathrm{Id}_K$. Let $\psi_0 : E \to Q \oplus K$ be the associated isomorphism, $\psi_0(e) = (\varphi_+(e), \rho(e))$, and define $\psi_+ : N \to N' \oplus K$ by $\psi_+(n) = (\varphi_+(n), \rho \circ i(n))$. The following commutative diagram

$$0 \to N \xrightarrow{i} E \xrightarrow{p} M \to 0$$
$$\downarrow \psi_+ \quad \downarrow \psi_0 \quad \downarrow \varphi$$
$$0 \to N' \xrightarrow{\sigma(j)} Q \xrightarrow{\rho(\pi)} M' \to 0$$

defines a morphism $\psi : \mathcal{E} \to \Sigma_+^K(\mathcal{E}')$. Since φ and ψ_0 are isomorphisms, it follows easily that $\psi_+ : N \to N' \oplus K$ is also an isomorphism, so that $\psi : \mathcal{E} \to \Sigma_+^K(\mathcal{E}')$ is an isomorphism over φ as claimed. $\quad\square$

In the following discussion, $M, M', N, N' \in \mathcal{F}(\Lambda)$. We denote by $\nu(M)$ the smallest integer $n \geq 1$ for which there exists a surjective Λ-homomorphism $\varphi : \Lambda^n \to M$.

Proposition 51.2: Suppose that $\mathcal{E} \in \mathbf{Stab}^1(M, N)$ and $\mathcal{E}' \in \mathbf{Stab}^1(M', N')$; if $\varphi : M \to M'$ is an isomorphism, then for each $\mu \geq \nu(N')$, there exists a finitely generated stably free module S such that φ lifts to an isomorphism of extensions $\psi : \Sigma_+^\mu(\mathcal{E}) \to \Sigma_+^S(\mathcal{E}')$.

Proof: Write $\mathcal{E} = (0 \to N \xrightarrow{i} F \xrightarrow{p} M \to 0)$ and $\mathcal{E}' = (0 \to N' \xrightarrow{j} F' \xrightarrow{\pi} M' \to 0)$. Since F is projective, $\varphi : M \to M'$ lifts to a morphism of extensions

$$\begin{matrix} \mathcal{E} \\ \downarrow \hat{\varphi} = \\ \mathcal{E}' \end{matrix} \begin{pmatrix} 0 \to N \xrightarrow{i} F \xrightarrow{p} M \to 0 \\ \downarrow \varphi_N \quad \downarrow \widehat{\varphi_0} \quad \downarrow \varphi \\ 0 \to N' \xrightarrow{j} F' \xrightarrow{\pi} M' \to 0 \end{pmatrix}$$

Since $\mu \geq \nu(N')$ we may choose a surjective homomorphism $\eta : \Lambda^\mu \to N'$. Let $\varphi_+ : N \oplus \Lambda^\mu \to N'$ be the homomorphism

$$\varphi_+(x, y) = \widehat{\varphi_0}(n, y) = \varphi_N(n) + \eta(y)$$

and let $\varphi_0 : F \oplus \Lambda^\mu \to F'$ be the homomorphism

$$\varphi_0(x, y) = \widehat{\varphi_0}(x) + j \circ \eta(y)$$

The following commutative diagram defines a morphism of extensions $\tilde{\varphi} : \Sigma_+^\nu(\mathcal{E}) \to \mathcal{E}'$

$$0 \to N \oplus \Lambda^\mu \xrightarrow{\sigma(i)} F \oplus \Lambda^\mu \xrightarrow{\rho(p)} M \to 0$$
$$\downarrow \varphi_+ \qquad \downarrow \varphi_0 \qquad \downarrow \varphi$$
$$0 \to \quad N' \xrightarrow{j} \quad F' \xrightarrow{\pi} M' \to 0$$

Since φ_+ is surjective, then by (51.1), φ lifts to an isomorphism of extensions $\psi : \Sigma_+^\mu(\mathcal{E}) \to \Sigma_+^S(\mathcal{E}')$ where $S = \text{Ker}(\varphi_0 : F \oplus \Lambda^\mu \to F')$. However, as noted in (51.1), since φ_+ is surjective, so is φ_0, and hence S is stably free. □

This generalizes to extensions of arbitrary length:

Theorem 51.3: Suppose that $\mathcal{E} \in \textbf{Stab}^{n+1}(M, N)$, $\mathcal{E}' \in \textbf{Stab}^{n+1}(M', N')$ where $n \geq 0$. If $\varphi : M \to M'$ is an isomorphism, then for each $\mu \geq v(N')$ there exist

  (i) a finitely generated stably free module S;
 (ii) an internal stabilization $\mathcal{P}$ of $\mathcal{E}$; and
(iii) a internal quasi-stabilization Q of $\mathcal{E}'$

with the property that φ lifts to an isomorphism of extensions

$$\psi : \Sigma_+^\mu(\mathcal{P}) \xrightarrow{\approx} \Sigma_+^S(Q)$$

Proof: By induction on n; note that the case $n = 0$ follows directly from (51.2), since we may take $\mathcal{P} = \mathcal{E}$ and $Q = \mathcal{E}'$. Thus suppose that $n \geq 1$ and that the statement is true for $n - 1$, and decompose $\mathcal{E}$ and $\mathcal{E}'$ as Yoneda products

$$\mathcal{E} = \mathcal{E}_1 \circ \mathcal{E}_n; \quad \mathcal{E}' = \mathcal{E}'_1 \circ \mathcal{E}'_n$$

where

$$\mathcal{E}_1 = (0 \to N \to F_n \to \Omega \to 0);$$
$$\mathcal{E}_n = (0 \to \Omega \to F_{n-1} \to \cdots \to F_0 \to M \to 0)$$

and

$$\mathcal{E}'_1 = (0 \to N' \to F'_n \to \Omega' \to 0);$$
$$\mathcal{E}'_n = (0 \to \Omega' \to F'_{n-1} \to \cdots \to F'_0 \to M' \to 0)$$

Put $\xi = v(\Omega')$. By induction, there exists a stably free module T, an internal stabilization $\mathcal{P}_n$ of $\mathcal{E}_n$, and a internal quasi-stabilization Q_n of $\mathcal{E}'_n$ such that φ lifts to an isomorphism of extensions

$$\psi_n : \Sigma_+^\xi(\mathcal{P}_n) \xrightarrow{\approx} \Sigma_+^T(Q_n)$$

Put $\mathcal{F} = \Sigma_-^\xi(\mathcal{E}_1)$ and $\mathcal{G} = \Sigma_-^T(\mathcal{E}'_1)$ and put $\mathcal{P} = \Sigma_-^\xi(\mathcal{E}_1) \circ \Sigma_+^\xi(\mathcal{P}_n) = \mathcal{F} \circ \Sigma_+^\mu(\mathcal{P}_n)$ and $Q = \Sigma_-^T(\mathcal{E}'_1 \circ \Sigma_+^T(Q_n) = \mathcal{G} \circ \Sigma_+^E(Q_n)$. Then $\mathcal{P}$ is an internal stabilization of $\mathcal{E}$, and Q is an internal quasi-stabilization of $\mathcal{E}'$. Now $\mathcal{F}, \mathcal{G}$ take the following forms

$$\mathcal{F} = (0 \to N \to F_n \oplus \Lambda^\nu \to \Omega \oplus \Lambda^\nu \to 0);$$
$$\mathcal{G} = (0 \to N' \to F'_n \oplus T \to \Omega' \oplus T \to 0)$$

The isomorphism $\psi_n : \Sigma_+^\xi(\mathcal{P}_n) \to \Sigma_+^T(\mathcal{Q}_n)$ restricts at *the left-hand end* to an isomorphism

$$\psi_+ : \Omega \oplus \Lambda^\xi \to \Omega' \oplus T$$

By (51.2), when $\mu \geq \nu(N')$, ψ_+ lifts to an isomorphism of extensions

$$\psi_1 : \Sigma_+^\mu(\mathcal{F}) \to \Sigma_+^S(\mathcal{G})$$

for some stably free module S. ψ_1 and ψ_n glue together by Yoneda product to give an isomorphism of extensions

$$\psi = \psi_1 \circ \psi_n : \Sigma_+^\mu(\mathcal{F}) \circ \Sigma_+^\xi(\mathcal{P}_n) \to \Sigma_+^S(\mathcal{G}) \circ \Sigma_+^T(\mathcal{Q}_n)$$

However

$$\Sigma_+^\mu(\mathcal{P}) = \Sigma_+^\mu(\mathcal{F}) \circ \Sigma_+^\xi(\mathcal{P}_n)$$

and

$$\Sigma_+^S(\mathcal{Q}) = \Sigma_+^S(\mathcal{G}) \circ \Sigma_+^T(\mathcal{Q}_n)$$

so that φ lifts to an isomorphism of extensions

$$\psi : \Sigma_+^\mu(\mathcal{P}) \to \Sigma_+^S(\mathcal{Q})$$

as claimed. $\square$

52 Eventual stability of homotopy types

We fix a finite group G and put $\Lambda = \mathbf{Z}[G]$; $\sigma(G)$ will denote the greatest integer s for which there is a stably free module $S \in \mathcal{F}(\Lambda)$ such that $S \otimes_{\mathbf{Z}} \mathbf{Q} \cong \mathbf{Q}[G]^s$ and $S \not\cong \mathbf{Z}[G]^s$. It is known from the Swan–Jacobinski Theorem that $\sigma(G)$ is either 0 or 1, and that $\sigma(G) = 0$ in the case where G satisfies the Eichler condition.

Fix a module $M \in \mathcal{F}(\Lambda)$, which, to avoid trivial cases, we assume is not projective. For any integer n, we define

$$\nu_n(M) = \min\{\nu(N) : N \in \Omega_n(M)\}$$

Clearly $\nu_n(M)$ is realized by some module N of *minimal height* in $\Omega_n(M)$. Here one should beware that, although $\mathrm{rk}_{\mathbf{Z}}$ is the same for all modules at the minimal level of $\Omega_n(M)$, it is *not generally true* that $\nu(N) = \nu_n(M)$ for *any* module N of *minimal height* in $\Omega_n(M)$. For example, over the generalized quaternion groups $Q(2^k)$ for $k \geq 5$, $\Omega_3(\mathbf{Z})$ contains at least two distinct minimal modules. One of these is $\mathbf{I}^*$, the dual of the augmentation ideal, and since $\mathbf{I}^*$ is monogenic it follows that $\nu_3(\mathbf{Z}) = 1$; however, all other minimal modules $J \in \Omega_3(\mathbf{Z})$ have

$\nu(J) = 2$. Moreover, in this case, $\text{rk}_Z(J) = \text{rk}_Z(\mathbf{I}^*) = |G| - 1$ for all minimal modules $J \in \Omega_3(\mathbf{Z})$.

Choose $N, N' \in \Omega_{n+1}(M)$, and let $\mathcal{E} \in \mathbf{Stab}^{n+1}(M, N)$, $\mathcal{E}' \in \mathbf{Stab}^{n+1}(M', N')$. In particular, neither N nor N' is projective. In the special case $n = 0$, weak homotopy equivalence is equivalent, by the Five Lemma, to 'isomorphism over Id_M':

Proposition 52.1: Let $N, N' \in \Omega_1(M)$ be minimal modules, and let $\mathcal{E} \in \mathbf{Stab}^1(M, N)$, $\mathcal{E}' \in \mathbf{Stab}^1(M, N')$; then

$$\Sigma^{\mu}_+(\mathcal{E}) \cong_{\text{Id}_M} \Sigma^{\mu}_+(\mathcal{E}')$$

when $\mu \geq \max\{\sigma(G) + 1, \nu_1(M)\}$. In particular, if $\mathbf{Z}[G]$ has the property that stably free modules are free, then

$$\Sigma^{\mu}_+(\mathcal{E}) \cong_{\text{Id}_M} \Sigma^{\mu}_+(\mathcal{E}')$$

provided $\mu \geq \nu_1(M)$.

Proof: Put $\nu = \nu_1(M)$. Since isomorphism over Id_M is an equivalence relation, we may, without loss of generality, suppose that $\nu(N') = \nu$. Write

$$\mathcal{E} = \left(0 \to N \xrightarrow{i} F \xrightarrow{p} M \to 0\right)$$

and

$$\mathcal{E}' = \left(0 \to N' \xrightarrow{j} F' \xrightarrow{\pi} M \to 0\right)$$

By (51.2), Id_M lifts to an isomorphism of extensions

$$\psi : \Sigma^{\nu}_+(\mathcal{E}) \to \Sigma^{S}_+(\mathcal{E}')$$

for some stably free module $S \in \mathcal{F}(\Lambda)$. Since N, N' are both minimal, $\text{rk}_Z(N) = \text{rk}_Z(N')$, and a straightforward rank calculation using the exactness of $\mathcal{E}, \mathcal{E}'$ shows that $F \cong F'$. Exactness of $\Sigma^{\nu}_+(\mathcal{E})$ and $\Sigma^{S}_+(\mathcal{E}')$ now gives $F \oplus \Lambda^{\nu} \cong F \oplus S$. If $\nu = \nu_1(M) \geq 2$, it follows by the Swan–Jacobinski Theorem that $S \cong \Lambda^{\nu}$, and there is an isomorphism

$$\Sigma^{\nu}_+(\mathcal{E}) \to \Sigma^{\nu}_+(\mathcal{E}')$$

The result for $\mu > \nu(N)$ follows after further stabilization by $\Sigma^{\mu-\nu}_+$.

If $\nu = \nu_1(M) = 1$ and $\sigma(G) = 1$ then $S \oplus \Lambda^{\mu-\nu} \cong \Lambda^{\mu}$ for $\mu - \nu \geq 1$, and the result follows on applying $\Sigma^{\mu-\nu}_+$ to $\psi : \Sigma^{\mu}_+(\mathcal{E}) \to \Sigma^{S}_+(\mathcal{E}')$. $\square$

When $n > 0$, we can no longer equate weak homotopy equivalence with isomorphism over the identity. We begin by comparing homotopy types at the minimal level.

Proposition 52.2: Let $n \geq 0$ and let $N, N' \in \Omega_{n+1}(M)$ be minimal modules. If $\mathcal{E} \in \mathbf{Stab}^{n+1}(M, N)$, $\mathcal{E}' \in \mathbf{Stab}^{n+1}(M, N')$ then there is a homotopy equivalence

$$\Sigma_+^{\mu}(\mathcal{E}) \xrightarrow{\sim} \Sigma_+^{\mu}(\mathcal{E}')$$

provided $\mu \geq \max\{\sigma(G) + 1, \nu_{n+1}(M)\}$. In particular, if $\mathbf{Z}[G]$ has the property that stably free modules are free, then there is a homotopy equivalence

$$\Sigma_+^{\mu}(\mathcal{E}) \xrightarrow{\sim} \Sigma_+^{\mu}(\mathcal{E}')$$

provided $\mu \geq \nu_{n+1}(M)$.

Proof: By (51.3) there is an isomorphism of extensions over Id_M

$$\psi : \Sigma_+^{\nu}(\mathcal{P}) \xrightarrow{\cong} \Sigma_+^{S}(\mathcal{Q})$$

where $\nu = \nu(N')$, $\mathcal{P}$ is an internal stabilization of $\mathcal{E}$, $\mathcal{Q}$ is an internal quasi-stabilization of $\mathcal{E}'$, $\nu = \nu(N')$ and S is a stably free module with the property that $S \otimes \mathbf{Q} \cong \Lambda^{\nu}$. If $\nu \geq 2$ then, by the Swan–Jacobinski Theorem, $S \cong \Lambda^{\nu}$, so that

$$\Sigma_+^{\nu}(\mathcal{P}) \xrightarrow{\cong}_{\mathrm{Id}_M} \Sigma_+^{\nu}(\mathcal{Q})$$

However $\mathcal{E} \approx \mathcal{P}$ and $\mathcal{Q} \approx \mathcal{E}'$; since congruence ($\approx$) implies homotopy equivalence ($\simeq$), we see in particular for $\nu \geq 2$ that

$$\Sigma_+^{\nu}(\mathcal{E}) \simeq \Sigma_+^{\nu}(\mathcal{P}) \simeq \Sigma_+^{\nu}(\mathcal{Q}) \simeq \Sigma_+^{\nu}(\mathcal{E}')$$

The result for $\mu > \nu \geq 2$ now follows by further stabilization. It remains to consider the case $\nu = 1$. Then (51.3) gives an isomorphism over Id_M

$$\Sigma_+^{1}(\mathcal{P}) \cong_{\mathrm{Id}_M} \Sigma_+^{S}(\mathcal{Q})$$

where S is a stably free module such that $S \otimes \mathbf{Q} \cong \mathbf{Q}[G]$. In the case where $\sigma(G) = 0$, that is when stably free modules over Λ are free, $S \cong \Lambda$ and we have

$$\Sigma_+^{1}(\mathcal{E}) \approx \Sigma_+^{1}(\mathcal{P}) \cong \Sigma_+^{1}(\mathcal{Q}) \approx \Sigma_+^{1}(\mathcal{E}')$$

giving a homotopy equivalence

$$\Sigma_+^{1}(\mathcal{E}) \simeq \Sigma_+^{1}(\mathcal{E}')$$

whilst for $\mu > 1$, the homotopy equivalences

$$\Sigma_+^{\mu}(\mathcal{E}) \simeq \Sigma_+^{\mu}(\mathcal{E}')$$

are obtained by stabilization. In general, we cannot assume that Λ has the cancellation property for free modules. Nevertheless, we still have $S \oplus \Lambda \cong \Lambda^2$ and we obtain

$$\Sigma_+^2(\mathcal{E}) \approx \Sigma_+^2(\mathcal{P}) \cong \Sigma_+^2(\mathcal{Q}) \approx \Sigma_+^2(\mathcal{E}')$$

and hence homotopy equivalences

$$\Sigma_+^\mu(\mathcal{E}) \simeq \Sigma_+^\mu(\mathcal{E}')$$

for all $\mu \geq 2$. This completes the proof. $\qquad\square$

We obtain the following 'eventual stability' theorem:

Corollary 52.3: (Eventual stability of homotopy types) Let N, $N' \in \Omega_{n+1}(M)$ and $\mathcal{E} \in \mathbf{Stab}^{n+1}(M, N)$, $\mathcal{E}' \in \mathbf{Stab}^{n+1}(M, N')$ where $n \geq 0$; then there exist integers $a, b \geq 1$ such that

$$\Sigma_+^a(\mathcal{E}) \simeq \Sigma_+^b(\mathcal{E}')$$

Proof: By (51.3), there is an internal stabilization $\mathcal{P}$ of $\mathcal{E}$, an internal quasi-stabilization $\mathcal{Q}$ of $\mathcal{E}'$, a stably free module S, and an integer $\nu \geq \nu(N')$ such that Id_M lifts to an isomorphism

$$\psi : \Sigma_+^\nu(\mathcal{P}_n) \xrightarrow{\;\approx\;} \Sigma_+^S(\mathcal{Q}_n)$$

Since S is stably free then by the Swan–Jacobinski Theorem there exists an integer μ in the range $0 \leq \mu \leq 1$ such that $S \oplus \Lambda^\mu \cong \Lambda^b$. Hence

$$\Sigma_+^\mu(\psi) : \Sigma_+^{\mu+\nu}(\mathcal{P}_n) \to \Sigma_+^b(\mathcal{Q}_n)$$

is an isomorphism over Id_M, so that

$$\Sigma_+^{\mu+\nu}(\mathcal{E}) \approx \Sigma_+^{\mu+\nu}(\mathcal{P}_n) \cong \Sigma_+^b(\mathcal{Q}_n) \approx \Sigma_+^b(\mathcal{E}')$$

Putting $a = \mu + \nu$ we have a homotopy equivalence $\Sigma_+^a(\mathcal{E}) \simeq \Sigma_+^b(\mathcal{E}')$ as claimed. $\qquad\square$

53 The Swan map

Let $\mathbf{E}$ be a *stably free n*-stem

$$\mathbf{E} = (0 \to J \to E_{n-1} \to \cdots \to E_0 \to \mathbf{Z} \to 0)$$

we have a *unique* ring isomorphism $\kappa^J : \mathrm{End}_{\mathcal{D}er}(J) \to \mathbf{Z}/|G|$ and a bijection (26.5)

$$e_{\mathbf{E}} : \mathcal{A}(J) \to \mathrm{Proj}^n(\mathbf{Z}, J)$$

given by $e_E(\alpha) = \alpha_*(E)$. We define the *Swan map* $S^E : (\mathbf{Z}/|G|)^* \to \tilde{K}_0(\mathbf{Z}[G])$ by

$$S^E = \chi \circ e_E \circ (\kappa^J)^{-1}$$

where $\chi : \mathbf{Proj}^n(\mathbf{Z}, J) \to \tilde{K}_0(\mathbf{Z}[G])$ is the Wall obstruction

$$\chi(0 \to J \to P_{n-1} \to \cdots \to P_0 \to \mathbf{Z} \to 0) = \sum_j (-1)^j [P_j]$$

S^E depends ostensibly upon the particular stably free n-stem $\mathbf{E}$. In fact, we show, in Theorem (53.7) below, that all Swan mappings are the same.

As in Section 36, for any $t \in (\mathbf{Z}/|G|)^*$ we denote by

$$\mathcal{E}(t) = (0 \to \mathbf{I}(G) \to (I, t) \xrightarrow{\eta_t} \mathbf{Z} \to 0)$$

the defining extension of the (projective) Swan module (I, t). In particular

$$\mathcal{E}(1) = \left(0 \to \mathbf{I}(G) \xrightarrow{i} \mathbf{Z}[G] \xrightarrow{\epsilon} \mathbf{Z} \to 0\right)$$

is the defining extension of the augmentation ideal. We put

$$S = S^{\mathcal{E}(1)}$$

S is called the *canonical* Swan mapping (more properly, in view of the uniqueness property we shall establish, it should perhaps be called the *canonical form* of the Swan mapping). By a straightforward chase of definitions we get:

Proposition 53.1: $S : (\mathbf{Z}/|G|)^* \to \tilde{K}_0(\mathbf{Z}[G])$ is the mapping given by

$$S([r]) = [(\mathbf{I}, r)]$$

It follows immediately from (37.9) that:

(53.2) The canonical Swan map is a homomorphism $S : (\mathbf{Z}/|G|)^* \to \tilde{K}_0(\mathbf{Z}[G])$.

Example: The finiteness obstruction of a projective n-stem
Let G be a finite group and let $\mathcal{P}_1, \mathcal{P}_2$ be projective n-stems in $\mathbf{Proj}^n(\mathbf{Z}, J)$ where $J \in \mathbf{D}_n(\mathbf{Z})$; then

$$\chi(\mathcal{P}_1) - \chi(\mathcal{P}_2) \in \mathrm{Im}(S)$$

To see this observe that, by (38.3), we may suppose that $\mathcal{P}_i \approx \mathcal{Q} \circ \mathcal{E}(t_i)$ for some fixed projective $(n-1)$-stem $\mathcal{Q} \in \mathbf{Proj}^{n-1}(\mathbf{I}(G), J)$. However, the Wall obstruction χ is invariant under homotopy equivalence, and hence, for projective stems, also under congruence (46.8). Thus $\chi(\mathcal{P}_i) = [(I, t_i)] - \chi(\mathcal{Q})$, so that $\chi(\mathcal{P}_1) - \chi(\mathcal{P}_2) = [(I, t_1)] - [(I, t_2)] = S(t_1 t_2^{-1}) \in \mathrm{Im}(S)$ as claimed.

It follows that for $J \in \mathbf{D}_n(\mathbf{Z})$, the finiteness obstruction gives a well-defined mapping

$$\chi : \mathbf{Proj}^n(\mathbf{Z}, J) \to \tilde{K}_0(\mathbf{Z}[G])/\mathrm{Im}(S)$$

We proceed to show that $S^{\mathbf{E}} = S$ for any stably free n-stem $\mathbf{E}$.

Proposition 53.3: Let $\mathbf{E}$, $\mathbf{F}$ be stably free n-stems; if $\mathbf{E} \approx \mathbf{F}$ then $S^{\mathbf{E}} = S^{\mathbf{F}}$.

Proof: Write

$$\mathbf{E} = (0 \to J \to E_{n-1} \to \cdots \to E_0 \to \mathbf{Z} \to 0)$$

and

$$\mathbf{F} = (0 \to J \to F_{n-1} \to \cdots \to F_0 \to \mathbf{Z} \to 0)$$

and let $[r] \in (\mathbf{Z}/|G|)^*$. Since $\mathbf{E}$ is a projective n-stem, the $\mathbf{Z}[G]$-homomorphism $\mathbf{Z} \xrightarrow{\times r} \mathbf{Z}$ lifts to a self-morphism of $\mathbf{E}$, thus

$$
\begin{array}{c} \mathbf{E} \\ \downarrow \hat{\mathbf{r}} = \\ \mathbf{E} \end{array}
\begin{pmatrix}
0 \to J \to E_{n-1} \to \cdots \to E_0 \to \mathbf{Z} \to 0 \\
\quad\ \ \downarrow \alpha \quad\ \downarrow \qquad\qquad\ \downarrow \quad \downarrow \times r \\
0 \to J \to E_{n-1} \to \cdots \to E_0 \to \mathbf{Z} \to 0
\end{pmatrix}
$$

Moreover, since $\mathbf{Z} \xrightarrow{\times r} \mathbf{Z}$ is an isomorphism in the derived category, so also is α, and from the definition of the ring homomorphism κ we have

$$\kappa[\alpha] = [r]$$

Since α is an isomorphism in the derived category, $\alpha_*(\mathbf{E})$ is a projective n-stem by (26.3). It follows directly from the definition that $S^{\mathbf{E}}([r]) = \chi(\alpha_*(\mathbf{E}))$. Since projectives are relatively injective in $\mathcal{F}(\mathbf{Z}[G])$, $\alpha : J \to J$ extends to a self-morphism of $\mathbf{F}$, thus

$$
\begin{array}{c} \mathbf{F} \\ \downarrow \tilde{\alpha} = \\ \mathbf{F} \end{array}
\begin{pmatrix}
0 \to J \to F_{n-1} \to \cdots \to F_0 \to \mathbf{Z} \to 0 \\
\quad\ \ \downarrow \alpha \quad\ \downarrow \qquad\qquad\ \downarrow \quad \downarrow \times s \\
0 \to J \to F_{n-1} \to \cdots \to F_0 \to \mathbf{Z} \to 0
\end{pmatrix}
$$

so that, from the independence of κ upon the choice of projective n-stem

$$\kappa[\alpha] = [s]$$

In particular, $[s] = [r]$, so that $S^{\mathbf{F}}([r]) = \chi(\alpha_*(\mathbf{F}))$. It suffices to show that $\chi(\alpha_*(\mathbf{E})) = \chi(\alpha_*(\mathbf{F}))$.

However, by hypothesis, there exists an elementary congruence $\mathbf{E} \to \mathbf{F}$, so that by (38.1), there exists an an elementary congruence $\nu : \alpha_*(\mathbf{E}) \to \alpha_*(\mathbf{F})$.

Since α is an isomorphism in the derived category, $\alpha_*(\mathbf{E})$ and $\alpha_*(\mathbf{F})$ are both projective n-stems, so ν is a homotopy equivalence. Since the Wall obstruction χ is invariant under homotopy equivalence, we have

$$\chi(\alpha_*(\mathbf{E})) = \chi(\alpha_*(\mathbf{F}))$$

as desired, and this completes the proof.   □

As before, let $\mathcal{E}(t)$ denote the standard projective 1-stem

$$0 \to \mathbf{I} \to (\mathbf{I}, t) \to \mathbf{Z} \to 0$$

Proposition 53.4: If $(\mathbf{I}, t)$ is stably free, then $S^{\mathcal{E}(t)} = S$.

Proof: Let r be coprime to $|G|$, so that $\times r : \mathbf{Z} \to \mathbf{Z}$ is an isomorphism in the derived category. We may lift $\times r$ to a self morphism of $\mathcal{E}(t)$, thus

$$
\begin{array}{ccccccc}
0 \to & \mathbf{I} & \to & (\mathbf{I}, t) & \to & \mathbf{Z} & \to 0 \\
 & \downarrow \mathbf{r} & & \downarrow \lambda_r & & \downarrow \times r & \\
0 \to & \mathbf{I} & \to & (\mathbf{I}, t) & \to & \mathbf{Z} & \to 0
\end{array}
$$

where, to distinguish the mapping from the formula, we write $\lambda_r : \mathbf{Z}[G] \to \mathbf{Z}[G]$ for the mapping $\lambda_r(\mathbf{x}) = r\mathbf{x}$, and $\mathbf{r}$ for its restriction $\mathbf{r} = \lambda_{r|\mathbf{I}} : \mathbf{I} \to \mathbf{I}$. However, it is straightforward to verify that $\mathbf{r}_*(\mathcal{E}(t)) = \mathcal{E}(rt)$, so that

$$
\begin{aligned}
S^{\mathcal{E}(t)}([r]) &= \chi((\mathbf{I}, rt)) \\
&= \chi((\mathbf{I}, r)) + \chi((\mathbf{I}, t)) \\
&= S([r]) + \chi((I, t))
\end{aligned}
$$

However, $\chi((\mathbf{I}, t)) = 0$, by hypothesis, so that $S^{\mathcal{E}(t)}([r]) = S([r])$ as claimed.
                                                                                        □

Proposition 53.5: If $\mathbf{E}$ is a stably free 1-stem, then $S^{\Sigma_+(\mathbf{E})} = S^{\mathbf{E}}$.

Proof: Write $\mathcal{F} = (0 \to J \xrightarrow{j} F \xrightarrow{p} \mathbf{Z} \to 0)$, so that

$$\Sigma_+(\mathcal{F}) = \left(0 \to J \oplus \mathbf{Z}[G] \xrightarrow{\hat{j}} F \oplus \mathbf{Z}[G] \xrightarrow{\hat{p}} \mathbf{Z} \to 0 \right)$$

where

$$\hat{j} = \begin{pmatrix} j & 0 \\ 0 & \mathrm{Id} \end{pmatrix} \quad \text{and} \quad \hat{p} = (p, 0)$$

Let r be coprime to $|G|$, so that $\times r : \mathbf{Z} \to \mathbf{Z}$ is an isomorphism in the derived category, and lift $\times r$ to a self morphism of $\mathcal{F}$ thus

$$
\begin{array}{ccccccccc}
0 & \to & J & \to & F & \to & \mathbf{Z} & \to & 0 \\
& & \downarrow \alpha & & \downarrow \alpha_0 & & \downarrow \times r & & \\
0 & \to & J & \to & F & \to & \mathbf{Z} & \to & 0
\end{array}
$$

where, since $\times r : \mathbf{Z} \to \mathbf{Z}$ is an isomorphism in the derived category, so also is α. Then $\alpha_*(\mathcal{F})$ is a projective cover by (23.2), and moreover

$$S^{\mathcal{F}}([r]) = \chi(\alpha_*(\mathcal{F}))$$

However, $\times r$ lifts to $\Sigma_+(\mathcal{F})$, thus

$$
\begin{array}{ccccccc}
0 \to & J \oplus \mathbf{Z}[G] & \to & F \oplus \mathbf{Z}[G] & \to & \mathbf{Z} \to & 0 \\
& \downarrow \hat{\alpha} & & \downarrow \hat{\alpha}_0 & & \downarrow \times r & \\
0 \to & J \oplus \mathbf{Z}[G] & \to & F \oplus \mathbf{Z}[G] & \to & \mathbf{Z} \to & 0
\end{array}
$$

where, for $\beta = \alpha, \alpha_0$

$$\hat{\beta} = \begin{pmatrix} \beta & 0 \\ 0 & \mathrm{Id} \end{pmatrix}$$

We see easily that $\alpha_*(\Sigma_+(\mathcal{F})) \approx \Sigma_+(\alpha_*(\mathcal{F}))$, and so

$$
\begin{aligned}
S^{\Sigma_+(\mathcal{F})}([r]) &= \chi(\alpha_*(\Sigma_+(\mathcal{F}))) \\
&= \chi(\alpha_*(\mathcal{F})) \\
&= S^{\mathcal{F}}([r])
\end{aligned}
$$

which completes the proof. $\qquad\qquad\qquad\qquad\qquad\qquad\qquad\qquad\qquad\square$

Corollary 53.6: If $\mathbf{F}$ is a stably free 1-stem, then $S^{\mathbf{F}} = S$.

Proof: The proof is essentially a repetition of the proof of the 'Eventual Stability Theorem' in a very easy case. Write $\mathbf{F}$ in the form

$$\mathbf{F} = (0 \to J \to F \to \mathbf{Z} \to 0)$$

Comparing $\mathbf{F}$ with $\mathcal{E}(1)$ we get

$$J \oplus \mathbf{Z}[G] \cong \mathbf{I} \oplus F$$

Moreover, since F is stably free, we have

$$J \oplus \mathbf{Z}[G]^a \cong \mathbf{I} \oplus \mathbf{Z}[G]^b$$

for some a, b. Write $[t] = k(\Sigma_+^b(\mathcal{E}(1)) \to \Sigma_+^a(\mathbf{F}))$. However, it is clearly true that $[t] = k(\Sigma_+^b(\mathcal{E}(1)) \to \Sigma_+^b(\mathcal{E}(t)))$, so that

$$\Sigma_+^a(\mathbf{F}) \approx \Sigma_+^b(\mathcal{E}(t))$$

Hence

$$S^{\Sigma_+^a(\mathbf{F})} = S^{\Sigma_+^b(\mathcal{E}(t))}$$

by (53.3). However, by (53.5), we have

$$S^{\mathbf{F}} = S^{\Sigma_+^a(\mathbf{F})} \quad \text{and} \quad S^{\Sigma_+^b(\mathcal{E}(t))} = S^{\mathcal{E}(t)}$$

so that

$$S^{\mathbf{F}} = S^{\mathcal{E}(t)}$$

From the congruence $\Sigma_+^a(\mathbf{F}) \approx \Sigma_+^b(\mathcal{E}(t))$ and the fact that F is stably free we infer that $(\mathbf{I}, t)$ is also stably free. Thus

$$S^{\mathcal{E}(t)} = S$$

by (53.4), and the result follows.   □

Theorem 53.7: If $\mathbf{F}$ is a stably free n-stem, then $S^{\mathbf{F}} = S$.

Proof: From (53.6) it suffices to take $n \geq 2$. Write $\mathbf{F}$ in the form

$$\mathbf{F} = (0 \to J \to F_{n-1} \to \cdots \to F_0 \to \mathbf{Z} \to 0)$$

where each F_i is stably free. As $J \in \Omega_n(\mathbf{Z}) = \Omega_{n-1}(\mathbf{I}(G))$, we may choose a projective $(n-1)$-stem $\mathcal{Q} \in \mathbf{Proj}^{n-1}(\mathbf{I}(G), J)$

$$\mathcal{Q} = \left(0 \to J \to Q_{n-1} \to \cdots \to Q_1 \xrightarrow{\mu} \mathbf{I}(G) \to 0\right)$$

By (38.3), $\mathbf{F} \approx \mathcal{Q} \circ \mathcal{E}(t)$ for some $t \in (\mathbf{Z}/|G|)^*$. Suppose that $[r]$ also belongs to $(\mathbf{Z}/|G|)^*$, and lift the $\mathbf{Z}[G]$ homomorphism $\mathbf{Z} \xrightarrow{\times r} \mathbf{Z}$ to a self morphism of $\mathbf{F}$, thus

$$\begin{matrix} \mathbf{F} \\ \downarrow \hat{\mathbf{r}} = \\ \mathbf{F} \end{matrix} \begin{pmatrix} 0 \to J \to F_{n-1} \to \cdots \to F_0 \to \mathbf{Z} \to 0 \\ \quad \downarrow \alpha \quad \downarrow \qquad\qquad\quad \downarrow \quad \downarrow \times r \\ 0 \to J \to F_{n-1} \to \cdots \to F_0 \to \mathbf{Z} \to 0 \end{pmatrix}.$$

Since $\mathbf{Z} \xrightarrow{\times r} \mathbf{Z}$ is an isomorphism in the derived category, so also is α, and from the definition of the ring homomorphism κ we have

$$\kappa[\alpha] = [r]$$

Also $\alpha_*(\mathbf{F})$ is a projective n-stem, and by definition

$$S^{\mathbf{F}}([r]) = \chi(\alpha_*(\mathbf{F}))$$

However, by (38.3), $\alpha_*(\mathbf{F}) \approx \mathcal{Q} \circ \mathcal{E}(\kappa(\alpha)t)$, and thus

$$\begin{aligned}
S^{\mathbf{F}}([r]) &= \chi(\alpha_*(\mathbf{F})) \\
&= \chi(\mathcal{Q} \circ \mathbf{E}(rt)) \\
&= \chi(\mathcal{E}(rt) - \chi(\mathcal{Q})) \\
&= \chi(\mathcal{E}(r)) + \chi(\mathcal{E}(t)) - \chi(\mathcal{Q}).
\end{aligned}$$

Since the Wall obstruction is invariant under homotopy equivalence, and thereby under congruence, we have $\chi(\mathcal{E}(t) - \chi(\mathcal{Q})) = \chi(\mathbf{F}) = 0$, so that, finally, $S^{\mathbf{F}}([r]) = \chi(\mathcal{E}(r)) = S([r])$ as desired. □

The definition of 'Swan map' can be extended to all projective, rather than merely stably free, n-stems $\mathbf{P}$, by means of

$$S^{\mathbf{P}} = \chi \circ c_{\mathbf{P}} \circ \kappa^{-1}$$

However, $S^{\mathbf{P}}$ is, in general, no longer a homomorphism, but rather one has

(53.8) $$S^{\mathbf{P}}([r]) = S([r]) + \chi(\mathbf{P})$$

We leave the verification to the reader.

Finally, if $J \in \Omega_n(\mathbf{Z})$, we can define a Swan map S^J directly on $\mathcal{A}(J)$

$$S^J = S \circ \kappa^J$$

where, as usual, $\kappa^J : \mathcal{A}(J) \to (\mathbf{Z}/|G|)^*$ is the unique ring isomorphism. We can re-phrase (53.7) as follows:

Proposition 53.9: Let $J \in \Omega_n(\mathbf{Z})$; then $S^J(\alpha) = \chi(\alpha_*(\mathbf{F}))$ for any stably free n-stem $\mathbf{F} \in \mathbf{Stab}^n(\mathbf{Z}, J)$.

Corollary 53.10: If $J \in \Omega_n(\mathbf{Z})$, then under the action of $\mathcal{A}(J)$ on $\mathrm{Proj}^n(\mathbf{Z}, J)$, the stabilizer of $\mathbf{Stab}^n(\mathbf{Z}, J)$ is $\mathrm{Ker}(S)$.

More generally, by virtue of (53.8), we have:

Corollary 53.11: Let $c \in \tilde{K}_0(\mathbf{Z}[G])$; if $J \in \mathbf{D}_n^c(\mathbf{Z})$; then under the action of $\mathcal{A}(J)$ on $\mathrm{Proj}^n(\mathbf{Z}, J)$, the stabilizer of $\mathrm{Proj}^n(\mathbf{Z}, J; c)$ is $\mathrm{Ker}(S)$.

(53.10) is the special case of (53.11) obtained by taking $c = 0$.

54 Module automorphisms and k-invariants

Let $J \in \mathbf{D}_n(\mathbf{Z})$; we have seen, in Chapter 7, how the ring $\operatorname{End}_{\mathcal{D}er}(J)$ of endomorphisms of J *in the derived category*, and in particular the unit group $\mathcal{A}(J) = \operatorname{Aut}_{\mathcal{D}er}(J)$, acts on $\mathbf{Ext}^n(\mathbf{Z}, J)$, and thereby on k-invariants. Here, instead, we consider the action of the group $\operatorname{Aut}_{\mathbf{Z}[G]}(J)$ of *module automorphisms*. We denote by [ ] : $\operatorname{End}_{\mathbf{Z}[G]}(J) \to \operatorname{End}_{\mathcal{D}er}(J)$ the natural ring homomorphism, and by ν the composite $\nu = \kappa \circ [\,] : \operatorname{End}_{\mathbf{Z}[G]}(J) \to \operatorname{End}_{\mathcal{D}er}(\mathbf{Z})$; clearly both [ ] and ν are surjective. Regarding $\operatorname{Aut}_{\mathbf{Z}[G]}(J)$ as the group of units of $\operatorname{End}_{\mathbf{Z}[G]}(J)$, ν restricts to a group homomorphism

$$\nu : \operatorname{Aut}_{\mathbf{Z}[G]}(J) \to (\mathbf{Z}/|G|)^*$$

The significant change is that, whereas $\kappa : \mathcal{A}(J) \to (\mathbf{Z}/|G|)^*$ is a group isomorphism, and $\operatorname{Proj}^n(\mathbf{Z}, J)$ corresponds to a single orbit, $\nu : \operatorname{Aut}_{\mathbf{Z}[G]}(J) \to (\mathbf{Z}/|G|)^*$ is, in general, no longer surjective, and $\operatorname{Proj}^n(\mathbf{Z}, J)$ decomposes into several orbits.

It is helpful to describe the action of $\operatorname{Aut}_{\mathbf{Z}[G]}(J)$ on $\operatorname{Ext}^n(\mathbf{Z}, J)$ explicitly. Thus let $\mathbf{E} \in \mathbf{Ext}^n(\mathbf{Z}, J)$ be an n-fold extension

$$\mathbf{E} = \left(0 \to J \xrightarrow{j} E_{n-1} \xrightarrow{\partial_{n-1}} \cdots \xrightarrow{\partial_1} E_0 \xrightarrow{\eta} \mathbf{Z} \to 0 \right)$$

For $\alpha \in \operatorname{Aut}_{\mathbf{Z}[G]}(J)$ we define an extension $\alpha \bullet \mathbf{E}$

$$\alpha \bullet \mathbf{E} = \left(0 \to J \xrightarrow{j\alpha^{-1}} E_{n-1} \xrightarrow{\partial_{n-1}} \cdots \xrightarrow{\partial_1} E_0 \xrightarrow{\eta} \mathbf{Z} \to 0 \right)$$

We note that

$$\begin{array}{ccc} J & \xrightarrow{j} & E_{n-1} \\ \downarrow \alpha & & \downarrow \operatorname{Id} \\ J & \xrightarrow{j\alpha^{-1}} & E_{n-1} \end{array}$$

is a pushout diagram so that we may identify $\alpha \bullet \mathbf{E}$ with the 'pushout extension' $\alpha_*(\mathbf{E})$ (Section 24). Moreover, there is an obvious homomorphism of extensions

$$\begin{array}{c} \mathbf{E} \\ \downarrow \alpha_* = \\ \alpha \bullet \mathbf{E} \end{array} \left(\begin{array}{ccccccc} 0 \to & J & \xrightarrow{j} & E_{n-1} \to & \cdots \to & E_0 \to & \mathbf{Z} \to 0 \\ & \downarrow \alpha & & \downarrow \operatorname{Id} & & \downarrow \operatorname{Id} & \downarrow \operatorname{Id} \\ 0 \to & J & \xrightarrow{j\alpha^{-1}} & E_{n-1} \to & \cdots \to & E_0 \to & \mathbf{Z} \to 0 \end{array} \right)$$

It follows immediately from the results of Section 34 that

(54.1) $$k(\mathbf{P} \to \alpha \bullet \mathbf{E}) = \nu(\alpha)k(\mathbf{P} \to \mathbf{E})$$

(54.2) $$k(\alpha \bullet \mathbf{P} \to \alpha \bullet \mathbf{E}) = k(\mathbf{P} \to \mathbf{E})$$

where $\mathbf{P} \in \mathbf{Proj}^n(\mathbf{Z}, J)$ is a projective n-stem, $\mathbf{E} \in \mathbf{Ext}^n(\mathbf{Z}, J)$ and $\alpha \in$ $\mathrm{Aut}_{\mathbf{Z}[G]}(J)$. Though tautologous, the following observation is nevertheless invaluable:

Proposition 54.3: For $\alpha \in \mathrm{Aut}_{\mathbf{Z}[G]}(J)$ the natural homomorphism $\alpha_* : \mathbf{E} \to$ $\alpha \bullet \mathbf{E}$ is an isomorphism over $\mathrm{Id}_{\mathbf{Z}}$.

Let $\alpha : J \to J$ be a $\mathbf{Z}[G]$ homomorphism which defines an isomorphism in $\mathcal{D}er$, and let $\mathbf{P}$ be a projective n-stem; for each $\alpha \in \mathrm{Aut}_{\mathbf{Z}[G]}(J)$ the natural homomorphism $\alpha_* : \mathbf{P} \to \alpha \bullet \mathbf{P}$ is a homotopy equivalence. In fact, we have:

Proposition 54.4: If $\alpha : J \to J$ is a $\mathbf{Z}[G]$-homomorphism, then

$$\mathbf{P} \simeq \alpha_*(\mathbf{P}) \iff [\alpha] \in \mathrm{Im}(\nu^J)$$

Proof: Suppose that $\Psi : \mathbf{P} \to \alpha_*(\mathbf{P})$ is a homotopy equivalence, so that Ψ takes the form

$$
\begin{matrix}
\mathbf{P} \\
\downarrow \Psi \\
\alpha_*(\mathbf{P})
\end{matrix}
=
\begin{pmatrix}
0 \to J \to P_{n-1} \to \cdots \to P_0 \to \mathbf{Z} \to 0 \\
\quad\quad \downarrow \psi \; \downarrow \nu_{n-1} \quad\quad\quad \downarrow \mathrm{Id} \; \downarrow \mathrm{Id} \\
0 \to J \to Q_{n-1} \to \cdots \to P_0 \to \mathbf{Z} \to 0
\end{pmatrix}
$$

where $\psi : J \to J$ is an isomorphism of $\mathbf{Z}[G]$ modules and $\nu_2 : P_2 \to Q_2$ is the pushout map. Then $\kappa^J([\psi]) = k(\mathbf{P} \to \alpha_*(\mathbf{P}))$. However, tautologously, one has $\kappa^J([\alpha]) = k(\mathbf{P} \to \alpha_*(\mathbf{P}))$. Thus $\kappa^J([\alpha]) = \kappa^J([\psi])$, and so $[\alpha] = [\psi]$. But $[\psi]$ is, by definition, $\nu^J(\psi)$; that is, $[\alpha] \in \mathrm{Im}(\nu^J)$ as required.

Conversely, suppose that $[\alpha] \in \mathrm{Im}(\nu^J)$, and choose an isomorphism of $\mathbf{Z}[G]$ modules $\psi : J \to J$ such that $\alpha \approx \psi$. Then $\alpha_*(\mathbf{P})$ is congruent to $\psi_*(\mathbf{P})$. However, ψ defines a tautologous homotopy equivalence $\mathbf{P} \xrightarrow{\sim} \psi_*(\mathbf{P})$

$$
\begin{matrix}
\mathbf{P} \\
\downarrow \hat{\psi} \\
\psi_*(\mathbf{P})
\end{matrix}
=
\begin{pmatrix}
0 \to J \xrightarrow{j} P_{n-1} \to \cdots \to P_0 \to \mathbf{Z} \to 0 \\
\quad\quad \downarrow \psi \quad \downarrow \mathrm{Id} \quad\quad\quad \downarrow \mathrm{Id} \; \downarrow \mathrm{Id} \\
0 \to J \xrightarrow{j\psi^{-1}} P_{n-1} \to \cdots \to P_0 \to \mathbf{Z} \to 0
\end{pmatrix}
$$

Thus $\psi_*(\mathbf{P}) \simeq \mathbf{P}$. Since $\alpha_*(\mathbf{P})$ is congruent to, and hence homotopy equivalent, to $\psi_*(\mathbf{P})$, we see that $\alpha_*(\mathbf{P}) \simeq \mathbf{P}$, as claimed. This completes the proof. $\square$

In particular, we have:

Proposition 54.5: If $\mathbf{P}$ is a projective n-stem, then for each $\alpha \in \mathrm{Aut}_{\mathbf{Z}[G]}(J)$

$$\chi(\alpha \bullet \mathbf{P}) = \chi(\mathbf{P})$$

It follows that, if $J \in \Omega_n(\mathbf{Z})$ and $\mathbf{F}$ is a stably free n-stem, then, for each $\alpha \in \mathrm{Aut}_{\mathbf{Z}[G]}(J)$, $\chi(\alpha \bullet \mathbf{F}) = 0$. This may be re-phrased: if $J \in \Omega_n(\mathbf{Z})$, and $[\ ] : \mathrm{Aut}_{\mathbf{Z}[G]}(J) \to \mathcal{A}(J)$ is the identification map, then

$$\mathrm{Im}([\ ]) \subset \mathrm{Ker}(S^J)$$

moreover, from (53.10) and (54.4), we see directly that:

Theorem 54.6: If $J \in \Omega_n(\mathbf{Z})$ then extensions in **Stab**$^{n+1}(\mathbf{Z}, J)$ are classified up to homotopy equivalence by $\mathrm{Ker}(S^J)/\mathrm{Im}(\nu^J)$.

$J \in \Omega_n(\mathbf{Z})$ is said to be *full* when the natural map $[\ ] : \mathrm{Aut}_{\mathbf{Z}[G]}(J) \to \mathrm{Ker}(S^J)$ is surjective.

Theorem 54.7: Let $J \in \Omega_n(\mathbf{Z}[G])$; if J is full, then $J \oplus \mathbf{Z}[G]$ is also full.

Proof: There is a ring homomorphism $\Sigma_{\mathbf{Z}[G]} : \mathrm{End}_{\mathbf{Z}[G]}(J) \to \mathrm{End}_{\mathbf{Z}[G]}(J \oplus \mathbf{Z}[G])$ defined by means of stabilization, viz.

$$\alpha \mapsto \begin{pmatrix} \alpha & 0 \\ 0 & \mathrm{Id} \end{pmatrix}$$

This formula also gives a ring *isomorphism* $\Sigma_{\mathcal{D}er} : \mathrm{End}_{\mathcal{D}er}(J) \to \mathrm{End}_{\mathcal{D}er}(J \oplus \mathbf{Z}[G])$. By uniqueness of the ring isomorphisms κ^J, $\kappa^{J \oplus \mathbf{Z}[G]}$, it follows firstly that

$$\kappa^J = \kappa^{J \oplus \mathbf{Z}[G]} \circ \Sigma_{\mathcal{D}er}$$

and by composition with the Swan map, defined on the unit groups, also that

$$S^J = S^{J \oplus \mathbf{Z}[G]} \circ \Sigma_{\mathcal{D}er}$$

$\Sigma_{\mathcal{D}er}$ induces an isomorphism making the following diagram commute

$$
\begin{array}{ccc}
\mathrm{Aut}_{\mathbf{Z}[G]}(J) & \xrightarrow{\quad [\ ] \quad} & \mathrm{Ker}(S^J) \\
\Big\downarrow {\scriptstyle \Sigma_{\mathbf{Z}[G]}} & & \Big\downarrow {\scriptstyle \Sigma_{\mathcal{D}er}} \\
\mathrm{Aut}_{\mathbf{Z}[G]}(J \oplus \mathbf{Z}[G]) & \xrightarrow{\quad [\ ] \quad} & \mathrm{Ker}\big(S^{J \oplus \mathbf{Z}[G]}\big)
\end{array}
$$

It follows that, if $[\ ]: \mathrm{Aut}_{Z[G]}(J) \to \mathrm{Ker}(S^J)$ is surjective, then

$$[\ ]: \mathrm{Aut}_{Z[G]}(J \oplus \mathbf{Z}[G]) \to \mathrm{Ker}(S^{J \oplus \mathbf{Z}[G]})$$

is also surjective. This completes the proof. $\qquad\qquad\qquad\qquad\qquad$ □

55 Realization Theorems

There is a natural geometric stabilization process within CW_G^2 : for $X \in \mathrm{CW}_G$, we define $\mathcal{S}(X) = X \vee S^2$. More generally, we write

$$\mathcal{S}^m(X) = X \vee \underbrace{S^2 \vee \cdots \vee S^2}_{m}$$

The correspondence $X \mapsto \mathcal{S}(X)$ is functorial with respect to based continuous maps; moreover, the following functorial square commutes up to natural equivalence

$$
\begin{array}{ccc}
\mathrm{CW}_G & \xrightarrow{\ S\ } & \mathrm{CW}_G \\
C_* \downarrow & & C_* \downarrow \\
\mathbf{Stab}^3_{\mathbf{Z}[G]}(\mathbf{Z}, \pi_2(-)) & \xrightarrow{\ \Sigma_+^!\ } & \mathbf{Stab}^3_{\mathbf{Z}[G]}(\mathbf{Z}, \pi_2(-) \oplus \mathbf{Z}[G])
\end{array}
$$

Recall that an algebraic 2-complex $\mathbf{E}$ over $\mathbf{Z}[G]$ is said to be *geometrically realized* when there is a finite 2-complex $\mathcal{L}$ with $\pi_1(\mathcal{L}) = G$, and a homotopy equivalence $\varphi : C_*(\mathcal{L}) \xrightarrow{\sim} \mathbf{E}$; equivalently, when there is a finite presentation $\mathcal{G}$ of G and a homotopy equivalence $\varphi : C_*(\mathcal{G}) \xrightarrow{\sim} \mathbf{E}$. If $\mathbf{E} = (0 \to J \to F_2 \to F_1 \to F_0 \to \mathbf{Z} \to 0)$ is an algebraic 2-complex, then $\mathbf{E}$ is said to be *minimal* when $J \in \Omega_3(\mathbf{Z})$ is minimal.

Say that G has the *realization property* when each algebraic 2-complex over $\mathbf{Z}[G]$ has a geometric realization; likewise say that G has the *minimal realization property* when each *minimal* algebraic 2-complex over $\mathbf{Z}[G]$ has a geometric realization. We arrive at:

Theorem II: Let G be a finite group; then G has the realization property if and only if all minimal algebraic 2-complexes are realizable.

Proof: If G has the realization property, then clearly G has the minimal realization property. It suffices to show the converse.

Suppose that, for each $N \in \Omega_3(\mathbf{Z})$ of *minimal height* and each $\mathcal{E} \in \mathbf{Stab}^3(\mathbf{Z}, N)$, the homotopy type of $\mathcal{E}$ is geometrically realizable. We must show that, if $N' \in \Omega_3(\mathbf{Z})$ has height $h \geq 1$, and $\mathcal{E}' \in \mathbf{Stab}^3(\mathbf{Z}, N')$, then $\mathcal{E}'$ is geometrically realizable. By the Swan–Jacobinski Theorem we may write

$$N' \cong N \oplus \Lambda^h$$

where $N \in \Omega_3(\mathbf{Z})$ is *any* module of minimal height. Let $(\mathcal{E}(r))_r$ be a complete set of representatives of congruence classes of extensions in $\mathbf{Stab}^3(\mathbf{Z}, N)$, where r runs through the elements of $\mathrm{Ker}(S : (\mathbf{Z}/|G|)^* \to \widetilde{K}_0(\mathbf{Z}[G]))$. By Yoneda's Theorem, $\mathcal{E}'$ is congruent to $\Sigma_+^h \mathcal{E}(r)$ for some unique $r \in \mathrm{Ker}(S)$. By hypothesis, each $\mathcal{E}(r)$ is homotopy equivalent to some finite 2-complex $K(r)$. Thus there is a homotopy equivalence $\Sigma_+^h \mathcal{E}(r) \simeq S^h(K(r))$ where

$$S^h(K) = K \vee \underbrace{S^2 \vee \cdots \vee S^2}_{h}$$

for any finite 2-complex K. By Yoneda's Theorem, $\mathcal{E}'$ is congruent to $\Sigma_+^h \mathcal{E}(r)$ for some unique $r \in \mathrm{Ker}(S)$, and so

$$\mathcal{E}' \simeq S^h(K(r))$$

In particular, $\mathcal{E}'$ is geometrically realizable, and this completes the proof. $\square$

The 'minimal realization' theorem just proved is a general condition on homotopy types. It is more useful to have a criterion for realizability in terms of homotopy *groups*. Say that a module $J \in \Omega_3(\mathbf{Z})$ is *realizable* when there exists a finite presentation $\mathcal{G}$ such that $J \cong \pi_2(\mathcal{G})$. By means of the Tietze operation, which adds a redundant relation, we see that realizability is preserved under the stabilization operation $J \mapsto J \oplus \mathbf{Z}[G]$; that is:

Proposition 55.1: Let G be a finite group and suppose that $J \in \Omega_3(\mathbf{Z})$; if J is realizable, then $J \oplus \mathbf{Z}[G]$ is also realizable.

A module $J \in \Omega_3(\mathbf{Z})$ is said to be *strongly realizable* when it is both realizable and full. From (54.7) and (55.1), we obtain:

Theorem 55.2: Let G be a finite group and suppose that $J \in \Omega_3(\mathbf{Z})$; if J is strongly realizable, then $J \oplus \mathbf{Z}[G]$ is also strongly realizable.

Strongly realizable modules have the following property:

Lemma 55.3: Let G be a finite group and suppose that $J \in \Omega_3(\mathbf{Z})$ is strongly realizable; then any algebraic 2-complex of the form

$$0 \to J \to F_2 \to F_1 \to F_0 \to \mathbf{Z} \to 0$$

is geometrically realizable.

Proof: By hypothesis, J is realized in the form $J \cong \pi_2(\mathcal{G})$ for some finite presentation

$$\mathcal{G} = \langle X_1, \dots, X_g \mid W_1, \dots, W_r \rangle$$

of G. Let [1] denote the algebraic Cayley complex of $\mathcal{G}$

$$[1] = (0 \to J \to \mathbf{Z}[G]^r \to \mathbf{Z}[G]^g \to \mathbf{Z}[G] \to \mathbf{Z} \to 0)$$

If $\mathcal{E}$ is an algebraic 2-complex of the form

$$0 \to J \to F_2 \to F_1 \to F_0 \to \mathbf{Z} \to 0$$

then by (38.3), $\mathcal{E}$ is congruent to an extension of the form

$$\mathcal{E}(s) = (0 \to J \to \mathbf{Z}[G]^r \to \mathbf{Z}[G]^g \to (I, s) \to \mathbf{Z} \to 0)$$

where $s = k([1] \to \mathcal{E})) \in (\mathbf{Z}/|G|)^*$. Since $\mathcal{E}$ is stably free, $\chi(\mathcal{E}) = 0$. However, χ is invariant under congruence (46.8), so that $\chi(I, s) = 0$ and $s \in \mathrm{Ker}(S : (\mathbf{Z}/|G|)^* \to \tilde{K}_0(\mathbf{Z}[G]))$. By hypothesis, there exists $u \in \mathrm{Aut}_{\mathbf{Z}[G]}(J)$ such that $v(u) = s$ where $v : \mathrm{Aut}_{\mathbf{Z}[G]}(J) \to (\mathbf{Z}/|G|)^*$ is the natural map.

Let $w = u^{-1} \in \mathrm{Aut}_{\mathbf{Z}[G]}(J)$, so that $v(w) = s^{-1} \in (\mathbf{Z}[G]/|G|)^*$. The k-invariant of the transition $[1] \to w_*\mathcal{E}(s)$ is given by

$$k([1] \to w_*\mathcal{E}(s)) = 1$$

Thus $w_*\mathcal{E}(s)$ is congruent to [1], and hence $w_*\mathcal{E}(s) \simeq [1]$. However, $w_* : \mathcal{E}(s) \to w_*\mathcal{E}(s)$ is an isomorphism, as in (54.3), *a fortiori*, a homotopy equivalence. Since $\mathcal{E} \approx \mathcal{E}(s)$, then $\mathcal{E} \simeq \mathcal{E}(s)$, and we see finally that $\mathcal{E} \simeq [1]$. The result follows, since [1] is the algebraic Cayley complex of the presentation $\mathcal{G}$. $\qquad\square$

We say that the stable module $\Omega_3(\mathbf{Z})$ is strongly realizable when each $J \in \Omega_3(\mathbf{Z})$ is strongly realizable. From (55.1) we obtain:

Theorem 55.4: The stable module $\Omega_3(\mathbf{Z})$ is strongly realizable if and only if each minimal module $J \in \Omega_3(\mathbf{Z})$ is both realizable and full.

From (55.3) and (55.4) we now get:

Theorem III: If each minimal module $J \in \Omega_3(\mathbf{Z})$ is both realizable and full, then G has the realization property.

56 Augmentation sequences

Following Swan [63], for any finite group G we denote by $SF(\mathbf{Z}[G])$ the isomorphism types of finitely generated stably free projective modules over $\mathbf{Z}[G]$. When G is understood we write this as SF. Moreover, we write $SF_1(\mathbf{Z}[G])$ ($= SF_1$) for the class of isomorphism types of stably free modules of rank $= 1$. Also following Swan, we write $LF(\mathbf{Z}[G])$ ($= LF$) for the class of all finitely projective (= locally free) modules over $\mathbf{Z}[G]$.

In (16.1) we pointed out the result of Swan that, if P is a projective module over $\mathbf{Z}[G]$, then $P \otimes \mathbf{Q} \cong \mathbf{Q}[G]^k$ for some $k \geq 1$; the integer k is called the *rank* of P; it follows that there exists a surjective homomorphism $\varphi : P \otimes \mathbf{Q} \to \mathbf{Q}$ on to the trivial $\mathbf{Q}[G]$ module $\mathbf{Q}$. Letting $\eta : P \to \mathbf{Q}$ be the $\mathbf{Z}[G]$ homomorphism given by $\eta(\mathbf{x}) = \varphi(\mathbf{x} \otimes 1)$, the G action on the submodule $\mathrm{Im}(\eta)$ is trivial. However, $\mathrm{Im}(\eta)$ is a nonzero finitely generated additive subgroup of $\mathbf{Q}$ and so is isomorphic to $\mathbf{Z}$:

Proposition 56.1: Let P be a (non zero) finitely generated projective module over $\mathbf{Z}[G]$; then there exists a surjective $\mathbf{Z}[G]$-homomorphism $\eta : P \to \mathbf{Z}$.

By an *augmentation* we mean a $\mathbf{Z}[G]$-homomorphism of the form $\eta : F \to \mathbf{Z}$ where F is a stably free module of rank 1; a $\mathbf{Z}[G]$-module J is said to be an *augmentation module* when it has the form $J = \mathrm{Ker}(\eta)$ where η is an augmentation. We denote by $\Omega_1^{min}(\mathbf{Z})$ the subset of $\Omega_1(\mathbf{Z})$ consisting of elements at the minimal level. The following is straightforward.

Proposition 56.2: J is an augmentation module if and only $J \in \Omega_1^{min}(\mathbf{Z})$.

By an *augmentation sequence* we mean an exact sequence of $\mathbf{Z}[G]$-modules of the form

$$0 \to J \to F \xrightarrow{\eta} \mathbf{Z} \to 0$$

where η is an augmentation. Clearly any augmentation sequence is an exact sequence in $\mathcal{F}(\mathbf{Z}[G])$. The augmentation ideal $\mathbf{I} = \mathbf{I}(G)$ is defined by the *standard augmentation sequence*

$$0 \to \mathbf{I} \to \mathbf{Z}[G] \stackrel{\epsilon}{\to} \mathbf{Z} \to 0$$

where $\epsilon : \mathbf{Z}[G] \to \mathbf{Z}$ is the standard augmentation map given by $\epsilon(g) = 1$ for all $g \in G$.

Proposition 56.3: Let $\mathcal{E}_k = (0 \to J_k \stackrel{J_k}{\to} F_k \stackrel{p_k}{\to} \mathbf{Z} \to 0)$ be augmentation sequences for $k = 1, 2$; then the natural map $\mathrm{Hom}(\mathcal{E}_1, \mathcal{E}_2) \to \mathrm{Hom}(F_1, F_2)$ is surjective.

Proof: Let $\varphi : F_1 \to F_2$ be a homomorphism over $\mathbf{Z}[G]$. Then $(p_2 \circ \varphi \circ j_1) \otimes \mathrm{Id} : D_1 \otimes \mathbf{Q} \to \mathbf{Q}$ is trivial, so that, since J_1 is torsion free, $\varphi \circ j_1(D_1) \subset \mathrm{Ker}(p_2) = \mathrm{Im}(j_2)$. The existence of homomorphisms $\varphi_J, \varphi_\mathbf{Z}$ making the following diagram commute is now clear

$$0 \to J_1 \stackrel{j_1}{\to} F_1 \stackrel{p_1}{\to} \mathbf{Z} \to 0$$
$$\downarrow \varphi_J \quad \downarrow \varphi \quad \downarrow \varphi_\mathbf{Z}$$
$$0 \to J_2 \stackrel{j_2}{\to} F_2 \stackrel{p_2}{\to} \mathbf{Z} \to 0 \qquad\qquad \square$$

Recalling from Section 37 the notion of full invariance, we have:

Corollary 56.4: Any augmentation sequence is fully invariant.

With the same notation, we have:

Corollary 56.5: If $\mathcal{E}_1, \mathcal{E}_2$ are augmentation sequences then

$$\mathcal{E}_1 \cong \mathcal{E}_2 \iff F_1 \cong F_2$$

Proof: The implication ($\Longrightarrow$) is clear. To prove ($\Longleftarrow$), observe that, if $\varphi : F_1 \to F_2$ is an isomorphism, φ can, by (56.3), be completed to a commutative diagram of the form

$$0 \to J_1 \stackrel{j_1}{\to} F_1 \stackrel{p_1}{\to} \mathbf{Z} \to 0$$
$$\downarrow \varphi_J \quad \downarrow \varphi \quad \downarrow \varphi_\mathbf{Z}$$
$$0 \to J_2 \stackrel{j_2}{\to} F_2 \stackrel{p_2}{\to} \mathbf{Z} \to 0$$

$\varphi_\mathbf{Z} : \mathbf{Z} \to \mathbf{Z}$ is clearly surjective and so is an isomorphism. It follows that φ_J is an isomorphism. $\qquad\qquad \square$

If F is a stably free rank 1-projective, we write $\Psi([F]) = [J]$ when there exists a sequence of the form $0 \to J \to F \to \mathbf{Z} \to 0$. On the face of it, Ψ is only a multi-valued (surjective) relation with domain SF_1 and codomain $\Omega_1^{min}(\mathbf{Z})$. In fact, the situation is simpler:

Proposition 56.6: Ψ is a surjective mapping $\Psi : SF_1 \to \Omega_1^{min}(\mathbf{Z})$.

Proof: Given augmentation sequences

$$\mathcal{E}_i = \left(0 \to J_i \to F_i \xrightarrow{p_i} \mathbf{Z} \to 0\right)$$

for $i = 1, 2$, it suffices to show that $F_1 \cong F_2 \implies J_1 \cong J_2$. However, by (56.5), $F_1 \cong F_2 \implies \mathcal{E}_1 \cong \mathcal{E}_2$, whilst it is clear that $\mathcal{E}_1 \cong \mathcal{E}_2 \implies J_1 \cong J_2$. $\square$

For any augmentation module J, there is a unique ring isomorphism

$$\kappa = \kappa^J : \text{End}_{\mathcal{D}er}(J) \to \text{End}_{\mathcal{D}er}(\mathbf{Z}) \cong \mathbf{Z}/|G|$$

Theorem 56.7: If J is an augmentation module, then there is a $1 - 1$ correspondence

$$\Psi^{-1}(J) \longleftrightarrow \text{Ker}(S^J)/\text{Im}(\nu^J)$$

Proof: Augmentation sequences $0 \to J \to F \to \mathbf{Z} \to 0$ are simply elements of $\mathbf{Stab}^1(\mathbf{Z}, J)$. By the k-invariant classification they are classified up to congruence by $\text{Ker}(S^J)$. In general, from (54.6), homotopy types of extensions in $\mathbf{Stab}^{n+1}(\mathbf{Z}, J)$ are classified by $\text{Ker}(S^J)/\text{Im}(\nu^J)$; however, for $\mathbf{Stab}^1(\mathbf{Z}, J)$, homotopy equivalence of such extensions is, by the Five Lemma, identical with isomorphism over $\text{Id}_{\mathbf{Z}}$. In fact, the qualification can be dispensed with, since $\text{Aut}(\mathbf{Z}) = \{\pm \text{Id}_{\mathbf{Z}}\}$, and given an isomorphism of augmentation sequences over $-\text{Id}_{\mathbf{Z}}$, thus

$$
\begin{array}{ccccccc}
0 \to & J & \to & F_1 \to & \mathbf{Z} \to & 0 \\
 & \downarrow \varphi_J & & \downarrow \varphi_F & \downarrow -\text{Id}_{\mathbf{Z}} & \\
0 \to & J & \to & F_2 \to & \mathbf{Z} \to & 0
\end{array}
$$

the sequences become isomorphic over $\text{Id}_{\mathbf{Z}}$ on multiplying through by -1. Thus the isomorphism classes of extensions in $\mathbf{Stab}^1(\mathbf{Z}, J)$ are classified by $\text{Ker}(S^J)/\text{Im}(\nu^J)$, and the conclusion follows from (56.5) $\square$

It follows from the Swan–Jacobinski Theorem that:

Proposition 56.8: *SF* is a fork with $|SF_1|$ prongs.

The Swan–Jacobinski Theorem also shows that, if F is a stably free module of rank $r \geq 2$, then $F \cong \mathbf{Z}[G]^r$, and together with Schanuel's Lemma implies that, if $\eta : \mathbf{Z}[G]^r \to \mathbf{Z}$ is surjective for $r \geq 2$, then

$$\mathrm{Ker}(\eta) \cong \mathbf{I} \oplus \mathbf{Z}[G]^{r-1}$$

One sees that Ψ extends to give a mapping of directed graphs (in this case, both forks)

$$\Psi : SF \to \Omega_1(\mathbf{Z})$$

using the same definition, namely $\Psi([F]) = [J]$ when there exists a sequence of the form $0 \to J \to F \to \mathbf{Z} \to 0$. In fact, we have:

Theorem 56.9: For any finite group G, $\Psi : SF \to \Omega_1(\mathbf{Z})$ is a surjective level preserving mapping of directed graphs with the property that, for each $J \in \Omega_1(\mathbf{Z})$, the multiplicity of the fibre $\Psi^{-1}(J)$ is $|\mathrm{Ker}(S^J)|/|\mathrm{Im}(\nu^J)|$.

Proof: The only detail left to be checked is that, when $J \in \Omega_1(\mathbf{Z})$ is non-minimal, the formal multiplicity, $|\mathrm{Ker}(S^J)|/|\mathrm{Im}(\nu^J)|$, is the actual multiplicity in this case, namely 1. This follows from the k-invariant classification together with the Swan–Jacobinski Theorem and Schanuel's Lemma. $\square$

One can generalize the above discussion to arbitrary projectives: for $c \in \tilde{K}_0(\mathbf{Z}[G])$, we put

$$LF^c = \{P \in LF : [P] = c \in \tilde{K}_0(\mathbf{Z}[G])\}$$

and denote by LF_k^c the subset of LF^c consisting of projectives of rank k.

As a directed graph, with edges defined, as usual, by stabilization, LF is the disjoint union of its connected components LF^c, where c runs through $\tilde{K}_0(\mathbf{Z}[G])$, and each LF^c is a fork, in which LF_1^c is the set of minimal elements. By a *quasi-augmentation* we mean a $\mathbf{Z}[G]$-homomorphism of the form $\eta : P \to \mathbf{Z}$ where P is a projective module of rank 1, not assumed to be stably free; a $\mathbf{Z}[G]$-module J is said to be a *quasi-augmentation module* when it has the form

$$J = \mathrm{Ker}(\eta)$$

where η is an augmentation. The previous discussion for augmentation modules goes through almost unchanged for quasi-augmentation modules; Ψ extends to give a mapping of directed graphs

$$\Psi : LF^c \to \mathbf{D}_1^c(\mathbf{Z})$$

using the same definition as before, namely $\Psi([P]) = [J]$ when there exists a sequence of the form

$$0 \to J \to P \to \mathbf{Z} \to 0$$

In fact, we have:

Theorem 56.10: $\Psi : LF^c \to \mathbf{D}_1^c(\mathbf{Z})$ is a surjective level preserving mapping of directed graphs with the property that, for each $J \in \mathbf{D}_1^c(\mathbf{Z})$, the multiplicity of the fibre $\Psi^{-1}(J)$ is $|\mathrm{Ker}(S^J)|/|\mathrm{Im}(\nu^J)|$.

57 Classification over groups of period 4

For any finite group G, the correspondence $\mathbf{F} \mapsto \pi_2(\mathbf{F})$ determines a level preserving surjective mapping of directed graphs $\pi_2 : \widehat{\mathbf{Alg}}_G \to \Omega_3(\mathbf{Z})$. We begin by finding the size of the fibres of this map.

Consider the natural action of $\mathrm{Ker}(S^J)$ on $\mathbf{Stab}^3(\mathbf{Z}, J)$, where $J \in \Omega_3(\mathbf{Z})$. By (54.4), for $\mathbf{F} \in \mathbf{Stab}^3(\mathbf{Z}, J)$ the stabilizer of the homotopy type $[\mathbf{F}]$ is $\mathrm{Im}(\nu^J)$. Thus we get:

Theorem 57.1: If G is a finite group, then the set of homotopy types within $\mathbf{Stab}^3(\mathbf{Z}, J)$ for $J \in \Omega_3(\mathbf{Z})$ is in 1–1 correspondence with $\mathrm{Ker}(S^J)/\mathrm{Im}(\nu^J)$.

In particular, for any finite group G and any $J \in \Omega_3(\mathbf{Z})$:

(57.2) $\left|\pi_2^{-1}(J)\right| = |\mathrm{Ker}(S^J)|/|\mathrm{Im}(\nu^J)|$

We are now in a position to parametrize two-dimensional algebraic homotopy types over a finite group of free period 4.

Theorem IV: Let G be a finite group of free period 4; there is an isomorphism of directed graphs

$$\widehat{\mathbf{Alg}}_G \longleftrightarrow SF_1(\mathbf{Z}[G])$$

In particular, the set of minimal two-dimensional algebraic homotopy types over G is faithfully parametrized by the set $SF_1(\mathbf{Z}[G])$ of stably free modules of rank 1.

Proof: By (56.9) there is a level preserving map $\lambda : SF(\mathbf{Z}[G]) \to \Omega_1(\mathbf{Z})$ with the property $|\lambda^{-1}(J)| = |\mathrm{Ker}(S^J)|/\nu(J)$, where $\nu(J) = |\mathrm{Im}(\nu^J)|$. Moreover, there is an isomorphism of directed graphs given by duality $\delta : \Omega_1(\mathbf{Z}) \to \Omega_{-1}(\mathbf{Z}), \delta(J) = J^*$, and it is straightforward to check that $\nu(\delta(J)) = \nu(J)$. Thus putting $\mu_1 = \delta \circ \lambda$ we see that μ_1 is a level preserving map $\widehat{\mathbf{Alg}}_G \to \Omega_{-1}(\mathbf{Z})$ with the property that $\mu_1^{-1}(J) = |\mathrm{Ker}(S^J)|/\nu(J)$ for all $J \in \Omega_{-1}(\mathbf{Z})$.

However, as we have seen in (57.2), there is a surjective level preserving map $\pi_2 : \widehat{\mathbf{Alg}_G} \to \Omega_3(\mathbf{Z})$ with the property that $|\pi_2^{-1}(J)| = |\mathrm{Ker}(S^J)|/\nu(J)$. However, since G has free period 4, we have equality $\Omega_3(\mathbf{Z}) = \Omega_{-1}(\mathbf{Z})$, and it follows easily now that the tree structures on $\widehat{\mathbf{Alg}_G}$ and $SF_1(\mathbf{Z}[G])$ are isomorphic. $\square$

In particular, we have:

Proposition 57.3: If G has free period 4, $\widehat{\mathbf{Alg}_G}$ contains at least two minimal homotopy types precisely when $\mathbf{Z}[G]$ admits a stably free module which is not free.

Notice that this occurs for all the generalized quaternion groups $Q(2^n)$ with $n \geq 5$. [63], [29].

Now consider the more general case where the group G admits a projective resolution of period 4. As in Section 56, when $c \in \widetilde{K}_0(\mathbf{Z}[G])$, we put

$$LF^c = \{P \in LF : [P] = c\}$$

Again, by the Swan–Jacobinski Theorem each LF^c is a fork in which the minimal level is the set LF_1^c of isomorphism types of projective modules P of rank $=$ 1 with $[P] = c$. Suppose $\mathbf{Z}$ admits a complete resolution over $\mathbf{Z}[G]$ of the form

$$0 \to \mathbf{Z} \to P_3 \to P_2 \to P_1 \to P_0 \to \mathbf{Z} \to 0$$

where each P_i is finitely generated projective. By a sequence of elementary congruences, we may suppose the extension is congruent to one of the form

$$0 \to \mathbf{Z} \to P \to F_2 \to F_1 \to F_0 \to \mathbf{Z} \to 0$$

where P is finitely generated and each F_i is finitely generated. It was shown by Milgram ([40], [41]) that for some groups G (for example, $G = Q(8, 3, 11)$ [4]) P cannot be chosen to be stably free; that is, the finiteness obstruction is necessarily nonzero. By appealing to (56.10) instead of (56.9), Theorem IV generalizes as follows:

Theorem 57.4: Let G be a finite group which admits a finitely generated projective resolution $\mathbf{P}$ of period 4; there is an isomorphism of directed graphs

$$\widehat{\mathbf{Alg}_G} \longleftrightarrow LF^c(\mathbf{Z}[G])$$

where $c = \chi(\mathbf{P}) \in \widetilde{K}_0(\mathbf{Z}[G])$.

Swan [60] showed that the finiteness obstruction $\chi(\mathbf{P})$ is not an absolute invariant of G, but can vary arbitrarily within a coset of $\mathrm{Im}(S)$. This implies that the isomorphism type of the directed graph $LF^\gamma(\mathbf{Z}[G])$ remains constant as γ runs through $c + \mathrm{Im}(S)$. It follows that:

Corollary 57.5: Let G be a finite group which admits a finitely generated projective resolution **P** of period 4; then the set $\widehat{\mathbf{Alg}}_G$ of homotopy classes of algebraic 2-complexes over G is fork, and has a unique homotopy type at each non-minimal level.

The results just proved are rather stronger than a more direct application of the methods developed to achieve them might suggest. To see this, let G be an arbitrary finite group, and revisit the stability theorems in the case of extensions in $\mathbf{Stab}^3(\mathbf{Z}, J)$ for $J \in \Omega_3(\mathbf{Z})$.

We know, from Chapter 5, that $\Omega_3(\mathbf{Z})$ is a fork, with a unique isomorphism type J_h at each height $h \geq 1$. In general, there are many choices of module at height 0. We choose J_0 at height 0 to have the property that $v(J_0) = \min\{v(J) : \mathrm{height}(J) = 0\}$; that is, $v(J_0) = v_3(\mathbf{Z})$. Then, by the Swan–Jacobinski Theorem, $J_h \cong J_0 \oplus \mathbf{Z}[G]^h$ when $h \geq 1$. Moreover, we also make an arbitrary choice of a stably free 3-stem $\mathcal{E}(1) \in \mathbf{Stab}^3(\mathbf{Z}, J_0)$. Then we have:

Proposition 57.6: Let G be a finite group, and let $\mathcal{E} \in \mathbf{Stab}^3(\mathbf{Z}, J_h)$; then there is a homotopy equivalence

$$\Sigma_+^h(\mathcal{E}(1)) \xrightarrow{\sim} \mathcal{E}$$

provided $h \geq \max\{\sigma(G) + 1, v_3(\mathbf{Z})\}$.

Proof: Without loss we may write $J_h = J_0 \oplus \mathbf{Z}[G]^h$. We use $\mathcal{E}(1)$ as the reference extension to parametrize $\mathbf{Stab}^3(\mathbf{Z}, J_0)$. Let $(\mathcal{E}(r))_r$ be a complete set of representatives of congruence classes of extensions in $\mathbf{Stab}^3(\mathbf{Z}, J_0)$, where r runs through the elements of $\mathrm{Ker}(S : (\mathbf{Z}/|G|)^* \to \widetilde{K}_0(\mathbf{Z}[G]))$, so that $r = k(\mathcal{E}(1) \to \mathcal{E}(r))$. Put $\mathcal{S} = \Sigma_+^h(\mathcal{E}(1))$. Then we may use $\mathcal{S}$ as a reference extension to parametrize $\mathbf{Stab}^3(\mathbf{Z}, J_h)$. In particular, for some unique $r \in \mathrm{Ker}(S)$, $r = k(\mathcal{S} \to \mathcal{E})$. However it is straightforward to see that $r = k(\mathcal{S} \to \Sigma_+^h(\mathcal{E}(r)))$. It follows that there is a congruence $\Sigma_+^h(\mathcal{E}(r)) \approx \mathcal{E}$. Now by (52.2), when $h \geq \max\{\sigma(G) + 1, v_3(\mathbf{Z})\}$ there is an isomorphism

$$\Sigma_+^h(\mathcal{E}(1)) \xrightarrow{\sim} \Sigma_+^h(\mathcal{E}(r))$$

Composition gives the desired homotopy equivalence $\Sigma_+^h(\mathcal{E}(1)) \xrightarrow{\sim} \mathcal{E}$. $\square$

This is not best possible, and in his thesis [9], Browning proved:

Theorem 57.7: (Browning's Stability Theorem) Let G be a finite group, let $J_0 \in \Omega_3(\mathbf{Z})$ be any module of minimal height, and let $\mathcal{E}_0 \in \mathbf{Stab}^3(\mathbf{Z}, J_0)$ be *any* stably free 3-stem at the minimal level; then for each $h \geq 1$, $\mathbf{Stab}^3(\mathbf{Z}, J_h)$ contains a unique homotopy type, namely that of $\Sigma_+^h(\mathcal{E}_0)$.

Browning's proof is a reworking of the Swan-Jacobinski Theorem in the context of extensions, extending earlier work of Williams [81]; see also [20], [21], [33]. For groups of period 4, Corollary (57.5) gives the same conclusion as Browning by a different route.

Chapter 10
The D(2)-problem

We finally turn to consider the D(2)-problem. We state it thus:

D(2)-problem: Suppose that X is a finite three-dimensional cell complex such that $H_3(\tilde{X}; \mathbf{Z}) = H^3(X; \mathcal{B}) = 0$ for all local coefficient systems $\mathcal{B}$ on X. Is X homotopy equivalent to a finite two-dimensional complex?

We show that, when G is finite, the D(2)-problem is equivalent to the Realization Problem; that is:

Theorem I: The D(2) property holds for the finite group G if and only if each algebraic 2-complex over G is geometrically realizable.

It then follows that Theorem III, already proved in Chapter 9, gives a sufficient condition for the D(2)-property to hold.

Moreover, for groups of period 4 using the parametrization of homotopy types of algebraic 2-complexes given by Theorem IV and Theorem (57.4), and by applying the computations of Swan [63], we are able, in Section 62, to verify the D(2)-property in some specific cases, as well as to identify some potential counterexamples.

58 Cohomologically two-dimensional 3-complexes

We begin with an elementary observation:

Proposition 58.1: Let X be a finite 3-complex with $\pi_1(X) \cong G$, and put $K = X^{(2)}$; then there is a canonical exact sequence of $\mathbf{Z}[G]$-modules

$$0 \to H_3(\tilde{X}; \mathbf{Z}) \to C_3(\tilde{X}) \xrightarrow{\partial_3} \pi_2(K) \to \pi_2(X) \to 0$$

Proof: We have exact sequences

$$0 \to Z_3(\tilde{X}) \to C_3(\tilde{X}) \xrightarrow{\partial_3} \mathrm{Im}(\partial_3) \to 0$$

and

$$0 \to \text{Im}(\partial_3) \to Z_2(\tilde{X}) \to H_2(\tilde{X}; \mathbf{Z}) \to 0$$

Splicing these together gives

$$0 \to Z_3(\tilde{X}) \to C_3(\tilde{X}) \xrightarrow{\partial_3} Z_2(\tilde{X}) \to H_2(\tilde{X}; \mathbf{Z}) \to 0$$

However, since $K = X^{(2)}$ we have

$$H_2(\tilde{K}; \mathbf{Z}) = Z_2(\tilde{K}) = Z_2(\tilde{X})$$

whilst

$$Z_3(\tilde{X}) = H_3(\tilde{X}; \mathbf{Z})$$

so that we have an exact sequence

$$0 \to H_3(\tilde{X}; \mathbf{Z}) \to C_3(\tilde{X}) \xrightarrow{\partial_3} H_2(\tilde{K}; \mathbf{Z}) \to H_2(\tilde{X}; \mathbf{Z}) \to 0$$

The result follows from the Hurewicz Theorem, as $\pi_2(K) \cong \pi_2(\tilde{K}) = H_2(\tilde{K}; \mathbf{Z})$ and $\pi_2(X) \cong \pi_2(\tilde{X}) = H_2(\tilde{X}; \mathbf{Z})$. $\qquad\qquad\square$

Let X be a CW complex of dimension 3; we say that is cohomologically two-dimensional when $H_3(\tilde{X}; \mathbf{Z}) = H^3(X; \mathcal{B}) = 0$ for any local coefficient system $\mathcal{B}$ on X. In the case when $\pi_1(X) = G$ is finite, there is a useful simplification.

Recall that $C_n(X)$ and $C_n(\tilde{X})$ have the same underlying abelian group, namely $H_n(\tilde{X}^{(n)}, \tilde{X}^{(n-1)}; \mathbf{Z})$. We take $C_n(\tilde{X})$ simply to be an abelian group, whilst $C_n(X)$ is a module over $\mathbf{Z}[G]$, from the covering action of G on $\tilde{X}$. Since $C_n(\tilde{X})$ is a free abelian group of finite rank, the Eckmann–Shapiro Lemma gives an isomorphism

$$\text{Hom}_{\mathbf{Z}}(C_n(\tilde{X}); \mathbf{Z}) \cong \text{Hom}_{\mathbf{Z}[G]}(C_n(X); \mathbf{Z}[G])$$

and so we may interpret

$$H^n(\tilde{X}; \mathbf{Z}) = H^n(X; \mathbf{Z}[G])$$

Since $\tilde{X}$ is a finite complex of dimension ≤ 3, by the Universal Coefficient Theorem for $\mathbf{Z}$-homology ([53], Chapter 5)

$$\text{Tor}(H_3(\tilde{X}; \mathbf{Z})) = \text{Tor}(H^4(\tilde{X}; \mathbf{Z})) = 0$$

and $H_3(\tilde{X}; \mathbf{Z})$ is a free abelian group whose $\mathbf{Z}$-rank is the same as that of $H^3(\tilde{X}; \mathbf{Z}) = H^3(X; \mathbf{Z}[G]) = 0$. Thus, when $\pi_1(X) = G$ is finite, the condition '$H_3(\tilde{X}; \mathbf{Z}) = 0$' is redundant and we obtain:

Proposition 58.2: Let X be a finite 3-complex in which $\pi_1(X)$ is finite; then X is cohomologically two-dimensional if and only if $H^3(X; \mathcal{B}) = 0$ for any local coefficient system $\mathcal{B}$ on X.

Observe that the above argument fails in general, since then the Eckmann–Shapiro argument gives only an inclusion

$$\operatorname{Hom}_{\mathbf{Z}[G]}(C_n(X); \mathbf{Z}[G]) \subset \operatorname{Hom}_{\mathbf{Z}}(C_n(\tilde{X}); \mathbf{Z})$$

which is never an isomorphism when G is infinite (cf. [12], p. 358).
Henceforth, unless stated explicitly to the contrary, G will denote a finite group. As a consequence, we have:

Proposition 58.3: Let X be a finite 3-complex in which $\pi_1(X) \cong G$ is finite; if X is cohomologically two-dimensional, then:

  (i) $\operatorname{Im}(\partial_3)$ is isomorphic to the free $\mathbf{Z}[G]$-module $C_3(\tilde{X})$;
  (ii) $\operatorname{Im}(\partial_3)$ is a direct summand of $\pi_2(K)$; and
(iii) $\pi_2(X) \cong \pi_2(K)/\operatorname{Im}(\partial_3)$,
(iv) $\pi_2(X) \cong \pi_2(K) \oplus \operatorname{Im}(\partial_3)$.

Proof: By hypothesis $H_3(\tilde{X}; \mathbf{Z}) = 0$. Thus $\partial_3 : C_3(\tilde{X}) \to \operatorname{Im}(\partial_3)$ is an isomorphism. This proves (i).
Furthermore, the exact sequence of (58.1) reduces to

$$0 \to C_3(\tilde{X}) \xrightarrow{\partial_3} \pi_2(K) \to \pi_2(X) \to 0$$

However

$$\operatorname{Tor}(\pi_2(X)) = \operatorname{Tor}(H_2(\tilde{X}; \mathbf{Z})) = \operatorname{Tor}(H^3(\tilde{X}; \mathbf{Z}))$$

so that, again by the cohomology assumption on X, $\pi_2(X)$ is torsion free. Thus we have a short exact sequence in $\mathcal{F}(\mathbf{Z}[G])$

$$0 \to C_3(\tilde{X}) \xrightarrow{\partial_3} \pi_2(K) \to \pi_2(X) \to 0$$

in which $C_3(\tilde{X}) \cong \operatorname{Im}(\partial_3)$ is $\mathbf{Z}[G]$-free (of rank n say) and hence relatively injective. Thus

$$\pi_2(K) \cong \pi_2(X) \oplus \operatorname{Im}(\partial_3) \cong \pi_2(X) \oplus \mathbf{Z}[G]^n$$

This proves (ii), (iii) and (iv) simultaneously.   □

Since $\operatorname{Im}(\partial_3)$ is finitely generated and free over $\mathbf{Z}[G]$, it follows immediately from (58.3) that:

Corollary 58.4: Let X be a finite 3-complex with $\pi_1(X) \cong G$; if X is cohomologically two-dimensional, then $\pi_2(X) \in \Omega_3(\mathbf{Z})$.

59 The virtual 2-complex

Continuing with the notation above, G will denote a finite group, X will denote a cohomologically two-dimensional finite 3-complex with $\pi_1(X) \cong G$, and K will denote the 2-skeleton of X. Up to homotopy type, there is no loss of generality in assuming that X is also reduced. Note that

$$C_*(X) = (0 \to C_3(\tilde{X}) \to C_2(\tilde{X}) \to C_1(\tilde{X}) \to C_0(\tilde{X}) \to \mathbf{Z} \to 0)$$

the cellular chain complex of X, fails in general to be exact at C_2.

We may refine the analysis of (58.3) slightly. Firstly, since $H_3(\tilde{X}; \mathbf{Z}) = 0$ the boundary map $\partial_3 : C_3(\tilde{X}) \to C_2(\tilde{X})$ is injective. Moreover, we have:

Proposition 59.1

$$\pi_2(X) \cong \mathrm{Ker}(\partial_2 : C_2(\tilde{X})/\mathrm{Im}(\partial_3) \to C_1(\tilde{X}))$$

Proof: This follows directly from the Hurewicz Theorem given that $\pi_2(X) \cong H_2(\tilde{X}; \mathbf{Z})$. $\square$

Proposition 59.2: $\mathrm{Im}(\partial_3)$ is a direct summand of $C_2(\tilde{X})$ and $C_2(\tilde{X})/\mathrm{Im}(\partial_3)$ is stably free over $\mathbf{Z}[G]$.

Proof: We have an exact sequence

$$0 \to \pi_2(X) \to C_2(\tilde{X})/\mathrm{Im}(\partial_3) \xrightarrow{\partial_2} C_1(\tilde{X})$$

in which $\pi_2(X)$ being a representative of $\Omega_3(\mathbf{Z})$, is torsion free over $\mathbf{Z}$. However, $C_1(\tilde{X})$ is free over $\mathbf{Z}[G]$ and hence free over $\mathbf{Z}$. Thus $C_2(\tilde{X})/\mathrm{Im}(\partial_3)$ is also torsion free over $\mathbf{Z}$. The result follows from the injectivity of the free module $\mathrm{Im}(\partial_3)$ relative to $\mathcal{F}(\mathbf{Z}[G])$, given that $C_2(\tilde{X})$ is also free. $\square$

Let $j : \pi_2(X) \to C_2(\tilde{X})/\mathrm{Im}(\partial_3)$ be the inclusion obtained by making the identifications

$$\pi_2(X) = \pi_2(K)/\mathrm{Im}(\partial_3); \quad C_2(\tilde{X}) = C_2(\tilde{K}).$$

We obtain an algebraic 2-complex $\langle X \rangle \in \mathbf{Alg}_G$ by

$$0 \to \pi_2(X) \xrightarrow{j} C_2(\tilde{X})/\mathrm{Im}(\partial_3) \to C_1(\tilde{X}) \to C_0(\tilde{X}) \to \mathbf{Z} \to 0$$

Observe that $\langle X \rangle$ is functorial in the cell structure of X. We may think of $\langle X \rangle$ as being a 'virtual' two-dimensional homotopy type representing X. There is a natural $\mathbf{Z}[G]$-chain map $\varphi : C_*(X) \rightarrow \langle X \rangle$ obtained by collapsing $C_3(\tilde{X}) \cong$ Im (∂_3), thus

$$0 \rightarrow C_3(\tilde{X}) \rightarrow \quad C_2(\tilde{X}) \quad \rightarrow C_1(\tilde{X}) \rightarrow C_0(\tilde{X}) \rightarrow \mathbf{Z} \rightarrow 0$$
$$\downarrow \qquad\qquad \downarrow \varphi_2 \qquad\quad \downarrow Id \quad\; \downarrow Id \quad \downarrow Id$$
$$0 \quad \rightarrow C_2(\tilde{X})/\mathrm{Im}(\partial_3) \rightarrow C_1(\tilde{X}) \rightarrow C_0(\tilde{X}) \rightarrow \mathbf{Z} \rightarrow 0$$

where, since X is reduced, we may identify the epimorphism $C_0(\tilde{X}) \rightarrow \mathbf{Z}$ with the augmentation map.

The 'algebraic π_2' of $\langle X \rangle$ is simply $\pi_2(\langle X \rangle) = \pi_2(K)/\mathrm{Im}(\partial_3)$. Identifying $\pi_2(X)$ with $H_2(\tilde{X}; \mathbf{Z})$, by means of the Hurewicz Theorem, it is straightforward to check that:

Proposition 59.3: $\varphi : C_*(X) \rightarrow \langle X \rangle$ induces an isomorphism $\varphi : \pi_2(X) \xrightarrow{\cong} \pi_2(\langle X \rangle)$.

It is tempting to restate (59.3) by saying that $\varphi : C_*(X) \rightarrow \langle X \rangle$ is a weak homotopy equivalence. The difficulty is that $C_*(X)$ and $\langle X \rangle$ are slightly different sorts of things; $\langle X \rangle$ is both augmented (by $C_0(\tilde{X}) \rightarrow \mathbf{Z}$) and co-augmented (by $\pi_2(X) \rightarrow C_2(\tilde{X})$) whereas $C_*(X)$ is merely augmented. In comparing the two, we should, strictly speaking, drop the co-augmentation, but as remarked in Section 47, this should cause no real confusion.

Observe that there is a natural transformation of (augmented, co-augmented) chain complexes $\nu : C_*(K) \rightarrow \langle X \rangle$ described by the following diagram

$$0 \rightarrow \quad \pi_2(K) \quad \rightarrow \quad C_2(\tilde{K}) \quad \rightarrow C_1(\tilde{K}) \rightarrow C_0(\tilde{K}) \rightarrow \mathbf{Z} \rightarrow 0$$
$$\downarrow \nu \qquad\qquad \downarrow \nu_2 \qquad\qquad \downarrow Id \quad\; \downarrow Id \quad \downarrow Id$$
$$0 \rightarrow \pi_2(K)/\mathrm{Im}(\partial_3) \rightarrow C_2(\tilde{X})/\mathrm{Im}(\partial_3) \rightarrow C_1(\tilde{X}) \rightarrow C_0(\tilde{X}) \rightarrow \mathbf{Z} \rightarrow 0$$

Theorem 59.4: Let G be a finite group, and let X be a finite cohomologically two-dimensional 3-complex with $\pi_1(X) \cong G$; then X is homotopy equivalent (over the identity on $\pi_1 = G$) to a finite 2-complex if and only if the virtual 2-complex $\langle X \rangle$ is geometrically realizable.

Proof: Suppose that L is a finite 2-complex for which there exists a homotopy equivalence $\psi : L \rightarrow X$ over the identity on G, then $\psi : \pi_2(L) \rightarrow \pi_2(X)$ is an isomorphism. Hence $\varphi \circ \psi_* : \pi_2(L) \xrightarrow{\cong} \pi_2(\langle X \rangle)$ is an isomorphism; thus $\varphi \circ \psi_* : C_*(L) \rightarrow \langle X \rangle$ is a weak homotopy equivalence, and so $\langle X \rangle$ is geometrically realizable.

Conversely, suppose that $\langle X \rangle$ is geometrically realizable; let $f : \langle X \rangle \to C_*(L)$, be a weak homotopy equivalence, where L is a finite 2-complex. As usual, let K denote the 2-skeleton of X; without loss of generality, we may suppose that:

(i) $L^{(1)} = K^{(1)}$, and that
(ii) $f_r = \mathrm{Id} : C_r(\tilde{K}) \equiv C_r(\tilde{L})$ for $r \le 1$.

Furthermore, assuming X has N cells of dimension 3, we may write

$$X = K \cup_{\alpha_1} E_1^{(3)} \cup_{\alpha_2} E_2^{(3)} \cdots \cup_{\alpha_N} E_N^{(3)}$$

Observe that f induces an isomorphism

$$f_* : \pi_2(\langle X \rangle) = \pi_2(X) \xrightarrow{\cong} \pi_2(L)$$

The chain map $f \circ \nu : C_*(K) \to C_*(L)$ has the property that $(f \circ \nu)_r = \mathrm{Id} : C_r(\tilde{K}) \equiv C_r(\tilde{L})$ for $r \le 1$. By (49.4), there exists a cellular map $g : K \to L$ such that:

(i) $f_* = \mathrm{Id} : \pi_1(K) \to \pi_1(L)$, and
(ii) $g_* = f_* \circ \nu_* : \pi_2(K) \to \pi_2(L)$.

In particular, since f_* is an isomorphism, we see that

$$\mathrm{Ker}(g_* : \pi_2(K) \to \pi_2(L)) = \mathrm{Ker}(\nu_*) = \mathrm{Im}(\partial_3 : C_3(\tilde{X}) \to \pi_2(K)) \cong \mathbf{Z}[G]^N$$

The classes of the attaching maps α_j all belong to

$$\mathrm{Im}(\partial_3) = \mathrm{Ker}(gi_* : \pi_2(K) \to \pi_2(L))$$

so that $g : K \to L$ extends over the 3-cells of X to a map $h : X \to L$ such that:

(iii) $h_* = \mathrm{Id} : \pi_1(X) \to \pi_1(L)$ and
(iv) $h_* = f_* : \pi_2(X) \to \pi_2(L)$.

Since $H_r(\tilde{X}; \mathbf{Z}) = H_r(\tilde{L}; \mathbf{Z}) = 0$ for $2 < r$, it follows from Whitehead's Theorem that h is a homotopy equivalence as required. $\square$

60 Reduction to two-dimensional data

Theorem (59.4) allows us to trace the fate of an individual cohomologically two-dimensional 3-complex X with finite fundamental group G; it will be homotopy equivalent to a finite 2-complex L precisely when the two-dimensional chain complex $\langle X \rangle$ is homotopy equivalent to the two-dimensional chain complex $C_*(L)$. We proceed to prove a result which treats the D(2)-problem in its entirety.

We first give a criterion which allows us to construct 3-complexes which are cohomologically two-dimensional. We adopt the following notation:

K: a finite 2-complex with $\pi_1(K) \cong G$;

m: an integer ≥ 1;

α: a $\mathbf{Z}[G]$-homomorphism $\mathbf{Z}[G]^m \to \pi_2(K)$;

$\{\epsilon_j\}_{1 \leq j \leq m}$: the canonical $\mathbf{Z}[G]$-basis for $\mathbf{Z}[G]^m$;

α_j: a map $S^2 \to K$ in the homotopy class $\alpha(\epsilon_j)$.

Furthermore, $X(\alpha)$ will denote the 3-complex obtained by attaching 3-cells to K by means of $\alpha_1, \alpha_2, \dots, \alpha_m$; that is

$$X(\alpha) = K \cup_{\alpha_1} E_1^{(3)} \cup_{\alpha_2} E_2^{(3)} \cdots \cup_{\alpha_m} E_m^{(3)}$$

Proposition 60.1: Let K be a finite 2-complex with $\pi_1(K) \cong G$, and let α : $\mathbf{Z}[G]^m \to \pi_2(K)$ be an *injective* $\mathbf{Z}[G]$-homomorphism such that $\pi_2(K)/\mathrm{Im}(\alpha)$ is torsion free over $\mathbf{Z}$; then $H^3(X(\alpha); \mathcal{B}) = 0$ for all local coefficient systems $\mathcal{B}$ on $X(\alpha)$.

Proof: We can make the following identifications

$$C_3(\widetilde{X(\alpha)}) \longleftrightarrow \mathbf{Z}[G]^m$$
$$C_2(\widetilde{X(\alpha)}) \longleftrightarrow C_2(\tilde{K})$$
$$\pi_2(K) \longleftrightarrow Z_2(\tilde{K})$$

Moreover, if $i : \pi_2(K) = Z_2(\tilde{K}) \subset C_2(\tilde{K})$ denotes the inclusion, we may further make the identifications

$$\partial_3 : C_3(\widetilde{X(\alpha)}) \to C_2(\widetilde{X(\alpha)}) \longleftrightarrow i \circ \alpha : \mathbf{Z}[G]^m \to C_2(\tilde{K})$$
$$H^3(X(\alpha); \mathcal{B}) \longleftrightarrow (\mathbf{Z}[G]^m)^* \otimes \mathcal{B}/\mathrm{Im}((i \circ \alpha)^* \otimes 1_{\mathcal{B}})$$

Since α is injective and $\pi_2(K)/\mathrm{Im}(\alpha)$ is $\mathbf{Z}$-free, all modules in the exact sequence

$$0 \to \mathbf{Z}[G]^m \xrightarrow{\alpha} \pi_2(K) \to \pi_2(K)/\mathrm{Im}(\alpha) \to 0$$

are in $\mathcal{F}(\mathbf{Z})[G]$; hence its dual sequence is also exact, and $\alpha^* : \pi_2(K)^* \to (\mathbf{Z}[G]^m)^*$ is surjective. Similarly

$$0 \to \pi_2(K) \to C_2(\tilde{K}) \to C_1(\tilde{K}) \to C_0(\tilde{K}) \to \mathbf{Z} \to 0$$

is an exact sequence within $\mathcal{F}(\mathbf{Z})[G]$, and its dual sequence is likewise exact. Thus $\iota^* : C_2(\tilde{K})^* \to \pi_2(K)^*$ is surjective and $(\alpha \circ i)^* : C_2(\tilde{K})^* \to (\mathbf{Z}[G]^m)^*$

is surjective. Thus

$$(i \circ \alpha)^* \otimes 1_{\mathcal{B}} : C_2(\tilde{K})^* \otimes \mathcal{B} \longrightarrow (\mathbf{Z}[G]^m)^* \otimes \mathcal{B}$$

is surjective, by right exactness of $- \otimes_{\mathbf{Z}[G]} \mathcal{B}$. From the correspondence

$$H^3(X(\alpha); \mathcal{B}) \longleftrightarrow (\mathbf{Z}[G]^m)^* \otimes \mathcal{B}/\mathrm{Im}((i \circ \alpha)^* \otimes 1_{\mathcal{B}})$$

we see that $H^3(X(\alpha); \mathcal{B}) = 0$ for all local coefficient systems $\mathcal{B}$. $\qquad \square$

We can use this to construct cohomologically two-dimensional 3-complexes corresponding to any two-dimensional $\mathbf{Z}[G]$-free chain complex.

Theorem 60.2: Let $\mathcal{E} \in \mathbf{Alg}_G$; then there exists a finite 3-complex X such that

(i) $\pi_1(X) \cong G$;
(ii) $H^3(X; \mathcal{B}) = 0$ for all local coefficient systems $\mathcal{B}$; and
(iii) $\langle X \rangle$ is homotopy equivalent to $\mathcal{E}$.

Proof: Represent the weak homotopy type $\mathcal{E} \in \mathbf{Alg}_G$ by

$$\mathcal{E} = \left(0 \to J \to \mathbf{Z}[G]^\gamma \xrightarrow{\delta_2} \mathbf{Z}[G]^\beta \xrightarrow{\delta_1} \mathbf{Z}[G] \xrightarrow{\epsilon} \mathbf{Z} \to 0 \right)$$

Let L be a finite 2-complex with $\pi_1(L) \cong G$; then by (52.3)

$$\Sigma_+^n(\mathcal{E}) \simeq \Sigma_+^m(C_*(L))$$

for some $n, m \geq 1$. However, $\Sigma_+^m(C_*(L)) \cong C_*(L \vee mS^2)$. Put $K = L \vee mS^2$, and let $\varphi : C_*(K) \to \Sigma_+^n(\mathcal{E})$ be a weak homotopy equivalence. In particular, φ induces an isomorphism

$$\varphi : \pi_2(K) \xrightarrow{\sim} J \oplus \mathbf{Z}[G]^n$$

If $\psi : \Sigma_+^n(\mathcal{E}) \to \mathcal{E}$ is the natural chain projection, then $\psi \circ \varphi : C_*(K) \to \mathcal{E}$ has the property that $\mathrm{Ker}(\psi \circ \varphi : \pi_2(K) \to J$ is isomorphic to $\mathbf{Z}[G]^n$. Choose a $\mathbf{Z}[G]$ basis $\alpha_1, \alpha_2, \dots, \alpha_m$ for $\mathrm{Ker}(f)$, and form $X(\alpha)$ by attaching 3-cells $E_1^{(3)}, E_2^{(3)}, \dots, E_m^{(3)}$ to K by means of $\alpha_1, \alpha_2, \dots, \alpha_m$; that is

$$X(\alpha) = K \cup_{\alpha_1} E_1^{(3)} \cup_{\alpha_2} E_2^{(3)} \cdots \cup_{\alpha_m} E_m^{(3)}$$

It follows from (60.1) that $H^r(X(\alpha); \mathcal{B}) = 0$ for all $r \geq 3$, and it is straightforward to see that $\langle X \rangle$ is homotopy equivalent to $\mathcal{E}$. $\qquad \square$

Finally we obtain:

Theorem I: The D(2) property holds for the finite group G if and only if each algebraic 2-complex over G is geometrically realizable.

Proof: Suppose the D(2)-property holds for G. If $\mathbf{E}$ is an algebraic 2-complex over G, then by (60.2) there exists a finite 3-complex X which is cohomologically two-dimensional and for which $\mathbf{E} \simeq \langle X \rangle$. Since we are assuming that the D(2)-property holds for G, then, by (59.4), there is a finite 2-complex K with $\pi_1(K) \cong G$ such that $\langle X \rangle$ is homotopy equivalent to $C_*(\tilde{K})$. Thus $\mathbf{E} \simeq C_*(\tilde{K})$, and $\mathbf{E}$ is geometrically realized.

Conversely, if each algebraic 2-complex is geometrically realized, then, for any finite cohomologically two-dimensional complex X, $\langle X \rangle$ is geometrically realized. By (59.4), X is homotopy equivalent to a finite 2-complex, and the D(2)-property holds.   □

From Theorems I and III, we get:

Corollary 60.3: Let G be a finite group; if each minimal module $J \in \Omega_3(\mathbf{Z})$ is both realizable and full, then the $D(2)$-property holds for G.

61 Group presentations and $\Omega_3(\mathbf{Z})$

Theorem III shows that strong realizability of all modules $J \in \Omega_3(\mathbf{Z})$ is a sufficient condition for the D(2)-property to hold. Though we shall not pursue the point here, the case of finite abelian groups shows that it is not necessary for minimal modules to be strongly realizable in order for the homotopy tree $\widehat{\mathbf{Alg}}_G$ to be geometrically realizable. This is essentially a result of Browning generalizing earlier work of Metzler [39]. Latiolais [32] gives a rather more general account. Consequently, strong realizability is not a necessary condition for the D(2)-property to hold.

On the other hand, realizability *is* a necessary condition, since $\Omega_3(\mathbf{Z})$ is the stable class of any module of the form $\pi_2(\mathcal{G})$ where $\mathcal{G}$ is a finite presentation of G. Since any $J \in \Omega_3(\mathbf{Z})$ is the 'algebraic π_2' of some algebraic 2-complex, we see immediately from Theorem I that:

Proposition 61.1: If the finite group G satisfies the D(2) property, then $\Omega_3(\mathbf{Z})$ is realizable.

The question of the realizability of $\Omega_3(\mathbf{Z})$ is related to the classical question of minimal presentations of (finite) groups ([60]). We saw previously, (29.1), that the *rational* isomorphism type of the stable module $\Omega_n(\mathbf{Z})$ is given by

$$(61.2) \qquad \Omega_n(\mathbf{Z}) \otimes \mathbf{Q} = \Omega_n(\mathbf{Q}) = \begin{cases} [\mathbf{Q}] & \text{if } n \text{ is even} \\ [\mathbf{I}_{\mathbf{Q}}(G)] & \text{if } n \text{ is odd} \end{cases}$$

If $J \in \Omega_3(\mathbf{Z})$ we define the *module height* $h_{\mathrm{mod}}(J)$ of J

$$h_{\mathrm{mod}}(J) = k \iff J \otimes \mathbf{Q} \cong \mathbf{I}_{\mathbf{Q}}(G) \oplus \mathbf{Q}[G]^k$$

We define an integer valued invariant $\omega_3(G)$ of G, thus

$$\omega_3(G) = h_{\mathrm{mod}}(J)$$

for any minimal module $J \in \Omega_3(\mathbf{Z})$. Clearly $0 \leq \omega_3(G)$; moreover, for any $N \in \Omega_3(\mathbf{Z})$, $\omega_3(G) \leq h_{\mathrm{mod}}(N)$.

We can approach this from a more geometric point of view. In Chapter 8 (48.3), we saw, for any finite presentation of the finite group G

$$\mathcal{G} = \langle x_1, \ldots, x_g \mid W_1, \ldots, W_r \rangle$$

that $g \leq r$. The difference $r - g$ is called the *excess* of the presentation, written $\mathrm{exc}(\mathcal{G})$. We define $\mathrm{exc}(G)$, the *excess* of G, to be the minimum of $\mathrm{exc}(\mathcal{G})$ as $\mathcal{G}$ runs through all finite presentations of G. Let $\mathcal{K}_{\mathcal{G}}$ denote the Cayley complex associated with $\mathcal{G}$ and let $C_*(\mathcal{G})$ denote the the cellular chain complex of $\widetilde{\mathcal{K}_{\mathcal{G}}}$; then we get an exact sequence of $\mathbf{Z}[G]$-modules of the form

$$0 \to \pi_2(\mathcal{G}) \to C_2(\tilde{\mathcal{K}}) \overset{\partial}{\to} C_1(\tilde{\mathcal{K}}) \overset{\partial}{\to} C_0(\tilde{\mathcal{K}}) \to \mathbf{Z} \to 0$$

Making the obvious identifications

$$C_2(\tilde{\mathcal{K}}) \cong \mathbf{Z}[G]^r; \quad C_1(\tilde{\mathcal{K}}) \cong \mathbf{Z}[G]^g; \quad C_0(\tilde{\mathcal{K}}) \cong \mathbf{Z}[G]$$

we calculate that

(61.3) $$h_{\mathrm{mod}}(\pi_2(\mathcal{G})) = r - g = \mathrm{exc}(\mathcal{G})$$

It follows immediately that

(61.4) $$\omega_3(G) \leq \mathrm{exc}(G)$$

(61.4) is called *Swan's Inequality*. In different notation, it was observed by Swan in [60]. In all cases where both sides of the inequality are known, they are actually equal. As an illustration, we note that, if D_{2N} is the dihedral group of order $2N$, then $\omega_3(D_{2N}) = \mathrm{exc}(D_{2N}) = 0$ when N is odd, and $\omega_3(D_{2N}) = \mathrm{exc}(D_{2N}) = 1$ when N is even (see, for example, [78]).

The question of whether the inequality is *always* an equation is both fundamental and difficult; $\omega_3(G)$ is a 'linear invariant' and, for a particular finite group G, one might have reasonable hope that it can be calculated exactly. By contrast, the task of calculating $\mathrm{exc}(G)$, defined by minimizing over a non-recursive set, is exceedingly difficult if not hopeless. The answer is still not known for many quite small and familiar groups. As Swan points out in [60], in the absence of any other method, the only real hope, at present, of finding the exact value of $\mathrm{exc}(G)$ is to try to find a presentation with excess equal to the computed value of $\omega_3(G)$.

One may avoid the difficulty of Swan's Inequality by restricting attention to groups G where $\mathrm{exc}(G)$ takes the minimum value 0; that is, when G has a presentation

$$\mathcal{G} = \langle x_1, \ldots, x_g \mid W_1, \ldots, W_r \rangle$$

in which $r = g$. Such a presentation is said to be *balanced*. In this case, by Swan's Inequality, one has:

Proposition 61.5: Let G be a finite group having a balanced presentation

$$\mathcal{G} = \langle x_1, \ldots, x_g \mid W_1, \ldots, W_g \rangle$$

then $\pi_2(\mathcal{G})$ is a minimal element of $\Omega_3(\mathbf{Z})$.

62 Verifying the D(2)-property

By Theorem I, the finite group G has the D(2)-property if and only if the tree $\widetilde{\mathbf{Alg}}_{\mathbf{G}}$ of two-dimensional algebraic homotopy types is geometrically realizable. This is equivalent to requiring that every minimal algebraic homotopy 2-type is homotopy equivalent to the Cayley complex of a presentation. If G is a finite group having a free resolution of period 4, it is easy to see that G has a minimal algebraic 2-complex of the form

$$\mathbf{F} = (0 \to J \to F_2 \to F_1 \to F_0 \to \mathbf{Z} \to 0)$$

where $\mathrm{rk}_{\mathbf{Z}}(J) = |G| - 1$. One can see this by first observing that there is a minimal complete resolution of the form

$$0 \to \mathbf{Z} \to P \to F_2 \to F_1 \to F_0 \to \mathbf{Z} \to 0$$

where P is stably free of rank 1, and taking $J = \mathrm{Im}(P \to F_2)$. Thus, if $\mathbf{F}$ is realizable, it must be realizable by a balanced presentation. In view of Theorem IV, we see that, if G has cancellation property for free modules, then $\mathbf{F}$ is the unique minimal algebraic 2-type. We obtain:

Theorem 62.1: Let G be a finite group which admits a free resolution of period 4; if G has the free cancellation property, then

$$G \text{ satisfies the D(2)-property} \iff G \text{ admits a balanced presentation.}$$

We now consider some examples:

(i) Cyclic groups

The cyclic group C_n evidently admits a balanced presentation

$$C_n = \langle x \mid x^n \rangle$$

The cyclic groups are unique in admitting a free resolution of period 2

$$0 \to \mathbf{Z} \xrightarrow{\epsilon^*} \mathbf{Z}[C_n] \xrightarrow{x-1} \mathbf{Z}[C_n] \xrightarrow{\epsilon} \mathbf{Z} \to 0$$

which can be repeated to give a free resolution of period 4

$$0 \to \mathbf{Z} \xrightarrow{\epsilon^*} \mathbf{Z}[C_n] \xrightarrow{x-1} \mathbf{Z}[C_n] \xrightarrow{\Sigma} \mathbf{Z}[C_n] \xrightarrow{x-1} \mathbf{Z}[C_n] \xrightarrow{\epsilon} \mathbf{Z} \to 0$$

by taking $\Sigma = 1 + x + \cdots + x^{n-1}$. Since $\mathbf{Z}[C_n]$ has the Eichler property, it has the cancellation property for free modules. Again by Theorem IV we have

(62.2) The cyclic groups C_n satisfy the D(2)-property.

(ii) Dihedral groups of order $\equiv 2 \bmod(4)$

The dihedral group D_{4n+2} of order $4n + 2$ can be defined by means of the balanced presentation

$$D_{4n+2} = \langle x, y \mid x^{2n+1} = y^2 , yx^n = x^{n+1}y \rangle$$

and D_{4n+2} admits a free resolution of period 4 (this is false for the dihedral groups of order $4n$)

$$0 \to \mathbf{Z} \xrightarrow{\epsilon^*} \mathbf{Z}[D_{4n+2}] \xrightarrow{\delta} \mathbf{Z}[D_{4n+2}]^2 \xrightarrow{\partial_2} \mathbf{Z}[D_{4n+2}]^2 \xrightarrow{\partial_1} \mathbf{Z}[D_{4n+2}] \xrightarrow{\epsilon} \mathbf{Z} \to 0$$

Here

$$\delta = \begin{pmatrix} 1 + x - x^{n+1} - y \\ -x + x^n y \end{pmatrix} ; \quad \partial_2 \sim \begin{pmatrix} \Sigma_x & \theta_1 - \theta_2 y \\ -\Sigma_y & x^n - 1 \end{pmatrix} ; \quad \partial_1 \sim (x - 1, y - 1)$$

where

$$\Sigma_x = 1 + x + \cdots + x^{2n}; \quad \Sigma_y = 1 + y; \quad \theta_1 = 1 + x + \cdots + x^{n-1} \text{ and}$$
$$\theta_2 = 1 + x + \cdots + x^n = \theta_1 + x^n$$

whilst ϵ is the augmentation map, and ϵ^* is its dual. On general grounds, one can show that $\mathbf{Z}[D_{4n+2}]$ has the Eichler property, and so satisfies free cancellation. Thus, again from Theorem IV, we get:

(62.3) The dihedral groups D_{4n+2} satisfy the D(2)-property.

(iii) Direct products $D_{2n} \times C_m$ (n, m odd and coprime)

It can be shown along the lines of [46] that, when n, m are odd and coprime, the 'quasi-dihedral' groups $G = D_{2n} \times C_m$ admit balanced presentations with two generators and two-relations, and that the corresponding algebraic Cayley complexes can be extended to give a periodic resolution of the form

$$0 \to \mathbf{Z} \to \mathbf{Z}[G] \to \mathbf{Z}[G]^2 \to \mathbf{Z}[G]^2 \to \mathbf{Z}[G] \to \mathbf{Z} \to 0$$

Once again, by virtue of the Eichler property, $D_{2n} \times C_m$ has free cancellation and we have:

(62.4) The 'quasi- dihedral' groups $D_{2n} \times C_m$, n, m odd and coprime, satisfy the D(2)-property.

By a celebrated theorem of Milnor [42], no dihedral or quasi-dihedral group can be the fundamental group of a closed 3-manifold. The next few examples, however, do all occur as fundamental groups of closed 3-manifolds. In this connection, one should note on general grounds, using Morse Theory, that, if the finite group G is the fundamental group of a closed 3-manifold, then G admits a finite free resolution of period 4 of the form

$$0 \to \mathbf{Z} \to \mathbf{Z}[G] \to \mathbf{Z}[G]^g \to \mathbf{Z}[G]^g \to \mathbf{Z}[G] \to \mathbf{Z} \to 0$$

where g is the number of generators/relations. We will pursue this point further in Chapter 11.

(iv) The spin Euclidean groups

We consider first the symmetry groups of the standard Euclidean solids; that is, denote by $\mathbf{T}, \mathbf{O}, \mathbf{I} \subset SO(3)$ respectively the tetrahedral, octahedral, and icosahedral groups. These groups do *not* have periodic cohomology. However, their liftings $\tilde{\mathbf{T}}, \tilde{\mathbf{O}}, \tilde{\mathbf{I}} \subset \mathrm{Spin}(3)$ are fundamental groups of closed 3-manifolds, and so, by the general considerations above, do have both resolutions of free period 4, and balanced presentations. In fact, explicit balanced presentations were given by Kenne in his thesis [31] (see also [46]). These groups *do not* satisfy the Eichler property. Nevertheless, Swan [63] has shown they have free cancellation. From Theorem IV, we see that:

(62.5) The spin Euclidean groups $\tilde{\mathbf{T}}, \tilde{\mathbf{O}}, \tilde{\mathbf{I}}$ satisfy the D(2)-property.

(v) The exceptional quaternion groups $Q(4n)$ ($2 \leq n \leq 5$)

For each $n \geq 2$, the quaternion group $Q(4n)$ is defined by the balanced presentation

$$Q(4n) = \langle x, y \mid x^n = y^2, xyx = y \rangle$$

$Q(4n)$ can be regarded as the spin double cover of the dihedral group D_{2n}. We saw in Chapter 8 that $Q(4n)$ has finite free resolution of period 4

$$0 \to \mathbf{Z} \xrightarrow{\epsilon^*} \mathbf{Z}[Q(4n)] \xrightarrow{\delta} \mathbf{Z}[Q(4n)]^2 \xrightarrow{\partial_2} \mathbf{Z}[Q(4n)]^2 \xrightarrow{\partial_1} \mathbf{Z}[Q(4n)] \xrightarrow{\epsilon} \mathbf{Z} \to 0$$

obtained by extending the algebraic Cayley complex of the above presentation one step to the left.

As can be seen from the Wedderburn calculations in Section 12, the quaternion groups all fail to have the Eichler property. Nevertheless, the first four examples, $Q(8)$, $Q(12)$, $Q(16)$, and $Q(20)$ are exceptional in that they do possess the cancellation property for free modules; thus we get:

(62.6) The exceptional quaternion groups $Q(8)$, $Q(12)$, $Q(16)$, $Q(20)$ satisfy the D(2)-property.

A similar argument shows:

(62.7) The direct product $Q(8) \times C_3$ satisfies the D(2)-property.

(vi) The generic quaternion group $Q(4n)$ ($n \geq 6$)
Now consider the generic quaternion groups $Q(4n)$ ($n \geq 6$). By Theorem IV, the minimal homotopy types are parametrized by SF_1. As we have already pointed out, $Q(4n)$ possesses a balanced presentation and a free resolution of period 4. However, the situation is now much more problematic, since for all $n \geq 6$, $Q(4n)$ has stably free modules which are not free, so that $Q(4n)$ has at least two minimal algebraic homotopy 2-types, and, of these, only one, the Cayley complex of the standard presentation, is as yet known to be realizable.

In the first few cases, the precise numbers of stably free modules, and hence minimal 2-types, can be retrieved from the calculations of Swan ([63], Theorems III and IV). They are:

$$|SF_1(Q(24))| = 3;$$
$$|SF_1(Q(28))| = 2;$$
$$|SF_1(Q(32))| = 2;$$
$$|SF_1(Q(36))| = 4;$$
$$|SF_1(Q(40))| = 8.$$

(vii) The groups $Q(8n; p, q)$
The group $Q(8n; p, q)$ can be defined as the pull back extension

$$1 \to C_p \times C_q \to Q(8n; p, q) \to Q(8n) \to 1$$

arising from the canonical imbedding $C_2 \times C_2 \to \text{Aut}(C_p) \times \text{Aut}(C_q)$, by projecting $Q(8n)$ on to $C_2 \times C_2$. When n, p, q are mutually coprime, the groups have period 4. From the point of view of the D(2)-property, they are genuinely more complicated than the generic quaternion groups. Milgram has shown([40], [41], see also [4]) that in some cases (for example $Q(8; 3, 11)$) they *do not* have free period 4. In all except the degenerate case where $p = q = 1$, $Q(8, p, q)$ admits a generic quaternion group as quotient. Thus Swan's Theorem [63] shows that the cancellation property for free modules fails for all except the degenerate case. In fact, from Swan's results, one sees slightly more: cancellation fails in *every* stable class of $\tilde{K}_0$ except in the case $Q(24) = Q(8; 3, 1)$. It follows that cancellation fails in the stable class of the finiteness obstruction, in every nondegenerate case. This includes $Q(8; 3, 1)$, since there cancellation fails in the stably free class, which in this case is also the class of the (zero) finiteness obstruction. Hence we see:

(62.8) Over $Q(8, p, q)$, the fork of two-dimensional algebraic homotopy types has at least two prongs except in the degenerate case $p = q = 1$.

Chapter 11

Poincaré 3-complexes

A finite Poincaré complex M of dimension n is said to be of *standard form* when it can be described thus

$$M = K \cup_\alpha e^n (= M(\alpha))$$

where K is a finite complex of dimension $\leq n - 1$ and $\alpha : S^{n-1} \to K$ is a continuous map. As we described in the Introduction, Wall showed in [72] that, when $n \geq 4$, every finite Poincaré n-complex is homotopy equivalent to one in standard form. Since the question in dimension 3 was the genesis of the D(2)-problem, it is appropriate to conclude our investigation of the D(2)-problem for finite fundamental groups by asking how far it takes us in the direction of obtaining standard forms in dimension 3. We prove:

Theorem V: Let G be a finite group; then G has a standard Poincare 3-form if and only if there is a finite presentation $\mathcal{G}$ of G with $\pi_2(\mathcal{G}) \cong \mathbf{I}^*(G)$. Moreover, G then necessarily has free period 4, and the presentation $\mathcal{G}$ is automatically balanced.

If G is a finite group of free period 4, which also has the D(2)-property, it is straightforward to see that there exists a finite Poincare 3-complex M of standard form such that $\pi_1(M) \cong G$. However, the intransigence of the D(2)-problem suggests the possibility that Poincaré 3-complexes might need to have more than one 3-cell. To an extent, this is supported by a consideration of classical examples. On general grounds, it can be shown that smooth 3-manifolds admit cellular representations with just one top-dimensional cell. Nevertheless, their natural representations are often more complicated. For example, if we regard the 3-sphere S^3 as the group of unit quaternions, then S^3 admits a cell structure, given by the Schläfli symbol $\{3, 3, 4\}$, which is equivariant under the action of the integral quaternion group $Q(8)$, and which has two orbits of 3-cells under the Q_8-action [15]. In other words, the natural geometric cell structure on the quotient $S^3/Q(8)$ has *two* top-dimensional cells, and so, in Wall's scheme of things, is nonstandard. Moreover, the associated presentation of $Q(8)$ is unbalanced. Now of course, in this case, since S^3/Q_8 is a closed 3-manifold, it also has another cell structure, with a single top-dimensional cell, and for which the associated presentation is balanced. Yet, at the algebraic level, this is not immediately obvious, and it requires at least some effort to go from the unbalanced presentation to a balanced one. That being so, for certain finite groups of period 4, we show that the connection between the standard form problem and the D(2)-problem is one of equivalence.

Theorem VI: Let G be a finite group which admits a free resolution of period 4; if G has the free cancellation property, then

$$G \text{ satisfies the D(2)-property} \iff G \text{ admits a balanced presentation.}$$

These results were published in [26].

63 Attaching 3-cells to a presentation

Let G denote a finite group, and let M be a closed connected 3-manifold with $\pi_1(M) \cong G$. By Morse theory, M admits a handle structure with one 0-handle, one 3-handle, and equal numbers of 1- and 2-handles. Let M_0 denote the bounded manifold obtained by removing the 3-handle, so that

(63.1) $$M = M_0 \cup e^3$$

With this notation, we have:

Proposition 63.2: There is an isomorphism of $\mathbf{Z}[G]$-modules

$$\pi_2(M_0) \cong \mathbf{I}^*(G)$$

Proof: We denote by $\widetilde{M}_0$ the universal covering of M_0, and by $\partial \widetilde{M}_0$ the (disconnected) covering of the boundary ∂M_0 induced from the covering map $\widetilde{M}_0 \to M_0$. Taking $\mathbf{Z}$-coefficients throughout, we get the following exact sequence of relative cohomology

$$H^0(\widetilde{M}_0; \partial\widetilde{M}_0) \to H^0(\widetilde{M}_0) \to H^0(\partial\widetilde{M}_0) \to H^1(\widetilde{M}_0; \partial\widetilde{M}_0) \to H^1(\widetilde{M}_0)$$

However, $H^1(\widetilde{M}_0) \cong 0$, whilst Lefschetz Duality gives

$$H^0(\widetilde{M}_0; \partial\widetilde{M}_0) \cong H_3(\widetilde{M}_0) \cong 0.$$

Moreover $H^0(\widetilde{M}_0) \cong \mathbf{Z}$. Let $\epsilon^* : \mathbf{Z} \to \mathbf{Z}[G]$, denote the dual augmentation map $\epsilon^*(1) = \sum_g g$. It is clear that $\partial \widetilde{M}_0$ is a disjoint union of $|G|$ copies of S^2, and under the covering action of G, we may make the identification $H^0(\partial \widetilde{M}_0) \cong \mathbf{Z}[G]$. Moreover, the mapping $H^0(\widetilde{M}_0) \to H^0(\partial \widetilde{M}_0)$ then corresponds to ϵ^*, and the above exact sequence takes the form

$$0 \to \mathbf{Z} \xrightarrow{\epsilon^*} \mathbf{Z}[G] \to H^1(\widetilde{M}_0; \partial \widetilde{M}_0) \to 0$$

from which we make the identification

$$\mathbf{I}^*(G) \cong H^1(\widetilde{M}_0; \partial \widetilde{M}_0)$$

However Lefschetz Duality gives an isomorphism $H^1(\widetilde{M}_0; \partial \widetilde{M}_0) \cong H_2(\widetilde{M}_0)$, so that, finally, the result follows from the Hurewicz Theorem

$$H_2(\widetilde{M}_0) \cong \pi_2(\widetilde{M}_0) \cong \pi_2(M_0)$$

This completes the proof $\qquad\qquad\qquad\qquad\qquad\qquad\qquad\qquad\qquad\qquad\qquad \square$

Suppose that the finite group G is the fundamental group of a closed 3-manifold M. As in (63.1), we write $M = M_0 \cup e^3$, noting that the presentation of G given by this handle decomposition is balanced. However, M_0 collapses on to a two-dimensional complex K with just one 0-cell and equal numbers of 1- and 2-cells, so that, for some mapping $\alpha : \partial e^3 \to K^{(2)}$

(63.3) $\qquad\qquad\qquad\qquad M \simeq K^{(2)} \cup_\alpha e^3$

and $K^{(2)} \cup_\alpha e^3$ is a finite Poincaré 3-complex with $\pi_1(K^{(2)} \cup_\alpha e^3) \cong \pi_1(M) \cong G$. Moreover, $\pi_2(\mathcal{G}) = \pi_2(K) \cong \pi_2(M_0)$, so from (63.2) we see that

(63.4) $\qquad\qquad\qquad\qquad \pi_2(\mathcal{G}) = \mathbf{I}^*(G).$

We shall show, as a partial converse, that a finite group G has the property that $\pi_2(G) = [\mathbf{I}^*(G)]$ precisely when there exists a finite connected Poincaré complex X_G with $\pi_1(X_G) \cong G$. First we prove some preparatory results.

By the Hurewicz Theorem, any simply connected Poincaré 3-complex is homotopy equivalent to S^3. Thus a finite complex X_G with finite fundamental group G is a Poincaré complex if and only if its universal cover $\widetilde{X}_G$ is homotopy equivalent to S^3. Suppose that $\mathcal{G} = \langle x_1, \ldots, x_g; W_1, \ldots, W_r \rangle$ is a finite presentation of a finite group G, and consider the effect of attaching a collection of 3-cells to $K_\mathcal{G}$. Let $\alpha : \mathbf{Z}[G]^{(N+1)} \to \pi_2(\mathcal{G})$ be a homomorphism of $\mathbf{Z}[G]$-modules, and let $\alpha_j : S^2 \to K_\mathcal{G}$ be a map in the homotopy class $\alpha(\epsilon_j)$, where

$\{\epsilon_j\}_{1 \le j \le N+1}$ is the canonical $\mathbf{Z}[G]$-basis for $\mathbf{Z}[G]^{(N+1)}$. We denote by $M(\alpha)$ the 3-complex defined by

$$M(\alpha) = K_{\mathcal{G}} \cup_{\{\alpha_j\}_j} \coprod_{j=1}^{N+1} e_j^3$$

Choose liftings $\tilde{\alpha}_i : S^2 \to \tilde{K}$ of the α_j to the universal covering of K and let $\alpha_* : \mathbf{Z}[G] \to \pi_2(\mathcal{G})$ be the homomorphism of $\mathbf{Z}[G]^{N+1}$ determined by sending ϵ_j to the homotopy class $[\tilde{\alpha}_j]$. For each $g \in G$ and each index j, $1 \le j \le N+1$, let $S_{g,j}^2$ (resp. $e_{g,j}^3$) denote a copy of S^2 (resp. e^3), and let $g\tilde{\alpha}_j : S_{g,j}^2 \to \tilde{K}$ denote the translate of α_j by g. Put $\hat{\alpha} = \coprod_{g,j} g\tilde{\alpha}_j : \coprod_{g,j} S_{g,j}^2 \to \tilde{K}$. Then the universal covering $\widetilde{M(\alpha)}$ of $M(\alpha)$ can be described as a pushout

$$\widetilde{M(\alpha)} = \tilde{K} \cup_{\hat{\alpha}} \coprod_{g,j} e_{g,j}^3$$

The Mayer–Vietoris Theorem for the pushout

$$\tilde{M} = \widetilde{M(\alpha)} = \tilde{K} \cup_{\hat{\alpha}} \coprod_{g,j} e_{g,j}^3$$

gives an exact sequence, with $\mathbf{Z}$-coefficients

$$0 \to H_3(\tilde{M}) \to H_2\left(\coprod_{g,j} S_{g,j}^2\right) \overset{(\hat{\alpha},0)}{\to} H_2(\tilde{K}) \oplus H_2\left(\coprod_{g,j} e_{g,j}^3\right) \to H_2(\tilde{M}) \to 0$$

Proposition 63.5: If $\mathcal{G} = \langle x_1, \dots, x_g; W_1, \dots, W_r \rangle$ is a finite presentation of a finite group G, and $\alpha : \mathbf{Z}[G]^{(N+1)} \to \pi_2(\mathcal{G})$ is a homomorphism of $\mathbf{Z}[G]$-modules, then the integral homology of $\tilde{M} = \widetilde{M(\alpha)}$ is given by

$$H_k(\tilde{M}; \mathbf{Z}) = \begin{cases} \mathbf{Z}, & k = 0 \\ 0, & k = 1 \\ \operatorname{Coker}(\alpha), & k = 2 \\ \operatorname{Ker}(\alpha), & k = 3 \end{cases}$$

Proof: The proof is entirely straightforward once it is appreciated that we may identify $H_2(\coprod_{g,j} S_{g,j}^2)$ with $\mathbf{Z}[G]^{N+1}$, and $(\hat{\alpha}, 0) : H_2(\coprod_{g,j} S_{g,j}^2) \to H_2(\tilde{K}) \oplus H_2(\coprod_{g,j} e_{g,j}^3)$ with α. $\qquad\square$

Corollary 63.6: Let $\mathcal{G} = \langle x_1, \dots, x_g; W_1, \dots, W_r \rangle$ be a finite presentation of a finite group G, and $\alpha : \mathbf{Z}[G]^{(n+1)} \to \pi_2(\mathcal{G})$ be a homomorphism of

$\mathbf{Z}[G]$-modules; then $M(\alpha)$ is a Poincaré complex if and only if α is surjective and $n = r - g$.

Proof: $M(\alpha)$ is a Poincaré complex if and only if $\widetilde{M(\alpha)}$ is homotopy equivalent to S^3. However, $\widetilde{M(\alpha)}$ is a 1-connected complex of dimension 3, so that, by the Whitehead Theorem, $\widetilde{M(\alpha)}$ is homotopy equivalent to S^3 if and only if $H_2(\widetilde{M}; \mathbf{Z}) = 0$ and $H_3(\widetilde{M}; \mathbf{Z}) \cong \mathbf{Z}$. By (63.5), we see that $H_2(\widetilde{M}; \mathbf{Z}) = 0$ if and only if α is surjective. A straightforward rank calculation shows in this case that

$$\mathrm{rk}_{\mathbf{Z}}[\mathrm{Ker}(\alpha)] = |G|(n - r + g) + 1$$

so that, when α is surjective, $H_3(\widetilde{M}; \mathbf{Z}) \cong \mathbf{Z}$ if and only if $n = r - g$. In fact, in this case we have seen, in (63.5), that $\mathrm{Ker}(\alpha)$ is the trivial $\mathbf{Z}[G]$-module with $\mathbf{Z}$-rank equal to 1. $\qquad\square$

64 A characterization of groups of period 4

We can now prove:

Theorem 64.1: The following conditions on the finite group G are equivalent:

(i) $[\pi_2(G)] = [\mathbf{I}^*(G)] \in \mathrm{Stab}(\mathbf{Z}[G])$;
(ii) G has a (finite) free resolution of period 4;
(iii) G has a (finite) free resolution of period 4 of the following type;

$$0 \to \mathbf{Z} \to \mathbf{Z}[G]^N \to C_2(\widetilde{K}) \overset{\partial}{\to} C_1(\widetilde{K}) \overset{\partial}{\to} C_0(\widetilde{K}) \to \mathbf{Z} \to 0$$

where

$$C_2(\widetilde{K}) \overset{\partial}{\to} C_1(\widetilde{K}) \overset{\partial}{\to} C_0(\widetilde{K}) \to \mathbf{Z} \to 0$$

is the partial resolution afforded by some finite presentation $\mathcal{G}$ for G.
(iv) there exists a finite connected Poincaré complex X_G with $\pi_1(X_G) \cong G$.

Proof: (i) $\implies$ (iv)
Suppose that $\pi_2(G) = [\mathbf{I}^*(G)]$, and let $\mathcal{H} = \langle x_1, \dots, x_g \mid W_1, \dots, W_s \rangle$ be a finite presentation for G; then we have the following exact sequence involving $\pi_2(\mathcal{H})$

$$0 \to \pi_2(\mathcal{H}) \to \mathbf{Z}[G]^r \to \mathbf{Z}[G]^g \to \mathbf{Z}[G] \to \mathbf{Z} \to 0$$

The hypothesis $\pi_2(G) = [\mathbf{I}^*(G)]$ implies that, for some $m, n \geq 0$

$$\pi_2(K_{\mathcal{H}}) \oplus \mathbf{Z}[G]^m \cong \mathbf{I}^*(G) \oplus \mathbf{Z}[G]^n$$

Put $r = s+m$. If $m = 0$, we put $\mathcal{G} = \mathcal{H}$; if $m > 0$ we modify $\mathcal{H}$ by adding m trivial relations $W_{s+1}, \ldots, W_r$ to obtain a presentation $\mathcal{G} = \langle x_1, \ldots, x_g \mid W_1, \ldots, W_r \rangle$ with

$$\pi_2(\mathcal{G}) \cong \mathbf{I}^*(G) \oplus \mathbf{Z}[G]^n$$

and $n = r - g$. Either way, we produce an exact sequence of $\mathbf{Z}[G]$-modules

$$0 \to \pi_2 \to \mathbf{Z}[G]^r \to \mathbf{Z}[G]^g \to \mathbf{Z}[G] \to \mathbf{Z} \to 0$$

in which $\pi_2 = \pi_2(\mathcal{G}) \cong \mathbf{I}^*(G) \oplus \mathbf{Z}[G]^n$. Since $\mathbf{I}^*(G)$ is an epimorphic image of $\mathbf{Z}[G]$, we may choose a surjective homomorphism of $\mathbf{Z}[G]$-modules $\alpha : \mathbf{Z}[G]^{n+1} \to \pi_2$. Put

$$X_G = M(\alpha) = K_{\mathcal{G}} \cup_{\alpha_0} e_0^3 \cup_{\alpha_1} e_1^3 \cdots \cup_{\alpha_n} e_n^3$$

Then, since $n = r - g$ and α is an epimorphism, it follows, by (63.6), that X_G is a finite Poincaré 3-complex with $\pi_1(X_G) \cong G$.

(iv) $\Longrightarrow$ (iii)
If X_G is a finite Poincaré 3-complex with $\pi_1(X_G)$, then the cellular chain complex

$$0 \to \mathbf{Z} \to C_3(\widetilde{X_G}) \to C_2(\widetilde{X_G}) \xrightarrow{\partial} C_1(\widetilde{X_G}) \xrightarrow{\partial} C_0(\widetilde{X_G}) \to \mathbf{Z} \to 0$$

is a finite free resolution of period 4 in which

$$C_2(\widetilde{X_G}) \xrightarrow{\partial} C_1(\widetilde{X_G}) \xrightarrow{\partial} C_0(\widetilde{X_G}) \to \mathbf{Z} \to 0$$

is the partial resolution afforded by some finite presentation $\mathcal{G}$ for G, namely that given by the 2-skeleton of X_G.

(iii) $\Longrightarrow$ (ii) follows from (63.5) and (63.6).

(ii) $\Longrightarrow$ (i) As we observed in Chapter 7, the condition that G has free period 4 is equivalent to requiring that $\Omega_3(\mathbf{Z}) = \Omega_{-1}(\mathbf{Z})$. However, $\pi_2(\mathcal{G})$ is a representative of $\Omega_3(\mathbf{Z})$, and $\mathbf{I}^*(G)$ is a representative of $\Omega_{-1}(\mathbf{Z})$. $\square$

For each integer $n \geq 0$ introduce a class $\mathcal{P}(n)$ of finite groups as follows:
$\mathcal{P}(n)$: A finite group G belongs to $\mathcal{P}(n)$ when there exists a finite Poincaré 3-complex X_G for which $\pi_1(X_G) \cong G$ and which has exactly $n + 1$ cells of dimension 3; that is, X_G has a cell structure

$$X_G = K^{(2)} \cup e_0^3 \cup \cdots \cup e_n^3$$

Let $\mathcal{M}$ denote the class of finite groups G for which there exists a (smooth) closed connected 3-manifold M with $\pi_1(M) \cong G$; these classes are related by a sequence of inclusions

$$\mathcal{M} \subset \mathcal{P}(0) \subset \mathcal{P}(1) \subset \cdots \subset \mathcal{P}(n) \subset \mathcal{P}(n+1) \subset \cdots$$

The inclusion $\mathcal{P}(n) \subset \mathcal{P}(n+1)$ is obtained by taking the one point union $X \vee e^3$ of $X \in \mathcal{P}(n)$ with a 3-cell e^3, attached at a point $* \in \partial e^3$. The inclusion $\mathcal{M} \subset \mathcal{P}(0)$ is obtained in the usual way, by taking a handle decomposition of a closed connected 3-manifold with a unique 3-handle.

The condition for membership of $\mathcal{P}(n)$ can be formulated algebraically; recall that to any finite presentation $\mathcal{G} = \langle x_1, \ldots, x_g \mid W_1, \ldots, W_r \rangle$ of the group G we associate a canonical two-dimensional complex $K_{\mathcal{G}}$ with $\pi_1(K_{\mathcal{G}}) \cong G$ by regarding the generators x_i as 1-cells and the relators W_j as 2-cells.

If $\mathcal{P}$ denotes the class of finite groups of free period 4, then evidently $\mathcal{P} = \bigcup_{n \geq 0} \mathcal{P}(n)$ by Theorem B. The filtration on $\mathcal{P}$ can then expressed by observing that the finite group G belongs to $\mathcal{P}(n)$ precisely when the trivial module $\mathbf{Z}$ admits a finitely generated free resolution of period 4 over $\mathbf{Z}[G]$ of the type

$$0 \to \mathbf{Z} \to \mathbf{Z}[G]^{n+1} \to C_2(\widetilde{K}) \xrightarrow{\partial} C_1(\widetilde{K}) \xrightarrow{\partial} C_0(\widetilde{K}) \to \mathbf{Z} \to 0$$

and where

$$C_2(\widetilde{K}) \xrightarrow{\partial} C_1(\widetilde{K}) \xrightarrow{\partial} C_0(\widetilde{K}) \to \mathbf{Z} \to 0$$

is the partial resolution afforded by some finite presentation $\mathcal{G}$ for G.

65 Poincaré 3-complexes of standard form

We begin with an algebraic result:

Lemma 65.1: Let A be a finitely generated module over $\mathbf{Z}[G]$ which is stably equivalent to $\mathbf{I}^*(G)$; if there exists an epimorphism $\varphi : \mathbf{Z}[G] \to A$, then $A \cong_{\mathbf{Z}[G]} \mathbf{I}^*(G)$.

Proof: If $\varphi : \mathbf{Z}[G] \to A$ is an epimorphism, then $\mathrm{rk}_{\mathbf{Z}}(A) \leq |G|$. Since A is stably equivalent to $\mathbf{I}^*(G)$, then A is torsion free over $\mathbf{Z}$, and a straightforward calculation of $\mathbf{Z}$-ranks shows that $\mathrm{rk}_{\mathbf{Z}}(A) = \mathrm{rk}_{\mathbf{Z}}(\mathbf{I}^*(G)) = |G| - 1$.

From the fact that A and $\mathbf{I}^*(G)$ are stably equivalent, it follows from Wedderburn's Theorem for modules over $\mathbf{Q}[G]$, that $A \otimes \mathbf{Q} \cong_{\mathbf{Q}[G]} \mathbf{I}^*(G) \otimes \mathbf{Q}$. In particular, $A \otimes \mathbf{Q}$ contains no trivial summand, so that the kernel of

$$\varphi \otimes \mathrm{Id} : \mathbf{Q}[G] \to A \otimes \mathbf{Q}$$

is the trivial one-dimensional module over $\mathbf{Q}[G]$. Hence we have a short exact sequence of $\mathbf{Z}[G]$-modules

$$0 \to \mathbf{Z} \to \mathbf{Z}[G] \stackrel{\varphi}{\to} A \to 0$$

where $\mathbf{Z} = \mathrm{Ker}(\varphi : \mathbf{Z}[G] \to A)$ denotes the trivial $\mathbf{Z}[G]$-module of rank 1.
   Dualization gives a short exact sequence

(I) $0 \to A^* \to \mathbf{Z}[G] \to \mathbf{Z} \to 0$

We also have another short exact sequence

(II) $0 \to \mathbf{I}(G) \to \mathbf{Z}[G] \to \mathbf{Z} \to 0$

Applying Schanuel's Lemma to (I) and (II), we obtain an isomorphism of $\mathbf{Z}[G]$-modules

$$\Psi : A^* \oplus \mathbf{Z} \to \mathbf{I}(G) \oplus \mathbf{Z}$$

Let $\pi : \mathbf{I}(G) \oplus \mathbf{Z} \to \mathbf{Z}$ be the projection. Since A^* and $\mathbf{I}(G)$ are rationally equivalent, A^* admits no non-trivial mapping to the trivial one-dimensional $\mathbf{Q}[G]$ module. Hence $\pi \circ \Psi : A^* \to \mathbf{Z}$ is trivial, so that $\Psi(A^*) \subset \mathbf{I}(G)$. Reversing the argument, we see that $\Psi^{-1}(\mathbf{I}(G)) \subset A^*$. That is, $\Psi : A^* \oplus \mathbf{Z} \to \mathbf{I}(G) \oplus \mathbf{Z}$ restricts to an isomorphism $\Psi : A^* \to \mathbf{I}(G)$. Dualizing, we see that $A \cong_{\mathbf{Z}[G]} \mathbf{I}^*(G)$. $\square$

   Now suppose that the finite group G possesses a standard three-dimensional form, that is, a Poincaré 3-complex X_G with just one top-dimensional cell and which satisfies $\pi_1(X_G) \cong G$. Let $\mathcal{G}$ be the presentation of G afforded by the 2-skeleton of X_G. From Theorem B and the algebraic interpretation of the filtration on $\mathcal{P}$, we see that there is a finitely generated free resolution of period 4 for $\mathbf{Z}$ over $\mathbf{Z}[G]$ which takes the form

$$0 \to \mathbf{Z} \to \mathbf{Z}[G] \to C_2(\widetilde{K}) \stackrel{\partial}{\to} C_1(\widetilde{K}) \stackrel{\partial}{\to} C_0(\widetilde{K}) \to \mathbf{Z} \to 0$$

where

$$C_2(\widetilde{K}) \stackrel{\partial}{\to} C_1(\widetilde{K}) \stackrel{\partial}{\to} C_0(\widetilde{K}) \to \mathbf{Z} \to 0$$

is the partial resolution afforded by $\mathcal{G}$. In particular, we have an epimorphism of $\mathbf{Z}[G]$-modules

$$\mathbf{Z}[G] \to \pi_2(\mathcal{G}) \cong \mathrm{Ker}(C_2(\widetilde{K}) \stackrel{\partial}{\to} C_1(\widetilde{K}))$$

and we have seen, in (65.1), that this implies that $\pi_2(\mathcal{G})) \cong \mathbf{I}^*(G)$. Since

$$C_0(\widetilde{K}) \cong \mathbf{Z}[G] \quad C_1(\widetilde{K}) \cong \mathbf{Z}[G]^g \quad C_0(\widetilde{K}) \cong \mathbf{Z}[G]^r$$

then the resolution becomes

$$0 \to \mathbf{Z} \to \mathbf{Z}[G] \to \mathbf{Z}[G]^r \xrightarrow{\partial} \mathbf{Z}[G]^g \xrightarrow{\partial} \mathbf{Z}[G] \to \mathbf{Z} \to 0$$

from which it follows that $r = g$; that is, the presentation $\mathcal{G}$ is balanced.

Suppose, conversely, that $\mathcal{G} = \langle x_1, \ldots, x_g; W_1, \ldots, W_g \rangle$ is a balanced presentation for G such that $\pi_2(\mathcal{G}) \cong \mathbf{I}^*(G)$ as modules over $\mathbf{Z}[G]$. Choose an epimorphism of $\mathbf{Z}[G]$-modules, $\alpha : \mathbf{Z}[G] \to \pi_2(\mathcal{G})$. On applying the construction of Section 63 we see that $X_G = M(\alpha) = K^{(2)} \cup_\alpha e^3$ is the required Poincaré 3-complex of standard form. We have proved:

Theorem V: Let G be a finite group; then there exists a Poincaré 3-complex of standard form $X_G = K^{(2)} \cup_\alpha e^3$ with $\pi_1(M) \cong G$ if and only if G admits a presentation $\mathcal{G}$ such that $\pi_2(\mathcal{G}) \cong \mathbf{I}^*(G)$, the dual module of the augmentation ideal. Moreover, G then necessarily has free period 4, and the presentation $\mathcal{G}$ is automatically balanced.

We note that the existence of a balanced presentation is not by itself enough to ensure the existence of an isomorphism $\pi_2(\mathcal{G}) \cong \mathbf{I}^*(G)$ over $\mathbf{Z}[G]$: the following example of Mennicke [38]

$$\mathbf{H} = \langle x, y \mid x^9 = y^3, \quad y^{-1}xy = x^4 \rangle$$

is a finite group of order 27, but is not of free period 4, nor indeed, since it has order 3^3 but is not cyclic, does it have any finite cohomological period.

66 Relationship with the D(2)-problem

In this section we link the standard form problem with D(2)-problem. We say that a finite group G is Class I, II, III according to the following:

Class I: Every minimal module $J \in \Omega_3(\mathbf{Z})$ is realized.

Class II: $\Omega_3(\mathbf{Z})$ contains at least two isomorphically distinct minimal modules one of which is realized, the other not.

Class III: No minimal module $J \in \Omega_3(\mathbf{Z})$ is realized.

It is clear that the classes are mutually disjoint. Note that Class I is nonempty; in fact we have:

Proposition 66.1: Class I contains all finite groups which possess the cancellation property for free modules and occur as the fundamental group of a closed 3-manifold.

Proof: Let M be a closed connected 3-manifold with finite fundamental group G, and let M_0 be the open manifold obtained by removing a small closed 3-disc e^3. It follows immediately from (63.2) that $\mathbf{I}(G)^*$ is realized as $\pi_2(K)$ for any finite 2-complex K, which is a deformation retract of M_0. The hypothesis that G has the cancellation property for free modules now guarantees that $\mathbf{I}(G)^*$ is the unique minimal representative of $\Omega_3(\mathbf{Z})$. $\qquad\square$

By contrast, it is presently unknown whether either Class II or Class III is non-empty. On the basis of present knowledge, which is rather meagre, it is still reasonable to hope that both classes are empty. The question for Class III is essentially classical and is closely related to Swan's Inequality. It follows immediately from Theorem I that:

Proposition 66.2: Let G be a finite group; if G is either Class II or Class III, then the D(2) property fails for G.

To focus the discussion, we restrict our attention to the case where G has free period 4. If G further has the cancellation property for free modules, then $\Omega_3(\mathbf{Z})$ is straight by (35.1), so that G is either Class I or Class III. When G has free period 4, $\mathbf{I}(G)^*$ is a minimal module for $\Omega_3(\mathbf{Z})$, so that any minimal module J has $\mathrm{rk}_{\mathbf{Z}}(J) = |G| - 1$. A straightforward calculation of ranks shows that a finite group of free period 4 is of class III precisely when it has no balanced presentation.

For groups of free period 4, $\mathbf{I}(G)^*$ is a distinguished minimal module in $\Omega_3(\mathbf{Z})$ so we can split Case II into two further subcases.

Class II(a): $\mathbf{I}(G)^*$ is realized, but some other minimal module is not.
Class II(b): $\mathbf{I}(G)^*$ is not realized, but some other minimal module is realized.

It is evident from the above discussion that, if G is a finite group of free period 4, then G has a standard Poincaré 3-form if and only if G is Class I or Class II(a); that is:

Theorem VI: Let G be a finite group which admits a free resolution of period 4; if G has the free cancellation property then

G satisfies the D(2)−property $\Longleftrightarrow$ G admits a balanced presentation.

If the existence of groups in Class III seems very difficult, the question of the existence of groups in Class II is slightly more clear cut, since, from the results of Swan [63] outlined at the end of Chapter 10, there are many balanced groups of free period 4, for example, the generalized quaternion groups $Q(2^n)$ with $n \geq 5$, which *do possess more than one minimal module* in $\Omega_3(\mathbf{Z})$. The question is then whether all of them can be realized. If at least one is not realized, then the D(2)-property fails.

Interesting examples to study in this context are the groups $Q(8; p, q)$. It can be shown that, when p, q are distinct primes, $Q(8; p, q)$ has the balanced presentation

$$Q(8; p, q) = \langle x, y \mid yx^p y = x^p, \quad xyx = y^{2q-1}\rangle$$

As we have already observed, Milgram [40], [41], and also Madsen and Bentzsen [4] have shown that, whilst many of the groups $Q(8; p, q)$ have free period 4, some, at least, do not. By Swan's computational results [63], none of them has free cancellation. It is conjectured, but not yet proved, that $Q(8; p, q)$ cannot be the fundamental group of a closed 3-manifold when both $p \neq 1$ and $q \neq 1$.

Likewise, for $n \geq 4$ the groups $Q(2^n; p, q)$ are also of interest. They have cohomological period 4, though very little appears to be known on the question of when this can be improved to free period 4. However, Charles Thomas [65] has shown that except in trivial cases, they cannot occur as the fundamental group of any closed 3-manifold. Furthermore, except in the degenerate case $n = 4; p = q = 1$ they do not possess the free cancellation property.

67 Terminus: the limits of this book

To see how much remains to be done one only has to pose the fundamental question: Does the D(2)-property hold for all finitely presented groups? If one believes the answer to be 'Yes' (and on present evidence this still seems the more likely outcome), then, even in the 'easy' case of groups of period 4 ('easy' because one has an easily understood model for $\Omega_3(\mathbf{Z})$), the task still seems formidable. For example, in the case of the groups of type $Q(8; p, q)$ where one knows there is more than one minimal module in $\Omega_3(\mathbf{Z})$, one is faced with the task of finding enough balanced presentations to exhaust the minimal homotopy types. This seems like a very interesting problem, though perhapsat the borders of computational intractability (c.f. [4], [40], [41], [63]).

We note that, in the case of finite abelian groups, enough minimal presentations have been found to exhaust the minimal homotopy classes. This was done by Browning, generalizing Metzler [39]; see Latolais's clear account in [32]. Incidentally, it follows from this that the D(2)-property holds for finite abelian groups.

For a general finite group, one lacks an understanding of $\Omega_3(\mathbf{Z})$. Indeed, the only non-periodic non-abelian finite group that the author knows to possess the D(2)-property is the dihedral group of order 8, and unfortunately the proof of this seems, at present, to require cancellation properties somewhat stronger than the Swan-Jacobinski Theorem, and which do not hold in general.

Secondly, the original motivation for considering the D(2)-problem for finite groups of period 4 in isolation arose from a desire to examine the structure of Poincaré 3-complexes, and it may fairly be argued that we have not considered the structure of Poincaré 3-complexes either generally, allowing for infinite fundamental groups, or specifically, by relating general Poincaré 3-complexes to manifolds.

Both questions are vexed. In the case of finite fundamental groups, it is not known which Poincaré 3-complexes correspond to manifolds. The conjecture is that the linear spherical space forms (see [82]) are the only closed 3-manifolds with finite fundamental groups. At the time of writing this still seems to be unproved, although strenuous attempts have been made towards it in recent years.

For the question of the general structure, the literature seems rather sparse, the most relevant references in the literature being [17], [24], [67]. In [17], Epstein gave some necessary conditions for a group to be the fundamental group of a closed 3-manifold, the most obvious restriction, which we noted in the text using a Morse Theory argument, being that the group must possess a balanced presentation. This is also observed by Turaev [67]. The detail in Turaev's paper is rather meagre, but it contains a sketch of a duality theory for Fox ideals which seems to coincide with that of Hempel [24].

Finally, it will be pointed out that we have not considered any case of the D(2)-problem with an infinite fundamental group.

In general, aspects of the problem which seem clear in the case of finite groups become far more problematic in the case of infinite fundamental groups, whilst the fundamental problem of finding enough minimal presentations remains at least as intractable as before. However, in any particular case there may be compensating advantages, and to demonstrate this, and make some small amends for this neglect of infinite groups we show in an Appendix that the

D(2)-property holds for non-abelian free groups. Comparing the solution for free groups with those we have obtained in the finite case, it should be apparent, if it were not already so, that the details of a successful solution to any D(2)-problem will, at the very least, depend heavily on the module theory of the particular fundamental group under consideration.

Appendix A: The D(2)-property for free groups

We sketch the proof of the following:

Theorem: The D(2)-property holds for all non-abelian free groups of finite rank.

Throughout we denote by $G(n)$ the non-abelian free group of rank n; then the following is known [2] (also [3] p. 213):

Theorem A.1: (Bass–Seshadri) If P is a finitely generated projective module over $\mathbf{Z}[G(n)]$, then P is free.

Now fix $G = G(n)$, and consider an algebraic 2-complex over $\mathbf{Z}[G]$

$$\mathbf{E} = \left(0 \to J \overset{j}{\to} E_2 \overset{\partial_2}{\to} E_1 \overset{\partial_1}{\to} E_0 \overset{\epsilon}{\to} \mathbf{Z} \to 0\right)$$

that is, we are assuming that E_0, E_1, E_2 are finitely generated and stably free. Then by (A.1), E_0, E_1, E_2 are all free. Furthermore, we have:

Proposition A.2: $\operatorname{Im}(\partial_1)$ is free.

Proof: Consider the standard presentation $\mathcal{G}(n) = \langle x_1, \ldots, x_n \rangle$ for $G = G(n)$ with n generators and no relations; the algebraic Cayley complex is the following exact sequence

$$0 \to \mathbf{Z}[G]^n \overset{\delta}{\to} \mathbf{Z}[G] \to \mathbf{Z} \to 0$$

where, if $\{e_1, \ldots, e_n\}$ is the canonical basis for $\mathbf{Z}[G]^n$, $\delta(e_i) = x_i - 1$. Comparing this with the exact sequence $0 \to \operatorname{Im}(\partial_1) \to E_0 \overset{\epsilon}{\to} \mathbf{Z} \to 0$, we see from Schanuel's Lemma that

$$\operatorname{Im}(\partial_1) \oplus \mathbf{Z}[G] \cong \mathbf{Z}[G]^n \oplus E_0$$

It follows that $\operatorname{Im}(\partial_1)$ is stably free, and hence free, by (A.1).   □

It now follows that the exact sequence $0 \to \operatorname{Im}(\partial_2) \to E_1 \overset{\partial_1}{\to} \operatorname{Im}(\partial_1) \to 0$ splits, so that

$$\operatorname{Im}(\partial_1) \oplus \operatorname{Im}(\partial_2) \cong E_1$$

Thus $\operatorname{Im}(\partial_2)$ is projective and so, again by (A.1):

Proposition A.3: $\operatorname{Im}(\partial_2)$ is free.

Iterating the argument, the sequence $0 \to J \to E_2 \overset{\partial_2}{\to} \mathrm{Im}(\partial_2) \to 0$ splits, and so:

Proposition A.4: J is free.

It follows that:

Proposition A.5: If K is a finite 2-complex with $\pi_1(K) \cong G(n)$, then $\pi_2(K)$ is a free module over $\mathbf{Z}[G(n)]$.

Now suppose that X is a cohomologically two-dimensional finite 3-complex with $\pi_1(X) = G(n)$, and consider the cellular chain complex

$$C_*(X) = \left(C_3(X) \overset{\partial_3}{\to} C_2(X) \overset{\partial_2}{\to} C_1(X) \overset{\partial_1}{\to} C_0(X) \to \mathbf{Z} \to 0 \right)$$

Since $\dim(X) \le 3$ it follows that $H_3(\tilde{X}; \mathbf{Z}) = \mathrm{Ker}(\partial_3 : C_3(X) \to C_2(X))$, and from the hypothesis $H_3(\tilde{X}; \mathbf{Z}) = 0$ we see that:

Proposition A.6: $\partial_3 : C_3(X) \to C_2(X)$ is injective.

Thus we may form the virtual 2-complex

$$\langle X \rangle = \left(0 \to \pi_2(X) \to C_2(X)/\mathrm{Im}(\partial_3) \overset{\partial_2}{\to} C_1(X) \overset{\partial_1}{\to} C_0(X) \to \mathbf{Z} \to 0 \right)$$

It is not immediately apparent that this is a genuine algebraic 2-complex. To check that it is we must show that $C_2(X)/\mathrm{Im}(\partial_3)$ is a free module. We do this in three stages. Put $\Lambda = \mathbf{Z}[G(n)]$; first observe that:

Proposition A.7: $\mathrm{Ext}^1_\Lambda(\pi_2(X); \mathcal{B}) = 0$ for any Λ-module $\mathcal{B}$.

Proof: Put $B_2 = \mathrm{Im}(\partial_3 : C_3(X) \to C_2(X))$, and $Z_2 = \mathrm{Ker}(\partial_2 : C_2(X) \to C_1(X))$. Then $B_2 \subset Z_2 \subset C_2(X)$. Let

$$j : B_2 \subset Z_2 \quad i : Z_2 \subset C_2(X), \quad i \circ j : B_2 \subset C_2(X)$$

denote the various inclusions. Observe that, by the Hurewicz Theorem

$$\pi_2(X) = H_2(\tilde{X}) = Z_2/B_2$$

For any coefficient module $\mathcal{B}$, we have induced maps

$$(i \circ j)^* : \mathrm{Hom}_\Lambda(C_2(X), \mathcal{B}) \to \mathrm{Hom}_\Lambda(B_2, \mathcal{B}); \quad j^* : \mathrm{Hom}_\Lambda(Z_2, \mathcal{B}) \to \mathrm{Hom}_\Lambda(B_2, \mathcal{B})$$

and since $(i \circ j)^* = j^* \circ i^*$, we have a filtration $\mathrm{Im}((i \circ j)^*) \subset \mathrm{Im}(j^*) \subset \mathrm{Hom}_\Lambda(B_2, \mathcal{B})$, and a short exact sequence

$$(*) \qquad 0 \to \mathrm{Im}(j^*)/\mathrm{Im}((i \circ j)^*) \to B_2^\mathcal{B}/\mathrm{Im}((i \circ j)^*) \to B_2^\mathcal{B}/\mathrm{Im}(j^*) \to 0$$

where $B_2^\mathcal{B} = \mathrm{Hom}_\Lambda(B_2, \mathcal{B})$. Now B_2 is free, since $\partial_3 : C_3(X) \to B_2$ is an isomorphism. Moreover, putting $K = X^{(2)}$, we see by the Hurewicz Theorem that $Z_2 \cong \pi_2(K)$, so that Z_2 is free by (A.5). Hence applying $\mathrm{Hom}_\Lambda(-, \mathcal{B})$ to the exact sequence

$$0 \to B_2 \overset{j}{\to} Z_2 \to \pi_2(X) \to 0$$

we get an exact sequence in cohomology

$$0 \to \mathrm{Hom}_\Lambda(\pi_2(X), \mathcal{B}) \to \mathrm{Hom}_\Lambda(Z_2, \mathcal{B}) \overset{j^*}{\to} \mathrm{Hom}_\Lambda(B_2, \mathcal{B}) \to \mathrm{Ext}^1(\pi_2(X), \mathcal{B}) \to 0$$

which shows that $\mathrm{Hom}_\Lambda(B_2, \mathcal{B})/\mathrm{Im}(j^*) \cong \mathrm{Ext}^1(\pi_2(X), \mathcal{B})$. However, $\partial_3 : C_3(X) \to B_2$ is an isomorphism, and induces an isomorphism

$$\mathrm{Hom}_\Lambda(B_2, \mathcal{B})/\mathrm{Im}((i \circ j)^*) \cong \mathrm{Hom}_\Lambda(C_3(X), \mathcal{B})/\mathrm{Im}(\partial_3^*) = H^3(X, \mathcal{B})$$

Thus the exact sequence (*) becomes

(**) $0 \to \text{Im}(j^*)/\text{Im}((i \circ j)^*) \to H^3(X, \mathcal{B}) \to \text{Ext}^1(\pi_2(X), \mathcal{B}) \to 0$

Since $H^3(X, \mathcal{B}) = 0$ it follows that $\text{Ext}^1(\pi_2(X), \mathcal{B}) = 0$ as claimed.   □

Theorem A.8: $\pi_2(X)$ is a free module.

Proof: As above, put $K = X^{(2)}$; then we have an exact sequence

(I) $0 \to C_3(X) \overset{\partial_3}{\to} \pi_2(K) \overset{p}{\to} \pi_2(X) \to 0$

Here we are identifying $\pi_2(X)$ with $H_2(\tilde{X}; \mathbf{Z})$, and p is the natural surjection

$$p : \pi_2(K) \to \pi_2(K)/\text{Im}(\partial_3) = \text{Ker}(\partial_2)/\text{Im}(\partial_3) = H_2(\tilde{X}; \mathbf{Z})$$

In the first instance, (I) gives rise to an exact sequence in cohomology

$$0 \to \text{Hom}_\Lambda(\pi_2(X), \mathcal{B}) \to \text{Hom}_\Lambda(\pi_2(K), \mathcal{B}) \overset{\partial_3^*}{\to} \text{Hom}_\Lambda(C_3(X), \mathcal{B}) \to \text{Ext}_\Lambda^1(\pi_2(X); \mathcal{B})$$

However, by (A.7), $\text{Ext}_\Lambda^1(\pi_2(X); \mathcal{B}) = 0$, and we are reduced to a short exact sequence

(II) $0 \to \text{Hom}_\Lambda(\pi_2(X), \mathcal{B}) \to \text{Hom}_\Lambda(\pi_2(K), \mathcal{B}) \overset{\partial_3^*}{\to} \text{Hom}_\Lambda(C_3(X), \mathcal{B}) \to 0$

Taking $\mathcal{B} = C_3(X)$, if $r \in \text{Hom}_\Lambda(\pi_2(K), C_3(X))$ is such that $\partial_3^*(r) = \text{Id}_{C_3}$, we see that r splits (I) on the left. Thus

$$\pi_2(K) \cong C_3(X) \oplus \pi_2(X)$$

so that $\pi_2(X)$ is projective and hence free by (A.1).   □

As a consequence, we obtain:

Corollary A.9: $C_2(X)/\text{Im}(\partial_3)$ is a free module.

Proof: We have an exact sequence $0 \to \pi_2(X) \to C_2(X)/\text{Im}(\partial_3) \overset{\partial_2}{\to} \text{Im}(\partial_2) \to 0$ in which $\text{Im}(\partial_2)$ is free, by the argument of (A.3) applied to the cellular chain complex $C_*(K)$. Hence

$$C_2(X)/\text{Im}(\partial_3) \cong \pi_2(X) \oplus \text{Im}(\partial_2)$$

and the result follows, again by (A.1).   □

It follows that if X is a cohomologically two-dimensional finite 3-complex with $\pi_1(X) = G(n)$, the virtual 2-complex $\langle X \rangle$ is a *bona fide* algebraic 2-complex. Now the geometric Cayley complex $K(n)$ of the standard presentation $\mathcal{G}(n)$ is just a wedge of n-copies of the 1-sphere $K(n) = \underbrace{S^1 \vee \cdots \vee S^1}_{n}$. We put $K(n, 0) = K(n)$ and

$$K(n, m) = K(n) \vee \underbrace{S^2 \vee \cdots \vee S^2}_{m}$$

whenever $m \geq 1$. Let $C_*(n, m)$ denote the cellular chain complex of $K(n, m)$

$$C_*(n, m) = (0 \to \pi_2(K(n, m)) \overset{\cong}{\to} \mathbf{Z}[G(n)]^m \overset{0}{\to} \mathbf{Z}[G(n)]^n \overset{\partial_1}{\to} \mathbf{Z}[G(n)] \to \mathbf{Z} \to 0)$$

Clearly $\pi_2(K(n, m)) \cong \mathbf{Z}[G(n)]^m$. Now let

$$\mathbf{E} = (0 \to J \to E_2 \to E_1 \to E_0 \to \mathbf{Z} \to 0)$$

be an algebraic 2-complex over $\mathbf{Z}[G(n)]$. By (A.5), J is free. Suppose $J \cong \mathbf{Z}[G(n)]^m$. Since $G(n)$ has cohomological dimension 1

$$\mathrm{Ext}^3(\mathbf{Z}, J) = \mathrm{Ext}^3(\mathbf{Z}, \pi_2(K(n, m))) = 0$$

so that the extensions $C_*(n, m)$ and $\mathbf{E}$ are necessarily congruent, and, since they are projective complexes, also homotopy equivalent; that is:

Proposition A.10: Every algebraic 2-complex over $\mathbf{Z}[G(n)]$ is geometrically realizable.

It follows that, if X is a finite cohomologically two-dimensional 3-complex with $\pi_1(X) \cong G(n)$ then the virtual 2-complex $\langle X \rangle$ is geometrically realizable by some finite 2-complex $K(n, m)$. Hence X is homotopy equivalent to $K(n, m)$, as in the easy half of (59.4). We have shown:

Theorem A.11: The D(2)-property holds for $G(n)$.

Generalizations of this argument will be considered in [30].

Appendix B: The Realization Theorem

The proof of the Realization Theorem (Theorem I) given in the text relies heavily on the fact that, over a finite group G, finitely generated projectives are injective relative to the class of $\mathbf{Z}[G]$-lattices. In this appendix, we give a proof of the Realization Theorem which avoids this property.

Let G be a finitely presented group; by an algebraic 2-complex over G we mean an exact sequence

$$\mathbf{F} = \left(0 \to J \to F_2 \xrightarrow{\partial_2} F_1 \xrightarrow{\partial_1} F_0 \xrightarrow{\epsilon} \mathbf{Z} \to 0\right)$$

where F_0, F_1, F_2 are finitely generated stably free modules over $\mathbf{Z}[G]$. We denote by $\mathbf{Alg}_G$ the category of algebraic 2-complexes over $\mathbf{Z}[G]$. In general, the module $J = \pi_2(\mathbf{F})$ need not be finitely generated over $\mathbf{Z}[G]$. We say that a finitely presented group G is *of type FL(3)* when there is at least one algebraic 2-complex for which $\pi_2(\mathbf{F})$ is finitely generated. We prove:

Realization Theorem: Let G be a finitely presented group of type FL(3); then the D(2)-property holds for G if and only each algebraic 2-complex $\mathbf{E} \in \mathbf{Alg}_G$ admits a geometric realization.

Without the FL(3) hypothesis, the argument holds in one direction, namely:

Weak Realization Theorem: If each algebraic 2-complex over $\mathbf{Z}[G]$ admits a geometric realization then the D(2)-property holds for G.

The property FL(3) has a slightly stronger formulation, since it follows easily from Schanuel's Lemma that:

Proposition B.1: If the finitely presented group G is of type FL(3) then $\pi_2(\mathbf{F})$ is finitely generated for *every* algebraic 2-complex $\mathbf{F}$ over $\mathbf{Z}[G]$.

For groups of type FL(3), the proof of the Eventual Stability Theorem given in the text (52.3) extends easily to give:

Proposition B.2: Let G be a finitely presented group of type FL(3), and let $\mathbf{E}$, $\mathbf{E}'$ be algebraic 2-complexes over $\mathbf{Z}[G]$; then for some $\mu, \nu \geq 0$ there is a chain homotopy equivalence

$$\Sigma_+^{\mu}(\mathbf{E}) \xrightarrow{\sim} \Sigma_+^{\nu}(\mathbf{E}')$$

If $\mathbf{E}'$ is geometrically realizable, then so is $\Sigma_+^{\nu}(\mathbf{E}')$. Hence we see:

256

Proposition B.3: If G is a finitely presented group of type FL(3), then for each algebraic 2-complex $\mathbf{E} \in \mathbf{Alg}_G$ there exists $n \geq 0$ such that $\Sigma^n\mathbf{E}$ admits a geometric realization.

The following proposition compensates for the fact that, in general, projectives are not relatively injective.

Proposition B.4: Let Λ be a ring and let

$$S = \left(0 \to Q \xrightarrow{i} M \xrightarrow{p} N \to 0\right)$$

be an exact sequence of Λ-modules in which Q is projective; then $p^* : \mathrm{Ext}^1(N, \mathcal{B}) \xrightarrow{\sim}$ $\mathrm{Ext}^1(M, \mathcal{B})$ is surjective for all $\mathcal{B}$. Moreover, the following conditions are then equivalent:

(i) S splits;
(ii) $p^* : \mathrm{Ext}^1(N, \mathcal{B}) \xrightarrow{\sim} \mathrm{Ext}^1(M, \mathcal{B})$ is an isomorphism for all $\mathcal{B}$;
(iii) $p^* : \mathrm{Ext}^1(N, Q) \xrightarrow{\sim} \mathrm{Ext}^1(M, Q)$ is an isomorphism.

Proof: Writing $A^{\mathcal{B}} = \mathrm{Hom}(A, \mathcal{B})$, we have an exact sequence

$$0 \to N^{\mathcal{B}} \xrightarrow{p^*} M^{\mathcal{B}} \xrightarrow{i_*} Q^{\mathcal{B}} \xrightarrow{\delta} \mathrm{Ext}^1(N, \mathcal{B}) \xrightarrow{p^*} \mathrm{Ext}^1(M, \mathcal{B}) \to 0$$

the final '0' arising $\mathrm{Ext}^1(Q, \mathcal{B}) = 0$ since Q is projective. In particular

$$p^* : \mathrm{Ext}^1(N, \mathcal{B}) \to \mathrm{Ext}^1(M, \mathcal{B})$$

is surjective as claimed. We now show (i) $\Longrightarrow$ (ii) $\Longrightarrow$ (iii) $\Longrightarrow$ (i).

(i) $\Longrightarrow$ (ii): Suppose that S splits. Then there exists a Λ-homomorphism $r : M \to Q$ such that $r \circ i = \mathrm{Id}_Q$. Hence

$$i^* \circ r^* = \mathrm{Id}_{\mathrm{Hom}(Q,\mathcal{B})}$$

In particular, $i^* : M^{\mathcal{B}} \to Q^{\mathcal{B}}$ is surjective. Thus

$$\delta : Q^{\mathcal{B}} \to \mathrm{Ext}^1(N, \mathcal{B})$$

is identically zero, since $\mathrm{Ker}(\delta) = \mathrm{Im} i^* = \mathrm{Hom}(Q, \mathcal{B})$. Since $\mathrm{Ker}(p^*) = \mathrm{Im}\delta = 0$, we have an injective map $p^* : \mathrm{Ext}^1(N, \mathcal{B}) \to \mathrm{Ext}^1(M, \mathcal{B})$. Hence

$$p^* : \mathrm{Ext}^1(N, \mathcal{B}) \xrightarrow{\sim} \mathrm{Ext}^1(M, \mathcal{B})$$

is an isomorphism, since surjectivity is already established.
(ii) $\Longrightarrow$ (iii) is clear.
(iii) $\Longrightarrow$ (i): If $p^* : \mathrm{Ext}^1(N, \mathcal{B}) \xrightarrow{\sim} \mathrm{Ext}^1(M, \mathcal{B})$ is an isomorphism, then $\delta : \mathrm{Hom}(Q, Q) \to \mathrm{Ext}^1(N, \mathcal{B})$ is necessarily zero, by exactness. Then the sequence

$$0 \to \mathrm{Hom}(N, Q) \xrightarrow{p^*} \mathrm{Hom}(M, Q) \xrightarrow{i_*} \mathrm{Hom}(Q, Q) \to 0$$

is exact. Choosing, $r \in \mathrm{Hom}(M, Q)$ such that $i^*(r) = \mathrm{Id}_Q$, then $r \circ i = \mathrm{Id}_Q$, and r is a left splitting of S. Thus (iii) $\Longrightarrow$ (i) as claimed. $\square$

Now fix a finitely presented group G. Let X be finite, cohomologically two-dimensional 3-complex with $\pi_1(X) \cong G$. and put $K = X^{(2)}$. Then we can decompose the cellular chain complex $C_*(X)$ as a pair of exact sequences

$$0 \to C_3(X) \xrightarrow{j} \pi_2(K) \xrightarrow{p} \pi_2(X) \to 0$$

and

$$0 \to \pi_2(K) \xrightarrow{i} C_2(X) \xrightarrow{\partial_2} C_1(X) \xrightarrow{\partial_1} C_0(X) \xrightarrow{\epsilon} \mathbf{Z} \to 0$$

where $i \circ j = \partial_3$. We form a new exact sequence by taking the quotient by the images of $C_3(X)$ thus

$$\langle X \rangle = \left(0 \to \pi_2(X) \xrightarrow{\hat{i}} C_2(X)/\mathrm{Im}(\partial_3) \xrightarrow{\hat{\partial}_2} C_1(X) \xrightarrow{\partial_1} C_0(X) \xrightarrow{\epsilon} \mathbf{Z} \to 0\right)$$

where $\hat{i}$, $\hat{\partial}_2$ are the obvious homomorphisms induced by i and ∂_2. $\langle X \rangle$ is called the *virtual 2-complex* of X. We first prove:

Proposition B.5: The sequence

$$0 \to C_3(X) \xrightarrow{j} \pi_2(K) \xrightarrow{P} \pi_2(X) \to 0$$

splits; in consequence $\pi_2(X) \in \Omega_3(\mathbf{Z})$.

Proof: Since $C_3(X)$ is free, it is enough, by (B.4), to show that

$$p^* : \mathrm{Ext}^1(\pi_2(X), \mathcal{B}) \xrightarrow{\approx} \mathrm{Ext}^1(\pi_2(K), \mathcal{B})$$

is an isomorphism for all coefficient modules $\mathcal{B}$. Thus consider the exact sequence

(I) $\qquad \pi_2(K)^{\mathcal{B}} \xrightarrow{j_*} C_3(X)^{\mathcal{B}} \xrightarrow{\delta} \mathrm{Ext}^1(\pi_2(X), \mathcal{B}) \xrightarrow{P_*} \mathrm{Ext}^1(\pi_2(K), \mathcal{B}) \to 0.$

Clearly there are inclusions

$$\mathrm{Im}(i \circ j)^* \subset \mathrm{Im}\, j^* \subset C_3(X)^{\mathcal{B}}$$

since

$$(i \circ j)^* = j^* \circ i^* : C_2(X)^{\mathcal{B}} \to C_3(X)^{\mathcal{B}}$$

giving an exact sequence

(II) $\qquad 0 \to \mathrm{Im}\, j^*/\mathrm{Im}(i \circ j)^* \to C_3(X)^{\mathcal{B}}/\mathrm{Im}(i \circ j)^* \to C_3(X)^{\mathcal{B}}/\mathrm{Im}\, j^* \to 0.$

However, $i \circ j = \partial_3$ so that $C_3(X)^{\mathcal{B}}/\mathrm{Im}(i \circ j)^* = H^3(X, \mathcal{B})$. By hypothesis, $H^3(X, \mathcal{B}) = 0$. It follows from (II) that $C_3(X)^{\mathcal{B}}/\mathrm{Im}\, j^* = 0$; that is, $j^* : \pi_2(K)^{\mathcal{B}} \to C_3(X)^{\mathcal{B}}$ is surjective. Thus in the exact sequence (I) it follows that $\delta = 0$, and so

$$p^* : \mathrm{Ext}^1(\pi_2(X), \mathcal{B}) \xrightarrow{\approx} \mathrm{Ext}^1(\pi_2(K), \mathcal{B})$$

is an isomorphism as required. $\square$

Let X be a finite cohomologically two-dimensional 3-complex with $\pi_1(X) \cong G$. In addition to the exact sequence

(III) $\qquad\qquad 0 \to C_3(X) \xrightarrow{j} \pi_2(K) \xrightarrow{P} \pi_2(X) \to 0$

we also have

(IV) $\qquad\qquad 0 \to \pi_2(K) \xrightarrow{i} C_2(X) \xrightarrow{\partial_2} \mathrm{Im}(\partial_2) \to 0$

where $\partial_3 = i \circ j$. Together these give rise to an exact sequence

(V) $\qquad\qquad 0 \to \pi_2(X) \xrightarrow{\hat{i}} C_2(X)/\mathrm{Im}(\partial_3) \xrightarrow{\hat{\partial}_2} \mathrm{Im}(\partial_2) \to 0$

where $\hat{i}$, $\hat{\partial}_2$ are induced by i , ∂_2 respectively. Applying $\mathrm{Hom}(-, B)$ to (V) gives a long exact sequence

(VI) $0 \to \mathrm{Im}(\partial_2)^B \xrightarrow{\hat{\partial}_2} (C_2(X)/\mathrm{Im}(\partial_3))^B \xrightarrow{\hat{i}} \pi_2(X)^B \xrightarrow{\delta} \mathrm{Ext}^1(\mathrm{Im}(\partial_2), B) \xrightarrow{\hat{\partial}_2} \cdots$

$\cdots \xrightarrow{\delta} \mathrm{Ext}^n(\mathrm{Im}(\partial_2), B) \xrightarrow{\hat{\partial}_2} \mathrm{Ext}^n(C_2(X)/\mathrm{Im}(\partial_3), B) \xrightarrow{\hat{i}} \mathrm{Ext}^n(\pi_2(X), B) \xrightarrow{\delta} \cdots$

Proposition B.6: There is a natural surjection $\nu : \pi_2(K)^B \to \mathrm{Ext}^1(\mathrm{Im}(\partial_2), B)$ and natural isomorphisms

$$\nu : \mathrm{Ext}^n(\pi_2(K), B) \to \mathrm{Ext}^{n+1}(\mathrm{Im}(\partial_2), B)$$

for $n \geq 1$.

Proof: Consider the long exact sequence obtained by applying $\mathrm{Hom}(-, B)$ to (IV),

(VII) $\cdots \xrightarrow{i^*} \pi_2(K)^B \xrightarrow{\nu} \mathrm{Ext}^1(\mathrm{Im}(\partial_2), B) \xrightarrow{\partial_2^*} \mathrm{Ext}^1(C_2(X), B) \cdots$

$\cdots \xrightarrow{\nu} \mathrm{Ext}^n(\mathrm{Im}(\partial_2), B) \xrightarrow{\partial_2^*} \mathrm{Ext}^n(C_2(X), B) \xrightarrow{i^*} \mathrm{Ext}^n(\pi_2(K), B) \xrightarrow{\nu} \mathrm{Ext}^{n+1}(\mathrm{Im}(\partial_2), B) \cdots$

where ν denotes the natural coboundary operator. Since $C_2 = C_2(X)$ is $\mathbf{Z}[G]$-free, we see that $\mathrm{Ext}^n(C_2(X), B) = 0$ for all $n \geq 1$. The result follows immediately. $\square$

Proposition B.7: $\mathrm{Ext}^n(C_2(X)/\mathrm{Im}(\partial_3), B) = 0$ for $n \geq 2$ and all coefficient modules B.

Proof: For $n \geq 1$, the following triangle commutes by naturality

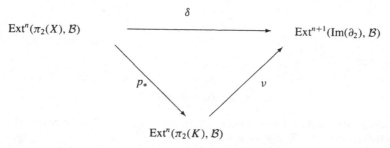

where δ is the coboundary map of the long exact sequence of (V), and ν is the coboundary map of the long exact sequence of (IV). In the proof of (B.4), we saw that $p^* : \mathrm{Ext}^1(\pi_2(X), B) \xrightarrow{\sim} \mathrm{Ext}^1(\pi_2(K), B)$ is an isomorphism for all B. Since $\nu : \mathrm{Ext}^n(\pi_2(K), B) \to \mathrm{Ext}^{n+1}(\mathrm{Im}(\partial_2), B)$ is an isomorphism for $n \geq 1$ (by (B.5)), we see that

$$\delta : \mathrm{Ext}^n(\pi_2(X), B) \to \mathrm{Ext}^{n+1}(\mathrm{Im}(\partial_2), B)$$

is an isomorphism for $n \geq 1$. In the exact sequence (VI)

$$\xrightarrow{\delta} \mathrm{Ext}^{n+1}(\mathrm{Im}(\partial_2), B) \xrightarrow{\hat{\partial}_2} \mathrm{Ext}^{n+1}(C_2(X)/\mathrm{Im}(\partial_3), B) \xrightarrow{\hat{i}} \mathrm{Ext}^{n+1}(\pi_2(X), B) \xrightarrow{\delta}$$

δ is an isomorphism; thus it follows that $\mathrm{Ext}^{n+1}(C_2(X)/\mathrm{Im}(\partial_3), B) = 0$ for $n \geq 1$; equivalently, $\mathrm{Ext}^n(C_2(X)/\mathrm{Im}(\partial_3), B) = 0$ for $n \geq 2$. $\square$

Slightly more delicate to prove is:

Proposition B.8: $\mathrm{Ext}^1(C_2(X)/\mathrm{Im}(\partial_3), \mathcal{B}) = 0$ for all coefficient modules $\mathcal{B}$.

Proof: The following triangle also commutes by naturality.

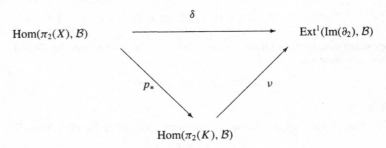

We have seen that the exact sequence

(III) $0 \to C_3(X) \overset{j}{\to} \pi_2(K) \overset{p}{\to} \pi_2(X) \to 0$

splits. If $r : \pi_2(K) \to C_3(X)$ is a splitting on the left of (III), then the mapping $(p^*, r^*) : \mathrm{Hom}(\pi_2(X), \mathcal{B}) \oplus \mathrm{Hom}(C_3(X), \mathcal{B}) \to \mathrm{Hom}(\pi_2(K), \mathcal{B})$ is an isomorphism. Thus δ factorizes as follows:

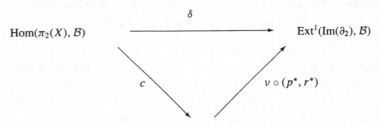

where $c : \mathrm{Hom}(\pi_2(X), \mathcal{B}) \to \mathrm{Hom}(\pi_2(X), \mathcal{B}) \oplus \mathrm{Hom}(C_3(X), \mathcal{B})$ is the canonical inclusion $c(\alpha) = (\alpha, 0)$. However, since

$$\partial_2 \circ i \circ j = \partial_2 \circ \partial_3 = 0$$

it is straightforward to check that $\nu \circ (p^*, r^*)_{|\mathrm{Hom}(C_3, \mathcal{B})} \equiv 0$. Since ν is surjective and (p^*, r^*) is an isomorphism, it follows that

$$\delta = \nu \circ (p^*, r^*)_{|\mathrm{Hom}(\pi_2(X), \mathcal{B})} : \mathrm{Hom}(\pi_2(X), \mathcal{B}) \to \mathrm{Ext}^1(\mathrm{Im}\partial_2, \mathcal{B})$$

is surjective. Now from the following portion of (VI)

$$\pi_2(X)^{\mathcal{B}} \overset{\delta}{\to} \mathrm{Ext}^1(\mathrm{Im}(\partial_2), \mathcal{B}) \overset{\hat{\partial_3}}{\to} \mathrm{Ext}^1(C_2(X)/\mathrm{Im}(\partial_3), \mathcal{B})$$

$$\to \mathrm{Ext}^1(\pi_2(X), \mathcal{B}) \overset{\delta}{\to} \mathrm{Ext}^2(\mathrm{Im}(\partial_2), \mathcal{B})$$

since $\delta : \pi_2(X)^{\mathcal{B}} \to \mathrm{Ext}^1(\mathrm{Im}(\partial_2), \mathcal{B})$ is surjective, and $\delta : \mathrm{Ext}^1(\pi_2(X), \mathcal{B}) \to \mathrm{Ext}^2(\mathrm{Im}(\partial_2), \mathcal{B})$ is an isomorphism, we see that $\mathrm{Ext}^1(C_2(X)/\mathrm{Im}(\partial_3), \mathcal{B}) = 0$ as claimed.   □

We can now proceed to:

Proof of Weak Realization Theorem: Let G be a finitely presented group, and let X be a finite cohomologically two-dimensional 3-complex with $\pi_1(X) \cong G$. We have an exact sequence

(VIII) $$0 \to \mathrm{Im}(\partial_3) \to C_2(X) \to C_2(X)/\mathrm{Im}(\partial_3) \to 0$$

Since $C_2(X)$ is free then $\mathrm{Ext}^n(C_2(X)), \mathcal{B}) = 0$ for all $n \geq 2$ and all coefficient modules $\mathcal{B}$. By (B.7) and (B.8), we see that $\mathrm{Ext}^n(C_2(X)/\mathrm{Im}(\partial_3), \mathcal{B}) = 0$ for all $n \geq 1$ and all coefficient modules $\mathcal{B}$. Thus the natural map $C_2(X) \to C_2(X)/\mathrm{Im}(\partial_3)$ induces isomorphisms

$$\mathrm{Ext}^n(C_2(X)/\mathrm{Im}(\partial_3), \mathcal{B}) \overset{\approx}{\to} \mathrm{Ext}^n(C_2(X), \mathcal{B}) = 0$$

for all $n \geq 1$. Since $\mathrm{Im}(\partial_3) \cong C_3(X)$ is free, hence projective, we see that (VIII) splits, by (B.4). Hence $C_2(X)/\mathrm{Im}(\partial_3)$ is stably free, since $\mathrm{Im}(\partial_3)$ and $C_3(X)$ are both free. In particular, the virtual 2-complex

$$\langle X \rangle = \left(0 \to \pi_2(X) \overset{i}{\to} C_2(X)/\mathrm{Im}(\partial_3) \overset{\partial_2}{\to} C_1(X) \overset{\partial_1}{\to} C_0(X) \overset{\epsilon}{\to} \mathbf{Z} \to 0 \right)$$

is a *bona fide* algebraic 2-complex. By hypothesis, there exists a finite 2-complex K with $\pi_1(K) \cong G$ such that $\langle X \rangle \simeq C_*(K)$. It then follows, as in the proof of (59.4), that X is homotopy equivalent to K, and so the D(2)-property holds for G. $\qquad\square$

Proof of Realization Theorem: Suppose that G is of type FL(3). By the Weak Realization Theorem, just proved, it suffices to show that, if the D(2)-property holds for G, then each algebraic 2-complex is geometrically realizable. Thus let $\mathbf{E} \in \mathbf{Alg}_G$

$$\mathbf{E} = \left(0 \to J \to \mathbf{Z}[G]^\gamma \overset{\delta_2}{\to} \mathbf{Z}[G]^\beta \overset{\delta_1}{\to} \mathbf{Z}[G] \overset{\epsilon}{\to} \mathbf{Z} \to 0 \right)$$

By the stable realization theorem, (B.3), if L is any finite 2-complex with $\pi_1(L) \cong G$

$$\Sigma_+^n(\mathbf{E}) \simeq \Sigma_+^m(C_*(L))$$

for some $n, m \geq 1$. However, $\Sigma_+^m(C_*(L)) \cong C_*(L \vee mS^2)$. Put $K = L \vee mS^2$, and let $\varphi : C_*(K) \to \Sigma_+^n(\mathbf{E})$ be a weak homotopy equivalence. In particular, φ induces an isomorphism

$$\varphi : \pi_2(K) \overset{\approx}{\to} J \oplus \mathbf{Z}[G]^n$$

If $\psi : \Sigma_+^n(\mathbf{E}) \to \mathbf{E}$ is the natural chain projection, then $\psi \circ \varphi : C_*(K) \to \mathbf{E}$ has the property that $\mathrm{Ker}(\psi \circ \varphi : \pi_2(K) \to J$ is isomorphic to $\mathbf{Z}[G]^n$. Choose a $\mathbf{Z}[G]$ basis $\alpha_1, \alpha_2, \ldots, \alpha_m$ for $\mathrm{Ker}(f)$, and form $X(\alpha)$ by attaching 3-cells $E_1^{(3)}, E_2^{(3)}, \ldots, E_m^{(3)}$ to K by means of $\alpha_1, \alpha_2, \ldots, \alpha_m$; that is

$$X(\alpha) = K \cup_{\alpha_1} E_1^{(3)} \cup_{\alpha_2} E_2^{(3)} \cdots \cup_{\alpha_m} E_m^{(3)}$$

Then it is easily checked that $H^r(X(\alpha); \mathcal{B}) = 0$ for all $r \geq 3$, and it is straightforward to see that $\langle X \rangle$ is homotopy equivalent to $\mathbf{E}$. However, by hypothesis, $X(\alpha)$ is homotopy equivalent to a finite 2-complex, K say, which thereby provides the required geometrical realization of $\mathbf{E}$. $\qquad\square$

References

[1] A. A. Albert, Involutorial simple algebras and real Riemann matrices. *Ann. of Math.* 36 (1935) 886–964.
[2] H. Bass, Projective modules over free groups are free. *J. of Algebra* 1 (1964) 367–373.
[3] H. Bass, *Algebraic K-Theory*, Benjamin, 1968.
[4] S. Bentzen and I. Madsen, On the Swan subgroup of certain periodic groups. *Math. Ann.* 264 (1983) 447–474.
[5] R. F. Beyl, M. P. Latiolais and N. Waller, Classification of 2-complexes whose finite fundamental group is that of a 3-manifold. *Proc. Edinburgh Math. Soc.* 40 (1997) 69–84.
[6] W. W. Boone, The word problem. *Ann. of Math.* 70 (1959) 265–269.
[7] W. Browder, *Surgery on Simply Connected Manifolds*, Springer-Verlag, 1969.
[8] E. H. Brown, Finite computability of Postnikov complexes. *Ann. of Math.* 65 (1957) 1–20.
[9] W. Browning, Homotopy types of certain finite C.W. complexes with finite fundamental group. Ph.D Thesis, Cornell University, 1978.
[10] W. Browning, Truncated projective resolutions over a finite group (unpublished notes). ETH, April 1979.
[11] J. F. Carlson, Modules and group algebras. *ETH Lecture Notes in Mathematics*, Birkhäuser Verlag, 1996.
[12] H. Cartan and S. Eilenberg, *Homological Algebra*, Princeton University Press, 1956.
[13] W. H. Cockroft and R. G. Swan, On the homotopy types of certain two-dimensional complexes. *Proc. London Math. Soc.* 11 (3) (1961) 194–202.
[14] C. W. Curtis and I. Reiner, *Methods of Representation Theory*, Volumes I and II, Wiley-Interscience, 1981 and 1987.
[15] P. Du Val, *Homographies, quaternions and rotations. Oxford Mathematical Monographs* (1964).
[16] M. N. Dyer and A. J. Sieradski. Trees of homotopy types of two-dimensional CW complexes. *Comment. Math. Helv.* 48 (1973) 31–44.
[17] D. B. A. Epstein, Finite presentations of groups and 3-manifolds, *Quart. J. Math.* 12 (1961) 205–212.
[18] D. K. Fadeev, An introduction to the multiplicative theory of modules of integral representations. *Proc. Steklov Inst. Math.* 80 (1965) 164–210.
[19] R. H. Fox, Free differential calculus V. *Ann. of Math.* 71 (1960) 408–422.

[20] K. W. Gruenberg, Homotopy classes of truncated projective resolutions. Comment. *Math. Helv.* 68 (1993) 579–598.

[21] M. Gutierrez and M. P. Latiolais, Partial homotopy type of finite two-complexes. *Math. Zeit.* 207 (1991) 359–378.

[22] D. Happel, *Triangulated Categories in the Representation Theory of Finite Dimensional Algebras*, Cambridge University Press, 1988.

[23] R. Hartshorne, Residues and duality. *Lecture Notes in Mathematics* 20, Springer-Verlag, 1966.

[24] J. Hempel, Intersection calculus on surfaces with applications to 3-manifolds. *Memoirs Amer. Math. Soc.* no. 282 (1983).

[25] H. Jacobinski, Genera and decompositions of lattices over orders. *Acta Math.* 121 (1968) 1–29.

[26] F. E. A. Johnson, Stable modules and the structure of Poincaré 3-complexes. *Geometry and Topology Aarhus: Proceedings of the 6th Aarhus Topology Conference. A.M.S. Contemporary Mathematics* 258 (2000) 227–248.

[27] F. E. A. Johnson, Explicit homotopy equivalences in dimension two. *Math. Proc. Camb. Phil. Soc.* 133 (2002) 411–430.

[28] F. E. A. Johnson, Stable modules and Wall's D(2)-problem. *Comment. Math. Helv.* 78 (2003) 18–44.

[29] F. E. A. Johnson, Minimal 2-complexes and the D(2)-problem (to appear in *Proc. Amer. Math. Soc.*).

[30] F. E. A. Johnson, The D(2)-property for certain infinite groups (in preparation).

[31] P. E. Kenne, Ph.D. Thesis, Australian National University (Canberra), 1984.

[32] M. P. Latiolais, The simple homotopy type of finite two-complexes with finite abelian fundamental group. *Trans. Amer. Math. Soc.* 293 (1986) 655–662.

[33] P. A. Linnell, Minimal free resolutions and (G, n)-complexes for finite abelian groups. *Proc. London Math. Soc.* 66 (1993) 303–326.

[34] S. MacLane, *Homology*. Springer-Verlag, 1963.

[35] S. MacLane and J. H. C. Whitehead, On the 3-type of a complex. *Proc. Nat. Acad. Sci.* 36 (1950) 41–48.

[36] I. Madsen, Reidemeister torsion, surgery invariants and spherical space forms. *Proc. London Math. Soc.* 46 (1983) 193–240.

[37] H. Maschke, *Math. Ann.* 51 (1899) 253–294.

[38] J. Mennicke, Einige endliche Gruppen mit drei Erzeugenden und drei Relationen. *Arch. Math.* 10 (1959) 400–418.

[39] W. Metzler, Uber den Homotopietyp zweidimensionaler CW-komplexe und Elementartransformationen bei Darstellungen von Gruppen durch Erzeugenden und definierende Relationen. *J. Reine Angew. Math.* 285 (1976) 7–23.

[40] R. J. Milgram, Evaluating the Swan finiteness obstruction for periodic groups, preprint, Stanford University, 1979.

[41] R. J. Milgram, Odd index subgroups of units in cyclotomic fields and applications. *Lecture Notes in Mathematics*, vol. 854, Springer-Verlag, 1981, 269–298.

[42] J. W. Milnor, Groups which act on S^n without fixed points. *Amer. J. Math.* 79 (1957) 623–630.

[43] J. W. Milnor, A procedure for killing the homotopy groups of differentiable manifolds. *Proc. Symp. Pure Math. 3 (Differential geometry). Amer. Math. Soc.* (1961) 39–51.

[44] J. W. Milnor, Whitehead torsion. *Bull. Amer. Math. Soc.* 72 (1966) 358–426.

[45] B. Mitchell, *Theory of Categories*, Academic Press, 1965.

264 *References*

[46] B. H. Neumann, Some finite groups with few defining relations. *J. Australian Math. Soc.* (Ser. A) 38 (1985) 230–240.
[47] P. S. Novikov, On the algorithmic unsolvability of the word problem in group theory. *Transl. Amer. Math. Soc.* 29 (1958) 1–122.
[48] S. P. Novikov, Homotopy equivalent smooth manifolds I. *Transl. Amer. Math. Soc.* 48 (1965) 271–396.
[49] R. S. Pierce, *Associative Algebras: Graduate Texts in Mathematics* 88, Springer-Verlag, 1982.
[50] K. W. Roggenkamp and V. Huber-Dyson, Lattices over orders: *Lecture Notes in Mathematics* 115, Springer Verlag, 1970.
[51] G. Shimura, On analytic families of polarised abelian varieties and automorphic functions. *Ann. of Math.* 78 (1963) 149–192.
[52] S. Smale, On the structure of manifolds. *Amer. J. Math.* 84 (1962) 387–399.
[53] E. H. Spanier, *Algebraic Topology*, McGraw-Hill, 1966.
[54] J. R. Stallings, On torsion free groups with infinitely many ends. *Ann. of Math.* 88 (1968) 312–334.
[55] N. Steenrod, *The Topology of Fibre Bundles*, Princeton University Press, 1951.
[56] M. Suzuki, On finite groups with cyclic Sylow subgroups for all odd primes. *Amer. J. Math.* 77 (1955) 657–691.
[57] R. G. Swan, Induced representations and projective modules. *Ann. of Math.* 71 (1960) 552–578.
[58] R. G. Swan, Periodic resolutions for finite groups. *Ann. of Math.* 72 (1960) 267–291.
[59] R. G. Swan, Projective modules over group rings and maximal orders. *Ann. of Math.* 76 (1962) 55–61.
[60] R. G. Swan, Minimal resolutions for finite groups. *Topology* 4 (1965) 193–208.
[61] R. G. Swan, Groups of cohomological dimension one. *J. of Algebra.* 12 (1969) 585–610.
[62] R. G. Swan, K-Theory of finite groups and orders (notes by E.G. Evans). *Lecture Notes in Mathematics* 149, Springer-Verlag, 1970.
[63] R. G. Swan, Projective modules over binary polyhedral groups. *Journal für die Reine und Angewandte Mathematik.* 342 (1983) 66–172.
[64] R. Thom, Sur quelques propriétés globales des variétés différentiables: Comment. *Math. Helv.* 28 (1954) 17–86.
[65] C. B. Thomas, A reduction theorem for free actions by the group $Q(8n, k, l)$ on S^3. *Bull. London Math. Soc.* 20 (1988) 65–67.
[66] H. Tietze, Über die topologischen Invarianten mehrdimensionaler Mannigfaltigkeiten. *Monatsh. Math. Phys.* 19 (1908) 1–118.
[67] V. G. Turaev, Fundamental groups of manifolds and Poincaré complexes. *Mat. Sbornik.* 100 (1979) 278–296 (in Russian).
[68] C. T. C. Wall, Finiteness conditions for CW Complexes. *Ann. of Math.* 81 (1965) 56–69.
[69] C. T. C. Wall, Finiteness conditions for CW complexes II. *Proc. Roy. Soc* A 295 (1966) 129–139.
[70] C. T. C. Wall, Formal deformations. *Proc. London Math. Soc.* 16 (3) (1966) 342–352.
[71] C. T. C. Wall, Surgery of non-simply connected manifolds. *Ann. of Math.* 84 (1966) 217–276.
[72] C. T. C. Wall, Poincaré complexes I. *Ann. of Math.* 86 (1967) 213–245.
[73] C. T. C. Wall, Surgery on compact manifolds. *London Math. Soc. Monographs*, Academic Press, 1970.

[74] C. T. C. Wall, Periodic projective resolutions. *Proc. London Math. Soc.* 39 (3) (1979) 509–553.

[75] A. H. Wallace, Modifications and co-bounding manifolds I. *Canadian J. Math.* 12 (1960) 503–528.

[76] A. H. Wallace, Modifications and co-bounding manifolds II. *J. Math. Mech.* 10 (1961) 773–809.

[77] A. H. Wallace, Modifications and co-bounding manifolds III. *J. Math. Mech.* 11 (78) 971–990.

[78] J. W. Wamsley, The deficiency of metacyclic groups. *Proc. Amer. Math. Soc.* 24 (1970) 724–726.

[79] J. H. M. Wedderburn, On hypercomplex numbers. *Proc. London Math. Soc.* (2) 6, 77–118.

[80] J. H. C. Whitehead, Combinatorial homotopy II. *Bull. Amer. Math. Soc.* 55 (1949) 45–496.

[81] J. S. Williams, Free presentations and relation modules of finite groups. *J. Pure Appl. Alg.* 3 (1973) 203–217.

[82] J. A. Wolf, *Spaces of constant curvature*, McGraw-Hill, 1967.

[83] N. Yoneda, On the homology theory of modules. *J. Fac. Sci. Tokyo*, Sec. I, 7 (1954) 193–227.

[84] H. Zassenhaus, Über endliche fastkörper. *Abhandlungen aus dem Mathematischen Seminar der Hamburgischen Universität*, 11 (1935) 187–220.

Index

Printed in the United States
By Bookmasters